Table of Critical Values of $\underline{t}$

For any given df, the table shows the values of t corresponding to various levels of probability. Obtained t is significant at a given level if it is equal to or *greater than* the value shown in the table.

	Level of significance for one-tailed test					
	.10	.05	.025	.01	.005	.0005
	Level of significance for two-tailed test					
df	.20	.10	.05	.02	.01	.001
1	3.078	6.314	12.706	31.821	63.657	636.619
2	1.886	2.920	4.303	6.965	9.925	31.598
3	1.638	2.353	3.182	4.541	5.841	12.941
4	1.533	2.132	2.776	3.747	4.604	8.610
5	1.476	2.015	2.571	3.365	4.032	6.859
6	1.440	1.943	2.447	3.143	3.707	5.959
7	1.415	1.895	2.365	2.998	3.499	5.405
8	1.397	1.860	2.306	2.896	3.355	5.041
9	1.383	1.833	2.262	2.821	3.250	4.781
10	1.372	1.812	2.228	2.764	3.169	4.587
11	1.363	1.796	2.201	2.718	3.106	4.437
12	1.356	1.782	2.179	2.681	3.055	4.318
13	1.350	1.771	2.160	2.650	3.012	4.221
14	1.345	1.761	2.145	2.624	2.977	4.140
15	1.341	1.753	2.131	2.602	2.947	4.073
16	1.337	1.746	2.120	2.583	2.921	4.015
17	1.333	1.740	2.110	2.567	2.898	3.965
18	1.330	1.734	2.101	2.552	2.878	3.922
19	1.328	1.729	2.093	2.539	2.861	3.883
20	1.325	1.725	2.086	2.528	2.845	3.850
21	1.323	1.721	2.080	2.518	2.831	3.819
22	1.321	1.717	2.074	2.508	2.819	3.792
23	1.319	1.714	2.069	2.500	2.807	3.767
24	1.318	1.711	2.064	2.492	2.797	3.745
25	1.316	1.708	2.060	2.485	2.787	3.725
26	1.315	1.706	2.056	2.479	2.779	3.707
27	1.314	1.703	2.052	2.473	2.771	3.690
28	1.313	1.701	2.048	2.467	2.763	3.674
29	1.311	1.699	2.045	2.462	2.756	3.659
30	1.310	1.697	2.042	2.457	2.750	3.646
40	1.303	1.684	2.021	2.423	2.704	3.551
60	1.296	1.671	2.000	2.390	2.660	3.460
120	1.289	1.658	1.980	2.358	2.617	3.373
∞	1.282	1.645	1.960	2.326	2.576	3.291

introductory statistics for psychology

introductory statistics for psychology

2nd Edition

Chester A. Insko
University of North Carolina

Douglas W. Schoeninger

Allyn and Bacon
Boston · London · Sydney · Toronto

Copyright © 1977, 1971 by Allyn and Bacon, Inc., 470 Atlantic Avenue, Boston, Massachusetts 02210. All rights reserved. Printed in the United States of America. No part of the material protected by this copyright notice may be reproduced or utilized in any form or by any means, electronic or mechanical, including photocopying, recording, or by any information storage and retrieval system, without written permission from the copyright owner.
Previous edition was published under the title *Introductory Statistics for the Behavioral Sciences* by Douglas W. Schoeninger and Chester A. Insko, Copyright © 1971 by Allyn and Bacon, Inc.

Library of Congress Cataloging in Publication Data

Insko, Chester A
    Introductory statistics for psychology.

        Bibliography: p.
        Includes index.
        1. Psychometrics. I. Schoeninger, Douglas W., joint author. II. Title.
BF39.I57 1977   150'.1'82   76-21814
ISBN 0-205-05575-3

Contents

Preface		**ix**
CHAPTER 1	**PRELIMINARY CONSIDERATIONS**	**3**

Problem Solving as a Goal • An Illustrative Research Report • Measurement • Scales of Measurement • Variables • Subfields of Statistics • Statistic, Statistics, and Statistics • Historical Perspective on Arabic Numerals • Exercises

CHAPTER 2	**FREQUENCY DISTRIBUTIONS**	**17**

Univariate Frequency Distributions • Bivariate Frequency Distributions • Beginnings of Descriptive Statistics • Exercises

CHAPTER 3	**CENTRAL TENDENCY**	**39**

Arithmetic Mean • Median • Mode • Comparison of the Mean, Median, and Mode • Geometric Mean • Tabling and Graphing Central Tendency • Graphs that Create False Impressions • Historical Perspective on the Indices of Central Tendency • Exercises

CHAPTER 4	**VARIABILITY**	**67**

Range • Variance and Standard Deviation • Semi-Interquartile Range • Historical Perspective on the Indices of Variability

CHAPTER 5	**CORRELATION**	**89**

Background Concepts • The Concept of Correlation and Slope • Coefficient of Determination • Prediction • Effect of Range upon Correlation • Multiple and Partial

Correlation • Correlation, Causation, and Experimentation • Historical Perspective on Correlation • Exercises

CHAPTER 6 PROBABILITY 137

Probability and Probability Numbers • Sampling and Probability • Computing Probabilities • Probability Distributions • Historical Perspective on Probability • Exercises

CHAPTER 7 SAMPLING DISTRIBUTIONS 175

The Concept of Sampling Distribution • Sampling Distributions of Proportions • Sampling Distributions of Means • Sampling Distributions and Statistical Inference • Exercises

CHAPTER 8 NORMAL DISTRIBUTION AND HYPOTHESIS TESTING 201

Normal Distribution • Ascertaining Probabilities for Intervals of a Normal Distribution • Using the Standard Normal Distribution to Test Hypotheses Concerning Population Means • Error and Power • Two Controversial Issues • Historical Perspective on the Normal Distribution • Exercises

CHAPTER 9 INFERENCES CONCERNING MEANS AND VARIANCES 239

Hypotheses Concerning a Single Population Mean • Hypotheses Concerning Differences Between Population Means • Recapitulation: Tails and Tests • Review of Formulas • Testing Equality of Two Population Variances • A Further Controversial Issue: Magnitude of Effect • Historical Perspective on the t Test • Exercises

CHAPTER 10 ANALYSIS OF VARIANCE 277

Overview of Analysis of Variance • One-Factor Analysis of Variance • Two-Factor Analysis of Variance • Three- or-More Factor Analysis of Variance • R. A. Fisher and Analysis of Variance • Exercises

CHAPTER 11 AN INTRODUCTION TO CHI SQUARE 323

Chi Square Statistic • Chi Square Applied to a Single Sample: Test of Goodness of Fit • Chi Square Applied to Two or More Samples: Test of Independence • Exercises

CHAPTER 12	**MEASURES OF CORRELATION AND TESTS OF SIGNIFICANCE**	**345**

Significance of the Pearson Product–Moment Correlation Coefficient, r • Spearman's Rank–Order Correlation Coefficient, r' • Point Biserial Coefficient, r_{pb} • Correlation Ratio, η • Exercises

CHAPTER 13	**DISTRIBUTION-FREE STATISTICAL TESTS**	**361**

Distribution-Free and Nonparametric Statistical Tests • Use of Nominal and Ordinal Characteristics of Obtained Data • Assumptions • Treatment of Zero Difference Scores and Tied Ranks • Sign Test for One Sample • Sign Test for Two Dependent Samples • Wilcoxon Rank–Sum Test • Wilcoxon Signed–Rank Test • Advantages and Disadvantages of Distribution-Free Tests • Historical Perspective on Distribution-Free Tests • Exercises

Appendix A	**Elementary Set Theory**	**385**
Appendix B	**An Introduction to Summation Algebra**	**395**
Appendix C	**Tables**	**401**

Table I. Proportions of Area Under the Normal Curve
Table II. Critical Values of t
Table III(a). Critical Values of F for α Equal to .05 and .01
Table III(b). Critical Values of F for α Equal to .025 and .005
Table IV. Critical Values of Chi Square
Table V. Critical Values of the Correlation Coefficient
Table VI. Critical Values of r', the Spearman Rank-Order Coefficient
Table VII. Table of Probabilities Associated With Values as Small as Observed Values of r in the Sign Test
Table VIII. Critical Lower-Tail Values of W_n for Wilcoxon's Rank-Sum Test
Table IX. Critical and Quasi-Critical Lower-Tail Values of W_+ (and Their Probability Levels) for Wilcoxon's Signed-Rank Test

References	**425**
Exercise Answers	**427**
Index	**445**

Preface

This book is designed to introduce psychology students to basic statistical concepts. An understanding of high school algebra, but not calculus, is assumed.

Five special features make this text unique. First, certain algebraic proofs and other more difficult materials are included in small print. The student who is not interested in this additional material can skip the small print without jeopardizing his or her understanding of subsequent material. On the other hand, the interested student can obtain additional information by reading it. This feature is included in recognition of the heterogeneous composition of most classes in introductory statistics. Students vary considerably regarding their interest in statistics and their background in mathematics. This feature also allows the instructor to select from among the more difficult material that which is considered essential, and assign these portions as he or she chooses. Continuity will not be sacrificed by adding or deleting any small print section.

The second feature is the inclusion of materials on set theory (Appendix A) and summation algebra (Appendix B). In the future, students will be familiar with set concepts by the time they enroll in introductory statistics. Set concepts provide a meaningful frame of reference for the discussion of measurement, probability, and sampling. Thus, an appendix introducing the student to set theory is included, and set concepts are used throughout the text wherever appropriate. The review of summation algebra should be particularly helpful to students interested in following the algebraic proofs.

A third feature of this book is the use of data from a single experiment—an interpersonal attraction experiment—to illustrate many of the concepts and techniques of statistics that are described in subsequent chapters. The data from this experiment are used to illustrate frequency distributions, graphs, central tendency, variability, correlation, t test for means, and analysis of variance. Thus, the student is given a concrete, in-depth example of the usefulness of various statistical procedures.

The fourth feature of the book is the inclusion of historical perspective sections at the ends of most chapters. It is perhaps this feature, more than any of the preceding, which makes the book truly unique. Most psychologists are relatively uninformed about the fascinating historical background of various statistical concepts and techniques. Some brief acquaintance with the cultural circumstances that contributed to the development of various techniques can make these techniques appear more human and more meaningful. Also the student is given a more accurate picture of the field of statistics as dynamic and changing, not static and fixed.

A fifth feature also gives the student a more accurate picture of statistics as dynamic and changing. This is the inclusion of discussion relating to various controversial issues. These include the arbitrariness of the significance level, one- versus two-tailed tests, significance of effect versus magnitude of effect, distribution-free versus classical tests. Although not specifically labeled as controversial, the correlation–causation issue is also discussed.

The text is roughly divided into two sections: the first deals with description of obtained data, the second is concerned with drawing inferences about population characteristics from samples of observations. Chapter 1 is an introduction that explains the focus of the text and discusses measurement. Chapters 2, 3, 4, and 5 cover frequency distributions, central tendency, variability, and correlation. These chapters comprise the section on descriptive statistics, although some material on inferential statistics is also included. Chapters 6 through 13 cover probability, sampling distributions, normal distributions and hypothesis testing, inferences concerning central tendency and variability, analysis of variance, chi-square analyses, additional measures of correlation, and distribution-free statistics. Chapters 6 through 13 comprise the section on inferential statistics, although they also contain material on descriptive statistics.

This book is based on an earlier book, *Introductory Statistics for the Behavioral Sciences*, by Douglas W. Schoeninger and Chester A. Insko. While the present book owes much to the earlier one, the content has been drastically altered and expanded.

Finally, I am indebted to a number of people for their help and assistance. Especially helpful were the numerous suggestions received from various readers of the earlier book. One in particular comes to mind, J. Lyn Elder, who, in a kindly way, expressed his disappointment regarding the lack of historical background material. I am further indebted to Susan F. Pearson for typing the manuscript and to my wife Verla for the many hours spent reading, criticizing, and correcting the manuscript.

CHESTER A. INSKO

introductory statistics for psychology

Prologue to Chapter 1

In many colleges and universities, statistics is one of the least appreciated courses in the entire psychology curriculum. Students study psychology for various reasons: to understand people, to learn how to help people, or to learn about themselves, but not to study some branch of mathematics. It is frequently very difficult to appreciate the necessity of mastering statistics. One of the purposes of Chapter 1 is to convince the reader that an understanding of statistics is an essential prerequisite to an in-depth study of psychology. If the study of psychology is taken seriously, then it is worth expending the effort to learn statistics, at least elementary statistics.

An additional purpose of Chapter 1 is to provide a short description of the superficially simple, but actually complex, problem of measurement. This sounds rather unimpressive. Recognize, however, that the numbers to which statistical techniques are applied arise from measurement. Recognize, further, that measurement in psychology is not a simple matter of "slapping down" a ruler. How, for example, could measurements even be taken of something as ephemeral as imagery? This chapter suggests one possible way.

1

Preliminary Considerations

Contrary to the notion held by many people, the field of statistics does not deal simply with quantities of compiled, tabulated, and numerical data. The field of statistics consists of the *mathematical theory and method useful in the collection, analysis, interpretation, and presentation of research data.* Since the field of statistics is concerned with both theory and method, an introduction to this field can place relatively more weight on theory or on method. This book places relatively more weight on method. At the same time, however, the intent has been to write something more than a "cookbook" or a mere listing of step-by-step procedures without any explanation as to why such procedures work. The emphasis is on the understanding and comprehension of methods that have been found most useful in psychology.

Beginning students in psychology are frequently puzzled about the necessity of mastering the subject matter of statistics. Such mastery is desirable for at least two main reasons. First, it is essential to understanding published research, and second, conducting psychological research typically requires facility in the use of statistical methods.

The authors' picture of you as students is diverse. We have both taken and taught statistics courses, and thus, are making judgments based on our memories as students and our more immediate experiences of communicating statistical concepts to students. Many students fear statistics, often because of past difficulties with mathematics. Other, more mathematically inclined students find introductory statistics easy in comparison to mathematics courses. This creates a dilemma as to how introductory statistics should be presented. If the subject matter is presented too simply, students with mathematical skills will be deprived and possibly bored. If it is presented at a more advanced level, students with less mathematical preparation will be denied the opportunity of mastering enough statistics to go on to advanced courses and training. In view of the fact that many of these latter students are quite capable of making contributions to the various subfields of psychology, this presents a real problem. This text makes use of two (explicit) procedures for handling the heterogeneous composition of most beginning statistics classes. First, some complex theoretical explanations and algebraic proofs have been included, but in small print separate from other material (with an easily noticeable vertical rule along side the text margin). These small print sections are interspersed throughout the text and vary in length from a brief paragraph to several pages. Students interested in this material can read it along with everything else; others can omit this material without jeopardizing their understanding of later chapters in the book. Second, since understanding the text requires some facility with set theory and summation algebra, brief presentations of these topics are included in Appendixes A and B, respectively. Students unfamiliar with set theory concepts should read Appendix A before completing Chapter 1. Set concepts are used throughout the book. The summation algebra in Appendix B should be studied in conjunction with Chapter 3.

PROBLEM SOLVING AS A GOAL

Statistical methods are important to psychologists because they are a means to an end, the solving of psychological problems. What is a problem? From a research perspective, a problem is a question proposed for solution. Here are some examples:

> What is the average IQ of students in the Clark School? Is it larger or smaller than the average IQ for the whole school system?
> Are opinions reached by group consensus more or less extreme than individual opinions?
> What is the relationship between degree of fear and amount of food eaten?

These types of questions are sometimes implied in the titles of journal articles and often directly stated in the articles themselves. The articles usually state a tentative answer to some question, preceded by a discussion of statistical procedures and results used to buttress or support the stated conclusion.

AN ILLUSTRATIVE RESEARCH REPORT

In the June 1972 issue of the *Journal of Abnormal Psychology* is an article entitled "Imagery Vividness, Reality Testing, and Schizophrenic Hallucinations." The authors are Sanford Mintz and Murray Alpert. One or more research questions are implied by this title. One question might be: "What is the relationship between the occurrence of hallucinations, vividness of imagery, and effectiveness of reality testing?" (Reality testing is a checking of assumptions regarding what is "out there.") Another question might be: "Is there any relationship between the occurrence of hallucinations, vividness of imagery, and effectiveness of reality testing?"

Such general questions do not indicate whether or not the investigators have a hypothesis regarding the properties to be investigated. Frequently, investigators do have a hypothesis and the present study is a case in point. Mintz and Alpert (1972, p. 310) set the stage for their hypothesis by reviewing some of the existing literature. (*Note*: The usage *S*s below refers to subjects, in this case, nurses.)

In a study of hypnotic-like experiences, Spanos and Barber (1968) gave the following instructions to 90 student nurses:

I want you to close your eyes and hear a phonograph record with words and music playing "White Christmas." Keep listening to the phonograph playing "White Christmas" until I ask you to stop [1968, p. 139]. Actually, no record or phonograph was present. The *S*s were then requested to check the one sentence among the following four choices that best described their experience:

A. "I heard a phonograph record of 'White Christmas' clearly and believed that the record was actually playing."
B. "I heard the phonograph record of 'White Christmas' clearly, but knew there was no record actually playing."
C. "I had a vague impression of hearing the record playing 'White Christmas.'"
D. "I did not hear the record" [p. 139].

Forty-three of these "normal" *S*s (48%) checked sentence B. The essentials of this design were replicated on several occasions with basically similar results (Barber and Calverly, 1964; Bowers, 1967).

Sir Francis Galton (Boismont, 1859) found that 10% of large groups of sampled adults admitted the "faculty of seeing pictures." These persons reported that "the mental image appears to correspond in all respects to reality," or "all objects in my mental picture are as bright as the actual scene," or "the images are brilliant and distinct."

The hypothesis is stated as follows: "...it is the combination of impaired reality testing function characteristic of the disorder plus the individual's predisposition to vivid imagery that results in hallucinations" (1972, p. 310). Mintz and Alpert also give us a statement of the problem: "We are concerned whether vivid auditory imagery may be a necessary (although not sufficient) prerequisite for hallucinations in schizophrenics..." (p. 311).

At the beginning of their "Results" section, Mintz and Alpert briefly summarize their procedure, which they had previously described in detail in a "Method" section.

> Thirty seconds after requesting S to imagine hearing a phonograph record of "White Christmas," he was presented four written statements and asked to choose the one that best described the experience he just had. If an S chose either statement A, "I heard a phonograph record of 'White Christmas' clearly and believed that the record was actually playing", or B, "I heard the phonograph record of 'White Christmas' clearly, but knew there was no record actually playing," he was considered to be a high vividness auditory imaginer, because in both cases the auditory image is described as being clearly heard. If S chose C, "I had a vague impression of hearing the record playing 'White Christmas'," or D, "I did not hear the record," he was considered to be a low vividness auditory imaginer (1972, pp. 312–313).

Application of this procedure to 20 hallucinating schizophrenics, 20 nonhallucinating schizophrenics, and 20 nonpsychotic control subjects produced the following results:

> Seventeen of the hallucinating schizophrenics checked A or B, the high vividness categories, and three checked C or D, the low vividness categories. Of the nonhallucinating schizophrenics, one checked B and 19 checked C or D. Among the controls, eight had high vividness scores and 12 had low vividness scores. These differences in frequencies of high and low vividness imaginers among the groups is highly significant ($\chi^2 = 26.89$, $df = 2$, and $p < .001$). As predicted, the hallucinating schizophrenic group contained a significantly greater number of vivid auditory imaginers, and the nonhallucinating schizophrenic group contained significantly more low vividness auditory imaginers than would be expected by a chance departure from null difference (1972, p. 313).

It is not expected that the above excerpt is totally comprehensible at this point. In fact, one of the objectives of this book is to make comprehensible such statements as the above. Another objective is to enable students to use statistical procedures as Mintz and Alpert have done. Here, the excerpt is included to illustrate the use of statistical procedures as a tool in problem solving.

MEASUREMENT

Before an investigator can make use of statistical procedures, he or she must first measure the properties that are of interest.

Properties

Vividness of imagery and effectiveness of reality testing are properties or attributes of human subjects. Funk and Wagnalls (dictionary) defines property as: "...a distinguishing quality, characteristic, or mode of any substance—any variable state." In psychology the substance is typically an organism, human or non-human.

Measurement Illustrated

Statistical procedures are applied to the results of measured properties. In the Mintz and Alpert study, the property, vividness of imagery, was measured by an already described procedure. Individually tested subjects were simply asked to close their eyes and imagine that they could hear a record of "White Christmas." After 30 seconds, the subjects were asked to check the following rating scale:

A. "I heard a phonograph record of 'White Christmas' clearly and believed that the record was actually playing."
B. "I heard the phonograph record of 'White Christmas' clearly, but knew there was no record actually playing."
C. "I had a vague impression of hearing the record playing 'White Christmas.'"
D. "I did not hear the record" (Spanos and Barber, 1968, p. 139).

Those subjects who rated themselves A or B were classified as high on vividness of imagery, those rating C or D, as low on vividness of imagery. Stated somewhat differently, the outcomes A or B were assigned to the high vividness of imagery subset and the outcomes C or D were assigned to the low vividness of imagery subset.

For purposes of measurement, outcomes assigned to the same subset are considered *alike* in the property being measured and *different* from those assigned to other subsets. Although Mintz and Alpert did not do so, measurement typically also involves the assignment of different numbers to the outcome subsets consistent with some rule. As an example, consider some measurements taken in an experiment by Stroebe, Insko, Thompson, and Layton (*Journal of Personality and Social Psychology*, 1971). The data resulting from the measurements taken in this experiment will be repeatedly used throughout the text to illustrate various statistical procedures. For present purposes the important aspect of the study is the quantitative measurement. Subjects from an introductory psychology course at the University of North Carolina initially spent some time examining materials regarding another University of North Carolina student. These materials included a picture of the other person and an attitude questionnaire supposedly filled out by this person. For female subjects the other person was male, and for male subjects, the other person was female. The subjects were told that the experiment was concerned with people's ability to make judgments about others on the basis of limited information. Thus, after examining the picture and attitude questionnaire, the subjects made a series of judgments regarding the other person's intelligence, knowledge of current events, and other similar properties. These judgments were followed by some measurements of various aspects of interpersonal attraction, the main interests of Stroebe *et al.* Three such measurements were taken: liking, preference as a date, and preference as a marital partner. The questions used to make these measurements are contained in Table 1.1. For each question the subject responded by checking one of seven

TABLE 1.1. *Three Questions Used by Stroebe et al. to Measure Different Aspects of Interpersonal Attraction.**

If you knew this boy, how much do you think you would like him? (*Check one.*)
_____ I feel that I would probably like this person very much.
_____ I feel that I would probably like this person.
_____ I feel that I would probably like this person to a slight degree.
_____ I feel that I would probably neither particularly like nor particularly dislike this person.
_____ I feel that I would probably dislike this person to a slight degree.
_____ I feel that I would probably dislike this person.
_____ I feel that I would probably dislike this person very much.

If you were looking for a date and you knew this boy, how likely would it be that you would consider him? (*Check one.*)
_____ I definitely would consider him.
_____ I very probably would consider him.
_____ I probably would consider him.
_____ I might or might not consider him.
_____ I probably would not consider him.
_____ I very probably would not consider him.
_____ I definitely would not consider him.

If you were thinking of getting married and you knew this boy, how likely would it be that you would consider him? (*Check one.*)
_____ I definitely would not consider him.
_____ I very probably would not consider him.
_____ I probably would not consider him.
_____ I might or might not consider him.
_____ I probably would consider him.
_____ I very probably would consider him.
_____ I definitely would consider him.

* NOTE: For male subjects "girl" was substituted for "boy", and "her" for "him."

alternatives that ranged from the lowest level of attraction to the highest level of attraction. Stroebe *et al.* subsequently assigned the numbers 1 through 7 to these alternatives such that the greater the attraction, the higher the number. Thus, for the measurement of liking: 1 was assigned to the alternative, "I feel that I would probably dislike this person very much;" 2 was assigned to the alternative, "I feel that I would probably dislike this person;" ...7 was assigned to the alternative, "I feel that I would probably like this person very much."

SCALES OF MEASUREMENT

It is helpful to distinguish among four scales of measurement; namely, these are the nominal, ordinal, interval, and ratio scales.

Nominal Scale

A nominal scale is a scale in which outcomes are assigned to subsets that are distinguished qualitatively, but not quantitatively. For example, to measure the property "eye color" as it occurs in a certain set of people, persons with blue eyes could be assigned to one subset, and brown to another. A rule to enable the assignment of every person would have to be developed. This rule would be some procedure for determining eye color. In the case of such simple measurement, the subsets differ qualitatively but not quantitatively, i.e., it is not assumed that blue eyes are in any sense greater than or less than brown eyes.

One of the most familiar examples of a nominal scale is the male–female classification. For some purposes it is convenient to assign numbers to the measurement subsets. Thus, 0 could be assigned to the male subset, and 1 to the female subset. Because, however, the subsets are not distinguished quantitatively, it is arbitrary which subset is assigned which number. Thus, a 1 (one) could be assigned to the male subset, and a 0 (zero) to the female subset. The numbers function simply as names.

Ordinal Scale

An ordinal scale is a scale in which the outcome subsets are ranked in order of magnitude. For example, in the Mintz and Alpert experiment, outcomes were assigned to high and low subsets that differ in the magnitude of the vividness of imagery property. As previously indicated, numbers are typically assigned to the outcome subsets. With an ordinal scale the numbers assigned to the outcome subsets indicate the relative amount of the property being measured. For example, in the Stroebe *et al.* experiment, the numbers 1 to 7 were used to convey ordinal information regarding the degree of interpersonal attraction. Thus, a 7 indicates more attraction than a 6, a 6 more than a 5, and so on. With an ordinal scale the order of the numbers corresponds to the relative amount of the property. The numbers, thus, do more than name the outcome subsets, as was the case with nominal scale subsets.

Interval Scale

An interval scale is a scale in which the numbers are assigned to the outcome subsets such that equal intervals between the numbers reflect equal amounts of the property. A common example of an interval scale is the Fahrenheit temperature scale. In the case of interval measurement, as in the case of ordinal measurement, the assigned numbers reflect relative amounts of the property. Interval measurement, however, has the additional characteristic that equal intervals between numbers reflect equal amounts of the property. For example, the temperature increase between 20 and 30 degrees Fahrenheit is equal to the increase between 30 and 40 degrees Fahrenheit, the increase between 30 and 40 degrees Fahrenheit equals the increase between 50 and 60 degrees Fahrenheit, and so forth. Equal differences between the numbers reflect equal differences in the property being measured (in this case temperature). The measurement of attraction in the Stroebe *et al.* study probably was not interval. Thus, while 7 indicates more attraction than 6, and 6 more attraction than 5, the increase in attraction from 5 to 6 probably is not equal to the increase in attraction from 6 to 7.

With an interval scale it is arbitrary which subset is assigned the score of zero. Stated differently, interval scales do not have absolute zeros. Zero degrees Fahrenheit, for example, is not as cold as it can get. Because a Fahrenheit temperature scale does not have an absolute zero, a temperature of 80 degrees Fahrenheit is not twice as hot as a temperature of 40 degrees. The ratio 80 / 40 does not express a true relationship. It is true, however, that the temperature difference between zero and 40 is equal to the difference between 40 and 80. The differences between numbers express true relationships even though the ratio of any two numbers does not.

Because the zero point of an interval scale is arbitrary, a constant may be added to or subtracted from all of the numbers without losing any valid information. Thus, for example, 32 could be subtracted from all of the numbers on the Fahrenheit temperature scale without losing any valid information.

Ratio Scale

A ratio scale is a scale that has all of the characteristics of an interval scale and, in addition, has a true zero point. Common examples of ratio scales are measurements of length in units of inches or feet, and measurements of weight in units of ounces or pounds. With a ratio scale the ratio between any two numbers on the scale does express a true relationship. Thus, it is true that a weight of 80 lbs. is twice as heavy as a weight of 40 lbs.

Each one of the last three scales is built upon the preceding. The ratio scale has all of the characteristics of an interval scale plus an absolute zero. An interval scale has all of the characteristics of an ordinal scale plus equality of the intervals. An ordinal scale has all of the characteristics of a nominal scale plus knowledge of the rank order relationship.

VARIABLES

In the psychological literature the term *variable* is typically used as a synonym for property. Thus, investigators may refer to attraction, for example, as a variable. In statistics, however, variable has a different meaning. *A variable is a symbol that can represent (stand in for) any of the values that result from measurement of a property.* The term variable is identified with the measurements of a property rather than the property itself. We will use this latter definition. Variables are usually identified by letters, such as X or Y. The different values of a variable may represent qualitative or quantitative measurements. Usually, however, a variable represents numerical values signifying quantitative measurements. Thus, in the equation,

$$Y = a + bX$$

the X and Y variables symbolize numerical values.

It will be helpful to distinguish between two types of quantitative variables, *discrete* and *continuous*. *A discrete variable represents isolated numerical values.* The most common examples of discrete variables are those that arise from enumeration. For example, we could measure the number of bar presses performed by a rat during a certain period of time. The variable representing "number of presses" is discrete. Only isolated numerical values can occur, e.g., 1, 2, 3, 4, and 5. On the other hand, most of the variables that concern psychologists are theoretically *continuous* in nature. Variables symbolizing height, conformity, anxiety, and intelligence are theoretically continuous. With a theoretically continuous variable, *there are no gaps, or values that cannot occur*, as there are with discrete variables. If the numbers 1, 2, 3, and 4 represent measures of a continuous variable, the infinite number of values between 1 and 2, 2 and 3, etc., then represent possible values of the variable.

In spite of the fact that most variables of interest are theoretically continuous, in practice the process of measurement always results in discrete variables. In the first place, no matter how refined the measurement process, an interval of numbers always is possible between any two numbers used as measurement subsets. The measurement process is always stopped after reaching a certain degree of precision, thus producing gaps in the measurement scale. Even when measurement procedures are highly refined and fine discriminations between different values of a property can be made, there are always gaps, or numbers not used as scores. For example, suppose a measurement procedure yields scores of 4.2953, 4.2954, 4.2955, 4.2956, etc. If these numbers represent quantities of a property, the researcher can, indeed, make very fine discriminations between values. Yet, these numbers are discrete. A gap exists between each pair of numbers and each gap contains an infinite set of possible values. Thus, a variable that represents numerical values resulting from actual measurements always will be discrete.

Further, even if it were possible to measure continuously, the actual populations from which samples are drawn do not contain instances of the infinite number of possible values of the variable in question. As an example, suppose conformity is measured using a procedure that results in scores between 1 and 7. No matter what population is measured, there will not be a conformity behavior having each of the infinite number of possible values between 1 and 7.

SUBFIELDS OF STATISTICS

Statistics has two subfields, *descriptive statistics* and *inferential statistics*. *Descriptive statistics is concerned with the description of research data.* Data or scores may be described by presenting them in a graph, by computing their average value, by giving an indication of how spread out the scores are, and so on. Chapters 2, 3, 4, and 5 will be devoted to discussion of the various ways in which obtained data or scores may be described.

Inferential statistics is concerned with inferring the properties of a population from the properties of a sample. Suppose it is necessary to know the average firing distance of the bullets produced by a given factory in a given period of time. The population would consist of all the bullets produced by this factory in that period of time. We could answer the question about average distance by firing the bullets and measuring the distance traveled by each bullet, and then computing the average of these distances. The question also could be answered by selecting at random a sample of bullets from the total population, and using the average distance traveled by bullets in the sample to make an inference about the average distance for the total population. The theory that enables such inferences defines the subfield of inferential statistics.

STATISTIC, STATISTICS, AND STATISTICS

Finally, the meanings of the terms statistic and statistics should be clarified. A *statistic is a summary measure of a set of scores.* For example, the average of a set of scores is a statistic. If a number of averages are computed on different sets of scores, it is appropriate to refer to these averages as statistics (using the plural).

At other times, the term statistics refers to the entire field or discipline that is the subject of this book. Thus, we say that the text is to be used as an introduction to statistics. The particular meaning implied is usually apparent from the context.

HISTORICAL PERSPECTIVE ON ARABIC NUMERALS

In this and most of the succeeding chapters, brief historical sketches are provided of the backgrounds of some of the described ideas, concepts, and procedures. These sections, like those describing more difficult mathematical treatments, are in small print, and thus may be omitted without jeopardizing understanding of later material. The intent of the small print sections containing mathematical derivations is to give the more mathematically inclined student a deeper understanding of statistics. Likewise, the intent of the small print sections containing historical material is to give the more scholarly, social-science inclined student an opportunity for deeper understanding of statistics. In this latter instance the "understanding" involves a knowledge of, and hopefully appreciation for, the rich cultural developments to which he or she has fallen heir.

The discipline of statistics is, of course, directly concerned with numbers or numerals. It is interesting to note that number systems based on ten are universal among literate cultures and very common among nonliterate cultures. It has been suggested plausibly that the reason for this lies in the fact that we have ten fingers and ten toes. The ancient Persians and Greeks had words for five that meant "hand;" and it is possible that the Roman symbol, V for five, represents the "vee" between the thumb and forefinger.

The Roman system of counting still exists as a part of our culture. The tables in Appendix C are, for example, numbered with Roman numerals. In addition to V for 5, the Romans used X for 10, L for 50, C for 100, D for 500, and M for 1,000. For very large numbers the Roman system is exceedingly

TABLE 1.2. *Multiplication of 27 "Times" 16 with Roman Numerals.*

XXVII (27)		
XVI (16)		
X "times" XXVII:		
X "times" X = XXXXXXXXXX or C		(100)
X "times" X = XXXXXXXXXX or C		(100)
X "times" V = VVVVVVVVVV or L		(50)
X "times" I = IIIIIIIII or X		(10)
X "times" I = IIIIIIIII or X		(10)
V "times" XXVII:		
V "times" X = XXXXX or L		(50)
V "times" X = XXXXX or L		(50)
V "times" X = VVVVV or XXV		(25)
V "times" I = IIII or V		(5)
V "times" I = IIII or V		(5)
I "times" XXVII:		
I "times" X = X or X		(10)
I "times" X = X or X		(10)
I "times" V = V or V		(5)
I "times" I = I or I		(1)
I "times" I = I or I		(1)
CC LLL XXXXXX VVVV II		(432)
or		
CDXXXII = 432		

awkward and difficult. For instance, in Roman numerals the mileage to the moon is written CCXXXMMMMMMMMDCCCLVII, instead of 238,857. The difficulty of performing a rather simple multiplication in Roman numerals (27 "times" 16) is illustrated in Table 1.2. After puzzling through this example, one should readily appreciate why such a cumbersome number system would impede cultural development in mathematics, and thus, in science and technology.

The Roman number system lacked a symbol for zero (0) and the concept of "place." In order to make use of the device of place, a symbol for 0 is, of course, required. Such a system allows for the re-use of the same symbols at different "places" and obviates the necessity of inventing new symbols for higher and higher numbers.

About A.D. 500, the Hindus started using a decimal notation in which each number's position showed what power of 10 it represented, while empty positions were designated by 0's. Thus, for example, three places to the left of the decimal became the 10^2 (or hundredths) position, two places to the left of the decimal, the 10^1 (or tens) position, and one place to the left of the decimal, the 10^0 (or units) position. Such a cultural development is as significant as the invention of the wheel. The person responsible for this remarkable development is, unfortunately, unknown.

The Hindu number system spread to the Moslem world, which, during the Middle Ages, was more culturally advanced than the Western world. Around A.D. 1100, the "arabic" numeral system reached the capitals of Christendom. Some of the diffusion came as a result of the crusades, but the most important connection was through trade. By caravan and ship the Arab traders brought silk, rugs, and spices to Europe —principally to the Italian city-states. The Italian merchants undoubtedly were more interested in the traders' goods than in their number system. In retrospect it is obvious that the number system was far more valuable than the silks and spices.

It was nearly four centuries later (around A.D. 1500) before arabic numerals were commonly used in Europe. It was claimed that the new numbers were a heathen invention and should not be used in Christian countries. Arabic numerals were, for example, banned by law in the city of Florence. Gradually, however, such ethnocentricism was overcome and arabic numerals were accepted throughout all the "civilized" or literate world.

EXERCISES

1. Define the following terms:
 (a) statistics (as a discipline)
 (b) nominal scale
 (c) ordinal scale
 (d) interval scale
 (e) ratio scale
 (f) property
 (g) variable
 (h) discrete variable
 (i) continuous variable
 (j) descriptive statistics
 (k) inferential statistics
 (l) statistic

2. Discuss the process of measurement using set theory concepts.

3. Some individuals have defined measurement so that the assignment of numbers is always involved. According to this perspective, did Mintz and Alpert measure vividness of imagery?

4. Design a procedure for measuring vividness of visual imagery that parallels the Mintz and Alpert procedure for measuring vividness of auditory imagery. Does this suggest a possible research project? Describe.

5. What does it mean to say that a variable is theoretically continuous?

6. What does it mean to say that a variable is discrete?

7. Indicate which of the following variables are discrete and which are continuous:
(a) height
(b) measurement of height
(c) weight

(d) measurement of weight
(e) number of students in a class
(f) measurement of number of students in a class
(g) number of trials required to learn a list of words
(h) measurement of number of trials required to learn a list of words
(i) annual income of professors at a university
(j) measurement of annual income of professors at a university

8. Why are all measurements discrete?

9. What are two possible meanings of the term "variable?"

10. Why is it necessary for a student of psychology to learn statistics?

11. Select a recent issue of any psychological or social-psychological journal: *Sociometry*, *Journal of Experimental Psychology*, or *Journal of Personality and Social Psychology*. Examine the titles of the various articles, and select a research report that looks interesting.
(a) What research questions are stated or implied in the introductory part of the article?
(b) What provisional conclusions to the research questions are stated in the "Discussion" section of the article?
(c) List by name the statistical techniques mentioned in the "Results" section.

12. Give two examples of each of the following scales: nominal, ordinal, interval, and ratio.

13. What is the role of numbers on a nominal scale?

14. What sense can be made of the term "ratio" when it is used to identify a scale?

15. John received a score of 100 on his arithmetic test. Carl received a score of 50. Did John perform twice as well as Carl? Why?

16. John received $5.00 for mowing a lawn. Carl received $2.50 for mowing a different lawn. Did John receive twice as much as Carl? Why?

17. John has a Stanford-Binet IQ of 150, and Carl has a Stanford-Binet IQ of 75. Is John twice as smart as Carl? Why?

18. Would the rank order of a set of scores be changed by any of the following operations:
(a) adding 5 to each score,
(b) subtracting 5 from each score,
(c) multiplying 5 times each score,
(d) dividing 5 into each score,
(e) or dividing each score into 1 (taking the reciprocal of each score)?

19. Does it make sense to talk about the absolute zero of imagery vividness, attraction, and IQ?

20. What kind of measurement is involved in the classification of individuals as non-smokers, light smokers, and heavy smokers? Is it possible to have an ordinal scale with an absolute zero? If so, what does this indicate about the classification scheme given in the text?

Prologue to Chapter 2

Chapter 2 begins the discussion of descriptive statistics by describing the procedures and conventions for ordering scores into what are called frequency distributions. The various types of frequency distributions are devices that enable someone to get a grasp or feel for a set of scores. The chapter also contains a discussion of the procedures and conventions for the construction of various types of graphs. Graphs have the unique advantage of providing a picture of a set of scores.

In view of the fact that topics treated in this chapter are relatively simple, students and statisticians alike have a tendency to underestimate their importance. We, however, should never lose sight of the advantage that is to be gained through the simple examination of frequency distributions and graphs.

2

Frequency Distributions

Consider a professor who wants to known more about the 64 students in her class. She has had difficulty explaining her ideas to them. Some students "catch-on" quickly, others she is "unable to reach." After pondering the situation, she asks herself, "Is there a wide range of intelligence in this class?" If there is, it might in part account for the communication difficulty.

Descriptive statistics helps answer questions of this nature. Such research questions do not ask for generalizations from certain persons to people in general, or to any people beyond the ones directly under observation. The concern is with the people, or organisms, that are measured. In the above example, the teacher measures her class by having each student complete an intelligence test. She then lists the resulting scores in rows according to the alphabetical order of the students' surnames.

```
 85,  81, 104,  86,  74,  80, 105,  75,  74,  89,  84,  79,  99,
101,  72,  77,  76,  80,  75,  73,  68,  95,  92,  63,  68,  98,
 84,  83,  81, 101,  81,  92,  61, 101,  70,  89,  91,  76,  82,
 74,  85,  61,  79,  84,  81,  81,  83,  68,  80,  64,  83,  75,
 87,  92,  70,  87,  79,  90,  93,  83,  97,  79,  69,  60.
```

When presented in this order, the scores do not appear very meaningful. It is very difficult to gain an intuitive feel for the distribution of scores.

A set of scores can be made more meaningful if the numbers are arranged in numerical order. For the above series we may start with 105, 104, 101, 101, 101,..., and end with 63, 61, 61, 60. For some purposes, such an ordering is sufficient. However, space can be saved and comprehension increased if the ordering process is carried one step further. All the possible, different values are listed in one column and the frequency with which each of these values occurs is

TABLE 2.1. Frequency Distribution.

X	f(X)	X	f(X)
105	1	82	1
104	1	81	5
103		80	3
102		79	4
101	3	78	
100		77	1
99	1	76	2
98	1	75	3
97	1	74	3
96		73	1
95	1	72	1
94		71	
93	1	70	2
92	3	69	1
91	1	68	3
90	1	67	
89	2	66	
88		65	
87	2	64	1
86	1	63	1
85	2	62	
84	3	61	2
83	4	60	1
			64

Ungrouped

listed in an adjacent column. The resultant set of pairings of numerical values with their respective frequencies of occurrence is a *frequency distribution*.

A frequency distribution constructed from the above scores is presented in Table 2.1. The symbol X is a variable that can represent any of the IQ scores. In Table 2.1, the symbols $f(X)$, read frequency of X, represent the frequency with which any X score occurs. The sum of the frequencies (in this case 64) is equal to the number of persons measured.

A frequency distribution so organizes scores as to reveal certain information. The highest and lowest scores, 105 and 60, are now apparent. The range of scores, highest minus lowest, can be quickly calculated ($105 - 60 = 45$). How many students received each score can be seen easily, and an intuitive grasp of the distribution of scores is facilitated. After examining the frequency of the distribution, the professor might decide to divide her students into sections of relatively homogeneous intelligence.

UNIVARIATE FREQUENCY DISTRIBUTIONS

Frequency distributions are of different types. Table 2.1 is a univariate frequency distribution. *Univariate* means *one variable*. *A univariate frequency distribution presents the values of one variable, each value paired with its frequency of occurrence.*

Univariate distributions are to be distinguished from bivariate distributions. *Bivariate* means *two variable*. If the 64 students in the above example were measured not only for intelligence but also for manual dexterity, each student would have a pair of scores. A bivariate frequency distribution presents frequencies associated with pairs of values. Univariate distributions will be discussed first, saving the discussion of bivariate distributions for a later section of this chapter.

Ungrouped Frequency Distributions

In Table 2.1, the values of variable X are *ungrouped*. They are the values assigned in the measurement process. There was no grouping of values into larger sets subsequent to assignment of the original IQ values. *An ungrouped frequency distribution is a function where frequencies are paired with the actual values that result from measurement.* This is not to imply, however, that all the observations assigned the same IQ value are exactly equal in the amount of intelligence. Thus, values representing amounts of a property are considered midpoints of intervals of possible values. In the ungrouped frequency distribution presented in Table 2.1, each value is the midpoint of an interval with the width of 1. For example, the score 105 is used to denote all intelligences between 104.5 and 105.5.

Grouped Frequency Distributions

Scores can be grouped into larger subsets. Table 2.2 contains a grouped frequency distribution for the ungrouped IQ scores in Table 2.1. In Table 2.2,

TABLE 2.2. *Grouped Frequency Distribution.*

CLASS INTERVALS	$f(X)$
105–109	1
100–104	4
95– 99	4
90– 94	6
85– 89	7
80– 84	16
75– 79	10
70– 74	7
65– 69	4
60– 64	5
	64

it is easier to perceive where in the total 45 score range the greatest number of scores occur. In this instance, grouping the scores into subsets facilitates perception or comprehension of how the scores are distributed across the total range.

Grouping frequently is a perceptual aid—particularly if the total number of scores is small in relation to the number of possible values, and many values occur with low frequency. By grouping scores close to each other, the areas of the measurement scale having a high or low frequency of occurrence become more visible. Such heightened visibility allows an investigator to answer more readily such questions as the following: Are there many students at the extremes of the distribution or only a few? Do most students fall in the middle or toward the upper end of the range?

PROCEDURE. The following instructions represent a typical set of conventions for constructing a grouped frequency distribution:

1. Round off the scores to the nearest whole number. Relative to the loss of information through grouping, the loss as a result of rounding error is inconsequential. The rounding procedure is as follows. The scores 1.1, 1.2, 1.3, and 1.4 are rounded to 1. The scores 1.6, 1.7, 1.8, and 1.9, are rounded to 2. The score 1.5 is rounded in the direction of an even number. Thus, 1.5 is rounded to 2. The score 2.5 would be rounded to 2.

2. Decide upon the number of desired classes; that is, make a decision about the number of classes into which the total range of measurement classes will be grouped. With too few classes the appearance of the data will be grossly distorted; with too many classes, the grouping will not aid comprehension. Frequently, between ten and twenty classes is optimum. However, in particular instances, fewer than ten or more than twenty may be required to get a "good picture" of the data.

3. Obtain the range of scores by subtracting the lowest score from the highest.

4. Divide the range by the number of desired classes. The resultant dividend, rounded off to the nearest whole number, will give the class interval width necessary to obtain *approximately* the number of desired classes.

5. Start the lowest class interval at or below the lowest score obtained, and, if possible, at a value that is a multiple of the class interval width. For example, if the width of the class interval is 5, the lowest class interval could be 5–9 or 60–64. Both 5 and 60 are multiples of 5. The lowest class interval is then chosen so that it includes the lowest score obtained. On the other hand, if in the above example the distribution of scores had included numbers between 0 and 4, it would not have been possible to start the lowest class interval at a value that is a multiple of 5. In this case the lowest class interval should begin with zero.

6. Starting with the lowest class at the bottom, arrange the intervals in order of increasing magnitude.

7. Indicate in an adjacent column the frequency with which scores fall within each class interval.

As previously indicated, the grouped frequency distribution presented in Table 2.2 was constructed using the scores contained in the ungrouped frequency distribution in Table 2.1. The procedure used to construct this grouped distribution follows directly the conventions described above.

1. The scores were whole numbers, so no rounding was necessary.
2. It was decided to have ten classes.
3. The range was obtained by subtracting the lowest score from the highest, $105 - 60 = 45$.
4. The range was divided by the number of desired classes, $45 \div 10 = 4.5$. The resultant dividend, 4.5, was rounded off to a whole number, 5.0, which gave the class interval width necessary to obtain the number of desired classes. Rounding to 4 would give more than 10 classes. (Note that the class interval width is 5 and not 4. The interval 60–64 contains 5 scores, 60, 61, 62, 63, and 64.)
5. The lowest class interval was started at a number, 60, which is a multiple of the class interval width, 5.
6. The class intervals were arranged in order of increasing magnitude.
7. The frequency of scores in each class interval was indicated in an adjacent column.

TERMINOLOGY. Certain technical terms are used to refer to various aspects of frequency distributions. The scores on either side of a class interval, for example 60 and 64, are referred to as the *apparent class limits*. In addition to the apparent class limits, however, are the *real class limits*. The real class limits, 59.5 and 64.5 for the above example, define the actual width of the class interval. A real class limit is always the point halfway between upper and lower apparent class limits for adjacent intervals. The difference between the upper and lower real limits is referred to as the *class interval width* and is symbolized by i. The width of the example class interval is 5 ($64.5 - 59.5 = 5$).

Finally, the *midpoint of the class interval* is the point half the distance between the upper and lower real limits. The midpoint of a given class interval can be obtained by averaging either the real class limits ($64.5 + 59.5 = 124$, $124 \div 2 = 62$) or the apparent class limits ($64 + 60 = 124$, $124 \div 2 = 62$). The apparent class limits, real class limits, and midpoints for the above grouped frequency distribution are presented in Table 2.3. All of the above concepts, except apparent class limits, apply to ungrouped as well as grouped frequency distributions. As noted above, in ungrouped frequency distributions the actual measurement values are regarded as midpoints of intervals of possible values. In the ungrouped frequency distribution contained in Table 2.1, the value 105, for example, is the midpoint of the interval that extends from 104.5 to 105.5. These latter values are, of course, the real class limits of the interval that contains 105 as its midpoint.

TABLE 2.3. *Grouped Frequency Distribution, Its Real Class Limits and Midpoints.*

APPARENT CLASS LIMITS	REAL CLASS LIMITS	MIDPOINTS	$f(X)$
105–109	104.5–109.5	107	1
100–104	99.5–104.5	102	4
95– 99	94.5– 99.5	97	4
90– 94	89.5– 94.5	92	6
85– 89	84.5– 89.5	87	7
80– 84	79.5– 84.5	82	16
75– 79	74.5– 79.5	77	10
70– 74	69.5– 74.5	72	7
65– 69	64.5– 69.5	67	4
60– 64	59.5– 64.5	62	5
			64

Graphing Univariate Frequency Distributions

Graphs of frequency distributions are pictorial descriptions using coordinate axes to represent the possible values and frequencies. By picturing frequency distributions, graphs aid in the description of frequency distributions. Three types of graphs will be discussed: *histograms, frequency polygons,* and *cumulative proportion graphs.*

HISTOGRAM. Figure 2–1 is a graph of the grouped frequency distribution in Table 2.2. It is a particular kind of graph—a *histogram. The histogram, also referred to as a bar graph, is a graph in which heights of bars are used to represent the frequencies with which scores fall into class intervals.*

A histogram is typically used to picture a grouped, as opposed to an ungrouped, frequency distribution. A grouped frequency distribution and a histogram summarize the same information. The key difference is that the histogram represents numerical frequencies with bars of differing heights. The relative height of the bars in Fig. 2–1 allows for a quick impression of the relative frequencies across the range of the distribution. It takes somewhat more careful examination of the histogram, however, to determine the exact frequency for each interval—information that is immediately available in a tabled frequency distribution (such as Table 2.2). Histograms are constructed so that the vertical lines forming the sides of each bar intersect the upper and lower real limits of the class intervals represented on the horizontal axis of the graph. The horizontal lines forming the tops of the bars are directly over the midpoints of the intervals and level with the frequencies of occurrence represented on the vertical axis. Typically, only the midpoints are indicated along the horizontal axis of the

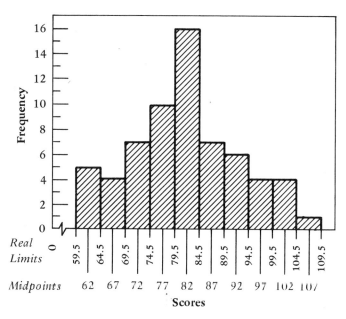

FIGURE 2-1. *Histogram.*

graph. In Fig. 2-1, the upper and lower real limits have also been indicated in order to make clear the manner in which histograms are constructed. A small break has been made in the horizontal axis as a visual reminder that the distance between 0 and 59.5 is not consistent with the metric (or scale) for the remainder of the axis.

Since ungrouped frequency distributions, like grouped frequency distributions, have midpoints, real limits, and frequencies, it would be possible to construct a histogram to represent an ungrouped frequency distribution. This is usually not done, however, since the individual bars would typically be exceedingly narrow and numerous.

FREQUENCY POLYGON. Figure 2-2 displays two different types of graphs—the histogram pictured in Fig. 2-1 and a frequency polygon for the same data. The histogram conveys a stronger impression of the relative differences in frequency among class intervals. On the other hand, the frequency polygon, because of its straight-line connections, conveys a stronger impression of the overall shape of the distribution. In both cases numerical information has been converted into pictorial shapes to give a visual–intuitive comprehension of the distribution. Exact frequencies, however, are somewhat more difficult to discern than in a grouped frequency distribution (Table 2.2).

Univariate Frequency Distributions

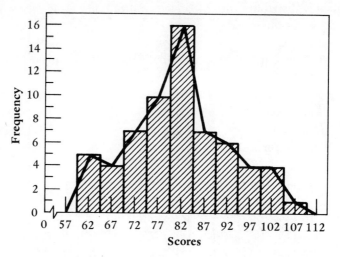

FIGURE 2-2. *Frequency polygon superimposed on a histogram.*

A frequency polygon is a graph in which points connected by lines represent the frequencies with which classes of scores occur. A frequency polygon is constructed by placing a dot mark directly above the midpoint of each class interval (represented on the horizontal axis) at a height that is level with the frequency (represented on the vertical axis) associated with that class interval. The dots are connected by a line, giving the graph the appearance of a polygon. A frequency polygon portraying the grouped frequency distribution contained in Table 2.2 is presented in Fig. 2-2. For purposes of illustration, this frequency polygon has been superimposed on the histogram constructed from the same scores. Note that the line defining the polygon begins with the midpoint of the lowest class interval having a frequency of zero and ends with the midpoint of the highest class interval having a frequency of zero.

As Fig. 2-2 implies, it is often arbitrary whether frequency distributions are pictured with frequency polygons or with histograms. However, in certain situations a histogram is typically preferable; in other situations, a frequency polygon is typically preferable. If the variable, or the property it symbolizes, has been measured with a nominal scale, a histogram rather than a frequency polygon is typically preferable. Suppose that someone counts the bushels of wheat, bushels of corn, and bushels of potatoes that are grown in a given year in a given country. If this information is to be pictured graphically, a histogram should be used. Because of its discontinuities, a histogram offers less of an impression of an increasing or decreasing continuum. In the case of a nominal scale, such an impression is, of course, incorrect.

A frequency polygon, rather than a histogram, typically should be used to portray frequency distributions with a large number (15 to 30, or more) of class intervals. A histogram representation of such frequency distributions would result in individual bars sufficiently narrow and numerous to blur the perceived discontinuities. Thus, it is somewhat simpler to use a frequency polygon.

What should be done with a nominal scale variable that has a large number of class intervals? In this situation, perhaps it is preferable to use a histogram—although there is no fixed rule. It might even be that a graph should not be used, and the data simply left in a frequency distribution. In this situation, as in many, there is no substitute for individual judgment.

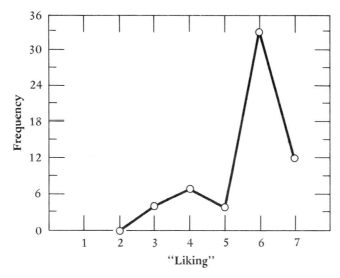

FIGURE 2-3. Frequency polygon of an ungrouped frequency distribution—liking in the Stroebe et al. high-similarity condition.

Figure 2-3 illustrates a frequency polygon constructed from an ungrouped frequency distribution. The data are from the high-similarity condition of the previously described Stroebe *et al.* experiment. Recall that in this experiment the subjects examined pictures and questionnaire responses of some other person, and then made judgments about that person. In the high-similarity condition, the questionnaire responses indicated that the other person had attitudes toward such things as black power and religion that were highly similar to the subjects'. The experimenter, in fact, individually so tailored the questionnaire for each subject as to produce such similarity. One of the judgments was an indication of how much the subject would like the other person. Responses were made by selecting one of seven alternatives ranging from "I feel that I would probably like this person very much" to "I feel that I would probably dislike this person very much." All seven alternatives are reproduced in Table 1.1. Numbers were assigned to the alternatives so that the higher the score, the greater the probable liking. As Fig. 2–3 indicates, probable liking in this high-similarity condition was toward the upper end of the scale. Over half of the subjects checked alternative 6.

Notice in Fig. 2–3 that the frequency polygon has been drawn so that it does not end with a zero frequency; the upper tail of the polygon does not touch the horizontal axis, as is the case in Fig. 2–2. Why not? In order for the polygon to end with a frequency of zero, the scale would have to extend beyond 7; and 7 is the end of the scale, the highest level of measured probable liking. Thus, dropping the line down to 8 would be a misrepresentation.

It is, of course, also possible to have a non-zero frequency in the *lowest* scale value. In this case the polygon will not begin with the line touching the horizontal axis. Whenever the lowest scale value has a non-zero frequency, the polygon will not begin on the horizontal axis; and whenever the highest scale value has a non-zero frequency, the polygon will not end on the horizontal axis. This is true for frequency polygons constructed from both grouped and ungrouped frequency distributions.

Figure 2–4 is an example of a frequency polygon whose lowest and highest scale values have non-zero frequencies. The lowest scale value, 1, has a frequency of 6; and highest scale value, 7, has a frequency of 1. The graph in Fig. 2–4 is of the liking scores for the Stroebe *et al.* low-similarity condition. In the low-similarity condition the questionnaire supposedly filled out by another person had, in fact, been individually tailored by the experimenter so as to contain belief and attitude responses highly dissimilar to that of each subject. Note the interesting differences between Fig. 2–4 for the low-similarity condition and Fig. 2–3 for the high-similarity condition.

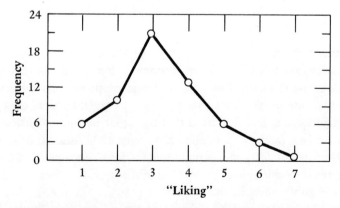

FIGURE 2–4. *Frequency polygon of an ungrouped frequency distribution—liking in the Stroebe et al. low-similarity condition.*

CUMULATIVE PROPORTION GRAPH. A cumulative proportion graph, constructed from the IQ scores we have been discussing, is illustrated in Fig. 2–5. Such a graph pictures the proportion of all the scores that are below any given value. Cumulative proportion graphs tend to have an "S" shape if the distribution of frequencies is greater in the middle of the range than at either end.

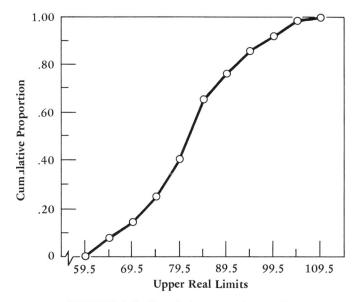

FIGURE 2–5. *Cumulative proportion graph.*

A cumulative proportion graph portrays the proportion of scores falling below the upper real limits of each class interval. Cumulative proportion graphs are used, in preference to frequency polygons and histograms, when the interest is in determining the relative position of a score in the distribution rather than the shape or form of the distribution.

In constructing a cumulative proportion graph, it first is necessary to compute the proportion of scores and cumulative proportion of scores for each class interval. These values for the grouped frequency distribution in Table 2.3 are contained in Table 2.4. In the $f(X)/n$ or proportion column is shown the proportion of scores falling within each class interval. These values are obtained either by dividing the total frequency (n), 64, into the frequency $f(X)$ for each class, or by multiplying 1/64th (.01562) "times" the frequency for each class. The values in the cp or cumulative proportion column are obtained by successively adding the separate proportions. Starting at the bottom with the proportion .07812, it is necessary to add the next proportion, .06250, to obtain .14062, and so

Univariate Frequency Distributions

TABLE 2.4. *Proportions and Cumulative Proportions for Grouped Frequency Distribution.*

REAL CLASS LIMITS	$f(X)$	$f(X)/n$	cp
104.5–109.5	1	.01562	1.00000
99.5–104.5	4	.06250	.98438
94.5– 99.5	4	.06250	.92188
89.5– 94.5	6	.09375	.85938
84.5– 89.5	7	.10938	.76563
79.5– 84.5	16	.25000	.65625
74.5– 79.5	10	.15625	.40625
69.5– 74.5	7	.10938	.25000
64.5– 69.5	4	.06250	.14062
59.5– 64.5	5	.07812	.07812
	64		

on. The highest cumulative proportion will always be within rounding error of 1.0.

In the cumulative proportion graph, the obtained cumulative proportions are associated with upper real limits of the class intervals, as illustrated in Fig. 2–5. Upper real limits rather than midpoints are represented along the horizontal axis of the graph because the cumulative proportions are accumulated up to and through each class interval. With the aid of Fig. 2–5, it is possible to obtain quickly the proportion of scores falling above and below any particular class interval.

CONSTRUCTION OF GRAPHS. Some general conventions regarding the construction of graphs follow:

1. Graphs, unlike tables, have their titles at the bottom.
2. Graphs of frequency distributions always represent values of a variable along the horizontal axis and frequencies along the vertical axis.
3. The lowest values are placed at the left end of the horizontal axis, and the smallest numbers (representing frequencies) are placed at the bottom of the vertical axis.
4. The zero points on both the vertical and horizontal axis are represented at the intersection of the two axes. If the first value is considerably above zero, a small break can be made in the axis between the zero point and the first value. Such a break is illustrated in Figs. 2–1, 2–2, and 2–5.
5. The size of the units along the two axes should be selected so that the ratio of the vertical axis length to the horizontal axis length is roughly 3 to 5. This particular ratio is, of course, arbitrary, but seems to have some aesthetic advantage.

Characteristics of Univariate Frequency Distributions

Descriptive statistics is concerned with the description of sets of scores. An initial step in describing a set of scores is the construction of a frequency distribution. The next two chapters are going to be concerned with univariate frequency distributions, and, in particular, with procedures for describing some of their characteristics. Four of these characteristics are: *central tendency*, *variability*, *skewness*, and *kurtosis*.

The central tendency of a frequency distribution refers to a single value representing the magnitude of scores in the distribution. Figure 2-6 graphically

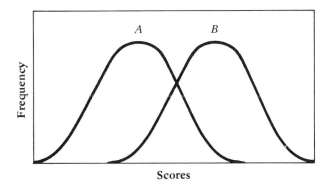

FIGURE 2-6. *Two frequency distributions differing in central tendency.*

presents two frequency distributions, A and B, which differ in central tendency. Distribution B has the greater central tendency. Various indices of central tendency, such as the arithmetic mean, will be discussed in Chapter 3.

The concept of central tendency is most meaningful when applied to frequency distributions like those in Fig. 2-6; i.e., to frequency distributions in which the values toward the center of the range occur more frequently than those toward either end of the distribution. The scores in distribution A appear to group around a central value, and the scores in distribution B appear to group around a central value. The less the tendency is for this to occur, the less meaningful is the concept of central tendency.

A measure of central tendency provides one way of comparing two frequency distributions. Suppose that the scores in Fig. 2-6 are for weight in pounds; distribution A is of smokers' weights and distribution B is of non-smokers' weights. A measure of central tendency would give a single value for smokers and a single value for non-smokers, indicating that non-smokers tend to weigh more.

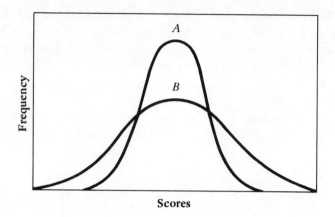

FIGURE 2–7. *Two frequency distributions differing in variability.*

Two frequency distributions that differ in *variability* are graphically presented in Fig. 2–7. *The variability of a frequency distribution refers to the extent to which the scores differ among themselves.* In Fig. 2–7, distribution B has greater variability than does distribution A. The scores in distribution B are more spread out; in other words, more scores appear at the extremes of the distribution.

Following an earlier example, suppose a professor measured the verbal IQ of students in two different classes. Suppose further that the results are represented in Fig. 2–7. Distribution A represents a section the professor found easy to teach, and distribution B shows a section found difficult to teach. The "difficult" section has greater variability in verbal intelligence. Various measures of variability will be discussed in Chapter 4.

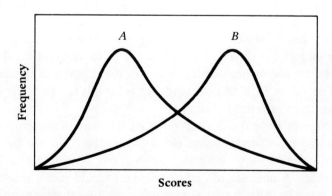

FIGURE 2–8. *Two frequency distributions differing in skewness.*

Figure 2–8 pictures two distributions differing in skewness. Distribution *A* with its "tail" to the right is said to be *positively* skewed, and distribution *B* with its "tail" to the left is said to be *negatively* skewed. *Skewness* refers to a distribution's *departure from symmetry*. Two frequency distributions differing in kurtosis are presented in Fig. 2–9. *Kurtosis refers to the flatness or peakedness*

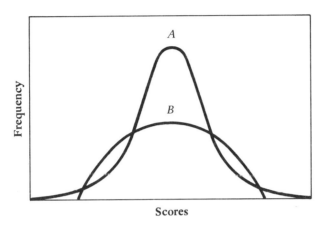

FIGURE 2–9. *Two frequency distributions differing in kurtosis.*

of a distribution. A markedly peaked distribution, such as distribution *A*, is referred to as *leptokurtic*; and a markedly flat distribution, such as distribution *B*, is referred to as *platykurtic.* A distribution that is neither markedly peaked nor flat is referred to as *mesokurtic.*

A complete discussion of the description of univariate frequency distributions would involve consideration of all four of the above characteristics. Psychologists have found less use for the indices of skewness and kurtosis than for the indices of central tendency and variability. For that reason, treatment of the description of frequency distributions will be limited to a discussion of central tendency and variability.

BIVARIATE FREQUENCY DISTRIBUTIONS

Bivariate frequency distributions describe the frequency of occurrence of measurement classes (grouped or ungrouped) defined with reference to two different properties. *A bivariate frequency distribution presents frequencies associated with pairs of values where each pair represents a unique measurement class (subset) defined by the combination of a value of one variable with a value of another variable.*

TABLE 2.5. *Bivariate Frequency Distribution Relating Grades on a History Exam to Grades on an English Exam.*

		HISTORY					
		F	D	C	B	A	f(X)
	A	0	0	0	1	2	3
	B	0	1	1	1	1	4
ENGLISH	C	0	2	3	1	0	6
	D	0	0	2	0	0	2
	F	1	0	0	0	0	1
	f(X)	1	3	6	3	3	

As an example, consider a class of sixteen students that took first an English and then a history exam. Each student was measured twice, once with reference to English performance and a second time with reference to history. The resultant bivariate frequency distribution is presented in tabular form in Table 2.5. The left vertical margin of the table indicates the various possible English scores while the upper horizontal margin indicates the various possible history scores. In a similar manner, the right vertical margin displays the frequency distribution of English scores; the lower horizontal margin displays the frequency of distribution of history scores. These two *marginal distributions* are, by themselves, univariate frequency distributions. The bivariate frequency distribution is described in the cells of the table, where the frequency of occurrence of each possible pair of values is presented. Examination of this bivariate distribution reveals, for example, that two of the three persons who received an *A* on the English exam also received an *A* on the history exam.

Although three dimensions are required, a bivariate frequency distribution of this kind can also be described in graphic form. In this instance the graph takes the form of a cube. The vertical and horizontal axes of the graph are used to represent values of the two variables as in Table 2.5. A third axis in a third (near–far) dimension is used to represent the various frequencies.

The set of all possible pairs of values represented by the cells in Table 2.5 is a *product set*, and the subset of pairs that actually occurred is a *relation*. Knowing an individual's score on one variable provides some information about his score on the other variable. As Table 2.5 illustrates, however, the relation between the two variables (history exam scores and English exam scores) suggests more than simply the occurrence or non-occurrence of pairs of values. The frequency with which pairs of history and English scores occur changes for different English scores (rows) and different history scores (columns). Thus, knowledge of a score on one variable (e.g., history) allows one to specify the differential frequency with which the various values occur on the other variable. Not only do certain pairs of values occur and others not occur, but certain pairs of values occur more frequently than others. In Table 2.5, for a history score of C, the most frequently associated English score is C. Given a history score of C, our best guess as to the paired English score is also C. Also, if the data are representative of future behavior, we can say that the best prediction of a future English score, given a history score of C, is C. It will be the most frequent outcome and is, therefore, the most likely to occur.

One method for describing and summarizing the relation between two variables in a set of data is called linear *correlation*. Careful examination of Table 2.5 will reveal a general tendency for the observations to scatter from the lower left-hand corner to the upper right-hand corner. This reflects the fact that students who made high grades on the English exam tended to make high grades on the history exam, and students who made low grades on the English exam tended to make low grades on the history exam. This relation could be summarized by stating that grades on the two exams are positively and linearly correlated; this might be accomplished by computing a correlation coefficient indicating the extent of linear correlation. The study of bivariate frequency distributions is intimately related to the study of correlation. The topic of linear correlation will be considered in Chapter 5; in the chapters preceding we will concentrate on the description of univariate frequency distributions.

BEGINNINGS OF DESCRIPTIVE STATISTICS

Ancient, Classic, and Medieval Worlds

The beginnings of descriptive statistics can be traced to the attempts of various ancient rulers to enumerate the resources of their state, including land, able-bodied men, roads, etc. In the ancient world such enumerations occurred in Egypt, Judea, China, and Persia. One of the oldest such recorded enumerations is of a census taken by King David of the number of "men who drew the sword." The estimated date for the following account from 2 Samuel, 24 is 2030 B.C.

>Again the anger of the Lord was kindled against Israel, and he incited David against them, saying, "Go number Israel and Judah." So the king said to Joab and the commanders of the army who were with him, "Go through all the tribes of Israel, from Dan to Beer-sheba, and number the people, that I may know the number of the people." ...They crossed the Jordan, and began from Aroer, and from the city that is in the middle of the valley, toward Gad and on to Jazar. Then they came to Gilead, and to Kadesh in the land of the Hittites; and they came to Dan, and from Dan they went around to Sidon, and came to the fortress of Tyre and to all the cities of the Hivites and Canaanites; and they went out to the Negeb of Judah at Beer-sheba. So when they had gone through all the land, they came to Jerusalem at the end of nine months and twenty days. Joab gave the sum of the numbering of the people to the king; in Israel there were eight hundred thousand valiant men who drew the sword; and the men of Judah were five hundred thousand.

This account is remarkable not only in that it is one of the earliest reported censuses, but, also, in that it gives the route taken by the census takers, as well as the time required to complete the census.

In the Greek city-states, numerous inquiries of

a statistical nature were taken. A census taken in Athens, for example, showed a population of 21,000 citizens, 10,000 alien residents, and 400,000 slaves. The statistical inquiries were motivated by a concern for the distribution of taxes, privileges, military service, and so forth. The Romans also had periodic enumerations of the population and the distribution of property—reputedly with great care and exactness. Some students will recall Luke's (Ch. 2) beginning of the Christmas story with the statement: "And it came to pass in those days, that there went out a decree from Caesar Augustus, that all the world should be taxed."*

During the middle ages, records of statistical enumerations are more scattered, although still existent. For example, Charlemagne ordered detailed descriptions of church lands, and records survive of the number of serfs attached to French estates, as well as later Church records of baptisms, marriages, and deaths.

Modern Descriptive Statistics until 1750

In his history of statistics, Meitzen (1886) labels the evolution of statistics in the modern world, through 1750, as the development of three semi-distinct traditions: comparative political statistics, official statistics, and population statistics. Each of these developments will be considered in turn.

Comparative Political Statistics

Toward the end of the fifteenth century, the medieval pattern of relatively autonomous estates had largely given way to modern states or monarchies. Concomitant with these, was an increasing concern with international relations and diplomacy. Diplomacy in those days, as in these, was associated with secrecy, duplicity, and an interest in accurate information relating to the resources of enemies and allies alike. In view of this historical context, it is not surprising that some individuals would desire to catalogue and compare the resources of various states. According to Meitzen, the first typical work in this area was *Cosmographia* by Sebastian Muenster, a professor at the Universities of Heidelberg and Basel. The first portion of the book appeared in 1536, and the remainder in 1544. It was not, however, until 1660 that comparative political statistics was introduced into a university curriculum. This was first done by Hermann Conring at the University of Helmstedt.

The most important name associated with comparative political statistics is that of Gottfried Achenwall. Achenwall (1719–1772) is typically hailed as the "father of statistical science." In 1748, he wrote an essay, "Vorbereitung zur Staatweissenschaft der europäischen Reiche," in which he first used the word, "Statistik." By this term Achenwall intended to convey a concern with the *state* and the practical politics thereof. While Achenwall was at the University of Göttingen, the essay was used as the introduction to a book on comparative political statistics. In the book Achenwall systematically discussed eight European states (Spain, Portugal, France, Great Britain, Netherlands, Russia, Denmark, Sweden) in terms of seven topics. Meitzen (1886, p. 24) describes these topics as follows:

1. The literature and sources of information. 2. The state, its territory, and the changes of the same. 3. The land, its climate, rivers, topography, divisions, and abundance or scarcity of products. 4. The inhabitants, numbers and character. 5. The rights of the rulers, the estates, the nobility, and the classes of the inhabitants. 6. The constitution of the court and the government, laws, and administration of churches, schools, and justice; industry, home and foreign commerce, currency, finances, debt, and army and navy. 7. The interests of national life and politics, as well as the outlook for the future.

Achenwall's lucidly written book found general recognition. It later was translated into numerous languages; thus, the term "statistics" passed into general usage. Achenwall founded a school of thought with the lofty purpose of describing and comparing political states in terms of everything that was worth knowing about them. In so far as this school had a methodology, it rarely went beyond the manner of grouping facts and a few principles of comparison. The sources of information were invariably existing literature, and statistics was conceived of as a largely verbal science. In fact, "Die Tablen Statistik" was regarded as a deviation from the true direction that the field of statistics should pursue.

Official Statistics

The second tradition that Meitzen describes is that of official statistics, or descriptive statistical inquiries

---

* "Taxed" meant counted or enrolled in this context.

carried out under the auspices of the state. Various early writers in the sixteenth century argued that governments ought to reinstate something akin to the Roman census, but such developments were slow in coming. One of the earliest official investigations was The Seventy-Five Questions of Philip II, which was directed to the prelates and corregidors of Spain concerning the state of their districts. In 1575, the answers were summarized for the king's use. Later and more noteworthy official investigations occurred in Prussia. In 1719, Frederik William I established a bureau for the semiannual construction of tables containing information relating to population, occupations, artisans, real estate holdings, taxes, etc. It was Frederick II, however, who placed the most emphasis upon such tabulation and took the greatest personal interest in the collection of the statistical data. In the mid-eighteenth century, the type of information was greatly expanded; and the results were supplied to the King in general outline tables, which he carried with him on his journeys.

The results of such official investigations, however, were not available to the general public until 1767 when Anton Frederich Büsching founded "das Magazin für Histioriographie und Geographie," which appeared in 23 issues until 1793. This first periodical journal on statistics collected the official statistical data from various German and other states—although most official data were not available to him. Büsching, unlike Achenwall, was less concerned with general reflections and more concerned with methodological details.

Population Statistics
By population statistics Meitzen means the enumeration of various facts relating to human life; for example, number of births, marriages, and deaths. The historical tradition here obviously overlaps with the two previously described ones. The overlap, however, is not total. The population statistics tradition was not concerned with national comparison, and the tradition existed to some extent outside official auspices.

The modern concern with population statistics had its beginning in the recording of events in church registers. According to Meitzen, the first regular and continuous registration of births, marriages, and deaths was begun in Augsburg in 1501. Thereafter, the tradition was spread throughout various church groups. Baptismal records started in London in 1550, and death registers in 1592, as a consequence of the plague. In 1629, women were employed to inspect the dead, estimate the probable age, and register the sex and probable cause of death. The results were published weekly in the Bills of Mortality for London.

Two of the most important men in this tradition are John Graunt and Edmund Halley. John Graunt (1620–1674) wrote an influential work criticizing and comparing the London Bills of Mortality for the period 1629 to 1661. In 1662, the work was presented to the newly-founded Royal Society. Graunt made the interesting finding that 14 boys are born for every 13 girls. He determined that of 100 persons born, 36 die in the first 6 years, 24 in the next 10 years, 15 in the following 10 years, and so forth. From these figures he demonstrated that the number of living persons could be calculated. The large population that Graunt ascribed to London aroused considerable jealously in Paris.

Edmund Halley (1656–1724) was an astronomer who is chiefly remembered for the comet bearing his name. In 1681, he published a report in the Philosophical Transactions of the Royal Society with the title, "An Estimate of the Degrees of Mortality of Mankind, drawn from curious Tables of Births and Funerals at the City of Breslau; with an Attempt to ascertain the Price of Annuities on Lives." Halley's mortality tables are the earliest ever developed apart from the figures given by Graunt. Halley was recognized by his contemporaries as the founder of the method for calculating mortality tables.

During the late seventeenth and early eighteenth century, numerous life insurance and annuity companies were founded. Such institutions had their historical roots in Medieval times when wagers were made on one's own or another's life prior to going on a sea voyage or a pilgrimage. In such developments there is an obvious concern with probability, and a merging of descriptive and inferential statistics.

Later Developments
By the mid-eighteenth century, Achenwall had numerous followers and students who produced a series of books and papers. Statistics, as Achenwall conceived of the field, was a "going concern." The most influential of Achenwall's students was August Ludwig Von Schlözer.

Official statistics also had come increasingly into vogue—although much of the collected information was still not generally available to the public. A particularly noteworthy date in the tradition of official statistics is 1790. It was at this time that the first decennial census was taken in the United States. The census was, and is, required by the U. S. Constitution to allow for the apportionment of seats in the House of Representatives. The census takers were federal marshalls—modern day descendents of Joab's army. The results were, of course, publicly available. After the French Revolution, men like Lavoisier were conscientious in making official statistical data available to the public and in gathering more such information. In 1806, Von Schlözer praised the new era with regard to publication, "How fortunate we statisticians of the new century, the disgraceful distinction between university and cabinet statistics has ceased to exist."

During the same period, however, Achenwall's followers felt increasingly threatened by the growing popularity of tabular and graphic presentations of statistical data. Von Schlözer and others felt that this was an aberration. Meitzen states: "In 1806 and 1807, a passionate controversy arose against the brainless bungling of the number statisticians, the slaves of the tables, and the skeleton-makers of statistics" (1886, p. 49). Within the field of statistics, the number statisticians, of course, eventually carried the day. Within the field of political science, though, amplified rumblings of the same controversy are still reverberating.

EXERCISES

1. Define the following terms:

(a) range
(b) frequency distribution
(c) univariate frequency distribution
(d) bivariate frequency distribution
(e) ungrouped frequency distribution
(f) grouped frequency distribution
(g) apparent class limits
(h) real class limits
(i) class interval width
(j) midpoint of the class interval
(k) histogram
(l) frequency polygon
(m) cumulative proportion graph

2. Name and define four characteristics of univariate frequency distributions.

3. Draw frequency polygons that are:

(a) positively skewed
(b) negatively skewed
(c) leptokurtic
(d) platykurtic
(e) mesokurtic

4. What is the difference between a univariate frequency distribution and a bivariate frequency distribution?

5. Briefly discuss and relate the following concepts: bivariate frequency distribution, product set, relation, and correlation.

6. What are some of the considerations determining the number of desired classes in a grouped frequency distribution?

7. Round the following numbers to whole numbers:

(a) 1.3 (b) 1.4 (c) 1.5 (d) 1.6 (e) 1.7 (f) 1.48
(g) 1.56 (h) 1.60 (i) 1.489 (j) 1.501 (k) 1.602

8. The following scores were obtained by a group of twenty students on an achievement test:

95	80	65	45
95	76	60	35
88	76	48	29
87	69	48	29
83	67	48	8

Using ten class intervals and starting the lowest class interval at 5, construct a grouped frequency distribution.

(a) Determine the midpoints for the ten class intervals above.

(b) What is the range for the above scores?

(c) Using the above data, construct a table similar to Table 2.4.

9. Using the grouped frequency distribution in

Exercise 8, construct a frequency polygon. Label both axes and title the figure.

10. Using the grouped frequency distribution in Exercise 8, construct a histogram. Label both axes and title the figure.

11. Using the grouped frequency distribution in Exercise 8, construct a cumulative proportion graph. Label both axes and title the figure.

12. Using the data in Table 2.5, construct a frequency polygon for the English scores. Label both axes and title the figure.

13. Using the data in Table 2.5, construct a frequency polygon for the history scores. Label both axes and title the figure.

14. Does it make sense to talk about the range of scores on a nominal scale?

15. Sketch the shape of a cumulative proportion graph for a frequency distribution in which all of the scores are of equal, non-zero frequency.

16. Sketch the shape of a cumulative proportion graph for a frequency distribution with positive skew.

17. Sketch the shape of a cumulative proportion graph for a frequency distribution with negative skew.

18. Sketch the shape of a cumulative proportion graph for a symmetrical, mesokurtic frequency distribution.

19. Using the measurement procedures described in Chapter 1, assess the imagery vividness of each person in your statistics class. Construct an ungrouped frequency distribution from the obtained data.

20. Construct a frequency polygon from the data collected in Exercise 19.

Prologue to Chapter 3

The most obvious way of describing univariate frequency distributions is in terms of their central tendency, or single most representative value. It is assumed that the student has previously learned to compute arithmetic means by summing a set of scores and then dividing by the number of scores. Chapter 3 contains an in-depth discussion of the arithmetic mean, as well as some additional measures of central tendency.

This is the first chapter in which the small print sections contain mathematical derivations. If these sections have been assigned, be reassured by the fact that the derivations are short and reasonably simple. Satisfaction can be gained by understanding the basis for a given formula. Be further reassured by the realization that a failure to follow these derivations will not hinder your understanding of the subsequent larger print material. If difficulty is encountered with these derivations, skip over them until after reading the entire chapter; then, study them carefully. If these sections still cannot be understood, ask for help from your instructor or from a fellow student. Do not feel bad about asking for help; the possession of superior mathematical skills is not *a prerequisite for being a psychologist.*

3

Central Tendency

This chapter will be concerned with various indices of central tendency. The three major indices are the *arithmetic mean*, the *median*, and the *mode*. In addition, one less widely used index, the *geometric mean*, will be presented.

ARITHMETIC MEAN

The arithmetic mean of a set of scores is the value obtained by summing the scores and then dividing the sum by the number of scores. The formula for the mean is:

$$\bar{X} = \frac{\sum X}{n} \qquad (3\text{–}1)$$

where $\bar{X}$ symbolizes the arithmetic mean, $\sum$ the operation of summing the set of n scores, X any score of the set of n scores, and n the number of scores in the set.

The use of formula 3–1 to calculate the arithmetic mean is illustrated in Table 3.1. The scores in Table 3.1 are from the Stroebe *et al.* high-similarity condition. In Chapter 2 these scores were used to illustrate a frequency polygon (Fig. 2–3). Inspection of Fig. 2–3 suggests that the central tendency is high. The calculations in Table 3.1 confirm this impression. The arithmetic mean is 5.70 on a scale in which the highest possible score is 7.

Table 3.2 contains the arithmetic mean calculation of the liking scores for the low-similarity condition (Stroebe *et al.*). Inspection of the graph of these scores (Fig. 2–4) suggests that the central tendency is relatively low. The calculations in Table 3.2 indicate that the arithmetic mean is, in fact, 3.27.

TABLE 3.1. *Calculation of the Arithmetic Mean for the Stroebe et al. High-Similarity Condition.*

X	
7	6
7	6
7	6
7	6
7	6
7	6
7	6
7	6
7	6
7	6
7	6
6	6
6	6
6	6
6	5
6	5
6	5
6	5
6	4
6	4
6	4
6	4
6	4
6	4
6	4
6	3
6	3
6	3
6	3
$\Sigma = 342$	

$$\bar{X} = \frac{\Sigma X}{n}$$

$$\bar{X} = \frac{\Sigma X}{60}$$

$$\bar{X} = \frac{342}{60}$$

$$\bar{X} = 5.70$$

Characteristics of the Arithmetic Mean

The arithmetic mean has four characteristics with which the student should be familiar.

1. *For any set of scores, the sum of the differences of scores from the arithmetic mean equals zero.* Symbolically, this first characteristic can be stated as:

$$\sum (X - \bar{X}) = 0 \qquad (3-2)$$

TABLE 3.2. *Calculation of the Arithmetic Mean for the Stroebe et al. Low-Similarity Condition.*

X	
7	3
6	3
6	3
6	3
5	3
5	3
5	3
5	3
5	3
5	3
4	3
4	3
4	3
4	3
4	2
4	2
4	2
4	2
4	2
4	2
4	2
4	2
4	2
4	2
3	2
3	1
3	1
3	1
3	1
3	1
3	1
$\Sigma = 196$	

$$\bar{X} = \frac{\Sigma X}{n}$$

$$\bar{X} = \frac{\Sigma X}{60}$$

$$\bar{X} = \frac{196}{60}$$

$$\bar{X} = 3.27$$

The expression $(X - \bar{X})$ occurs so frequently that it is typically symbolized x, and referred to as a *deviation score*. Thus we can write:

$$\Sigma x = 0 \qquad (3\text{--}3)$$

The scores 101, 105, 109, 110, and 115 will be used to illustrate the characteristics of the arithmetic mean. The arithmetic mean of these five scores is 108 ($\bar{X} = 540 / 5 = 108$). The calculation of Σx is illustrated in Table 3.3. Here it can be seen that Σx is zero, in agreement with formulas 3–2 and 3–3.

TABLE 3.3. *Calculation of $\sum x$.*

X	$\bar{X}$	(x) $X - \bar{X}$	
101	108	-7	
105	108	-3	
109	108	$+1$	$\sum x = \sum (X - \bar{X}) = 0$
110	108	$+2$	
115	108	$+7$	
		$\sum = 0$	

The fact that the sum of differences between the arithmetic mean and every score is zero can be intuitively grasped by picturing the relationship of individual scores to the mean as a balanced seesaw. The mean is at the fulcrum of the seesaw and the scores are at appropriate distances from the fulcrum, distances corresponding to the sizes of the differences between the individual scores and the mean. Such a seesaw is presented in Fig. 3–1. Balance results from the fact that every plus unit distance to the right of the fulcrum is matched by a minus unit distance to the left of the fulcrum.

> In order to prove that the sum of deviations from the arithmetic mean is zero, it is necessary to make use of rules 1 and 2 from summation algebra, see Appendix B. By definition:
>
> $$\sum x = \sum (X - \bar{X})$$
>
> Using rules 2 and 1 in that order:
>
> $$\sum x = \sum X - \sum \bar{X}$$
> $$= \sum X - n\bar{X}$$
> $$= \sum X - \frac{n \sum X}{n}$$
> $$= \sum X - \sum X$$
> $$= 0$$

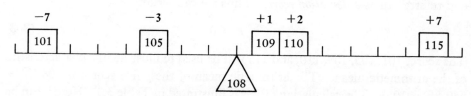

FIGURE 3–1. *Arithmetic mean as the balance point of a seesaw.*

TABLE 3.4. *Calculation of $\sum x^2$.*

X	$\bar{X}$	x	x^2	
101	108	−7	49	
105	108	−3	9	
109	108	+1	1	$\sum x^2 = 112$
110	108	+2	4	
115	108	+7	49	
			$\sum = 112$	

2. *For any set of scores, the sum of the squared differences from the arithmetic mean is less than the sum of squared differences from any other number.* This second characteristic of the arithmetic mean can be more simply expressed by stating that $\sum x^2$ is a minimum. The calculation of $\sum x^2$ is illustrated in Table 3.4. The above principle tells us that if we had subtracted any number other than the mean from all the scores, squared the differences, and then summed the squares, a number larger than 112 would have been obtained.

3. *If each score in a distribution is increased by a constant, the arithmetic mean is increased by that same constant, and if each score in a distribution is decreased by a constant, the mean is decreased by the same constant.* This third characteristic of the arithmetic mean can be symbolically expressed as:

$$\frac{\sum (X \pm a)}{n} = \bar{X} \pm a \qquad (3\text{--}4)$$

where the double sign ($\pm$) means plus or minus and *a* represents the constant.

In Table 3.5, a constant, 101, has been subtracted from the distribution of scores whose arithmetic mean had been determined as 108. The arithmetic mean of the $X - a$ scores is 7; and 7 plus the constant, 101, equals 108, the arithmetic mean of the original scores. This illustrates the subtraction part of the third characteristic. The student can be further convinced of the addition part by working through an example.

The algebraic proof of the third characteristic is very short. Using summation rule 2:

$$\frac{\sum (X \pm a)}{n} = \frac{\sum X \pm \sum a}{n}$$

Substituting *na* for $\sum a$ according to summation rule 1:

$$\frac{\sum (X \pm a)}{n} = \frac{\sum X \pm na}{n}$$

$$= \frac{\sum X}{n} \pm \frac{na}{n}$$

$$= \frac{\sum X}{n} \pm a$$

$$= \bar{X} \pm a$$

TABLE 3.5. *Subtraction of a Constant from the Arithmetic Mean.*

X	a	X − a	
101	101	0	
105	101	4	$\bar{X} = \dfrac{\Sigma X}{n}$
109	101	8	
110	101	9	
115	101	14	$\bar{X} = \dfrac{540}{5}$
$\Sigma = 540$		$\Sigma = 35$	$\bar{X} = 108$

Also,
$$\bar{X} = \frac{\Sigma(X - a)}{n} + 101$$

$$\bar{X} = \frac{\Sigma(X - 101)}{5} + 101$$

$$\bar{X} = \frac{35}{5} + 101$$

$$\bar{X} = 7 + 101$$

$$\bar{X} = 108$$

4. If each score in a distribution is multiplied by a constant, the arithmetic mean becomes multiplied by the same constant; if each score is divided by a constant, the arithmetic mean becomes divided by that constant. Symbolically, the two parts of this fourth characteristic can be written as:

$$\frac{\Sigma aX}{n} = a\bar{X} \qquad (3\text{–}5)$$

and,

$$\frac{\Sigma(X/a)}{n} = \frac{\bar{X}}{a} \qquad (3\text{–}6)$$

Formula 3–6 is illustrated in Table 3.6. With an a of 10, the arithmetic mean of the X/a scores is 10.8; and 10.8 multiplied by a (or 10) equals 108, the arithmetic mean of the original scores.

The division part of the fourth characteristic and the subtraction part of the third characteristic provide coding techniques for use in computing the arithmetic mean. When the scores comprising a distribution are large, labor may be saved by subtracting a constant from the scores, computing the arithmetic mean of the $X - a$ scores, and then adding the constant, a, to the arithmetic mean of the $X - a$ scores. This procedure results in the mean of the distribution of

TABLE 3.6. *Division of the Arithmetic Mean by a Constant.*

X	a	X / a	
101	10	10.1	
105	10	10.5	$\bar{X} = \dfrac{\Sigma X}{n}$
109	10	10.9	
110	10	11.0	$\bar{X} = \dfrac{540}{5}$
115	10	11.5	
$\Sigma = 540$		$\Sigma = 54.0$	$\bar{X} = 108$

Also,
$$\bar{X} = a\frac{\Sigma(X/a)}{n}$$

$$\bar{X} = 10\frac{54}{5}$$

$$\bar{X} = 108$$

X. In other instances, it may be helpful to divide the scores by a constant, compute the arithmetic mean of the X/a scores, and then multiply the constant, a, by the arithmetic mean of the X/a scores. This procedure also results in the mean of the distribution of X.

The algebraic proofs of first the multiplication and then the division part of the fourth characteristic follow. Using summation rule 3:

$$\frac{\Sigma aX}{n} = \frac{a\Sigma X}{n}$$

Substituting $\bar{X}$ for $\Sigma X/n$ (formula 3–1):

$$\frac{\Sigma aX}{n} = a\bar{X}$$

Likewise:

$$\frac{\Sigma(X/a)}{n} = \frac{(1/a)\Sigma X}{n}$$

$$\frac{\Sigma(X/a)}{n} = \frac{\bar{X}}{a}$$

Computing the Mean of a Set of Scores from the Means of Its Subsets

Sometimes one has the arithmetic means of a number of subsets of scores and wishes to obtain the mean of the set comprised of the union of all the subsets. Suppose, for example, the mean exam score obtained in each of the several sections of a statistics course is known and it is desirable to know the overall mean.

If all of the original scores for each section were known, one could, of course, sum all these scores and obtain the mean of the union in the usual manner. This procedure, however, is somewhat laborious; and in addition, the original scores upon which each subset (or section) mean is based may not be available. In order to calculate correctly the mean of the union without recourse to all the original scores, it is necessary to weight differentially the subset means according to the number of scores in each subset. Such differential weighting is accomplished by formula 3–7:

$$\bar{X} = \frac{n_1 \bar{X}_1 + n_2 \bar{X}_2 + \cdots + n_k \bar{X}_k}{n_1 + n_2 + \cdots + n_k} \tag{3-7}$$

The subscripts in this formula refer to the first subset, second subset, and last (or kth) subset—k corresponding to the number of subsets. Table 3.7 illustrates the use of formula 3–7 to calculate the mean of the union of three subsets.

> The algebraic proof of formula 3–7 involves only a few steps. The symbols with subscripts refer to the specific subsets while the symbols without subscripts refer to the total set of scores (the union). That is, X_1 represents any score in subset 1; X_2, any score in subset 2, etc., while X represents any score in the union of all subsets.
> Substituting $\sum X_1 / n_1$ for $\bar{X}_1$, etc.:
>
> $$\frac{n_1 \bar{X}_1 + n_2 \bar{X}_2 + \cdots + n_k \bar{X}_k}{n_1 + n_2 + \cdots + n_k} = \frac{\frac{n_1 \sum X_1}{n_1} + \frac{n_2 \sum X_2}{n_2} + \cdots + \frac{n_k \sum X_k}{n_k}}{n_1 + n_2 + \cdots + n_k}$$
>
> $$= \frac{\sum X_1 + \sum X_2 + \cdots + \sum X_k}{n_1 + n_2 + \cdots + n_k}$$
>
> Since $\sum X_1 + \sum X_2 + \cdots + \sum X_k$ is equivalent to the total sum of all scores across the k sets, and likewise, $n_1 + n_2 + \cdots + n_k$ is equivalent to the total number of scores in all k sets, then:
>
> $$\frac{\sum X_1 + \sum X_2 + \cdots + \sum X_k}{n_1 + n_2 + \cdots + n_k} = \frac{\sum X}{n} = \bar{X}$$

TABLE 3.7. *The Mean of a Set Computed from Means of Its Subsets.*

(1)	(2)	(3)	
5	8	10	$\bar{X} = \dfrac{n_1 \bar{X}_1 + n_2 \bar{X}_2 + n_3 \bar{X}_3}{n_1 + n_2 + n_3}$
7	11	4	
10	20	2 ⟋ 14	$\bar{X} = \dfrac{(5)(7) + (4)(10) + (2)(7)}{5 + 4 + 2}$
9	1	7	
4	4 ⟋ 40		$\bar{X} = \dfrac{35 + 40 + 14}{11}$
5 ⟋ 35	10		
7			$\bar{X} = 8.09$

Computing the Arithmetic Mean of an Ungrouped Frequency Distribution

The usual procedure for calculating the arithmetic mean involves the initial step of summing the scores to obtain $\sum X$; e.g., as is shown in Table 3.1, where $\sum X = 342$. Examination of Table 3.1, however, suggests an alternative way of obtaining $\sum X$. Notice that all of the score values occur with frequencies greater than one. The 7 score occurs with a frequency of 12, the score 6 occurs with a frequency of 33, and so forth. All of these frequencies are listed in the second column of Table 3.8, which contains an ungrouped frequency distribution for the X scores in Table 3.1. The alternative way of obtaining $\sum X$ is to multiply each score value by its frequency, $f(X)$, and then sum the products, $Xf(X)$'s. This procedure is illustrated in Table 3.8. The sum of the third $Xf(X)$ column equals 342, which, as indicated above, is $\sum X$.

When scores have been arranged in an ungrouped frequency distribution, as in Table 3.8, the arithmetic mean of those scores can be calculated by the following formula:

$$\bar{X} = \frac{\sum_1^k Xf(X)}{n} \qquad (3\text{-}8)$$

Where: $Xf(X)$ symbolizes the products of each score value by its frequency.
k symbolizes the number of different score values.
$\sum_1^k$ thus indicates summation over the k score values.*

TABLE 3.8. *The Arithmetic Mean of an Ungrouped Frequency Distribution for the Stroebe et al. High-Similarity Condition.*

X	f(X)	Xf(X)
7	12	84
6	33	198
5	4	20
4	7	28
3	4	12
	$\sum = 60$	$\sum = 342$

$$\bar{X} = \sum_1^k Xf(X) / n = 342 / 60 = 5.70$$

---

* Space limitations prevent writing $\sum_1^k$ as in formula 3-8, with k above and 1 below the summation symbol, $\sum$.

$\sum_{1}^{k}$ symbolizes the operation of beginning with the first score value and summing to the last or kth one. Most typically, summation is over the total number of scores, n. In this case, however, it is over the number of k score values. For the ungrouped frequency distribution in Table 3.8, $n = 60$, since there are 60 scores; and $k = 5$, since there are 5 score values (7, 6, 5, 4, and 3).

The use of formula 3-8 to calculate the mean of an ungrouped frequency distribution is illustrated in Table 3.8. Notice that the mean of 5.70 is the same one obtained in Table 3.1. Formula 3-8 for the arithmetic mean of a frequency distribution is equivalent to formula 3-1 for the arithmetic mean of a set of scores. The numerator in formula 3-8, which calls for the summation of products for each different score value times its frequency [$\sum_{1}^{k} Xf(X)$], is equivalent to the numerator of formula 3-1, which calls for the summation of all scores in the set ($\sum X$).

MEDIAN

A second very common measure of central tendency is the median. *The median is defined as that value on a scale of measurement at or below which 50 percent of a set of scores fall.* The median is also called the 50th percentile. *A percentile is that value at or below which a certain percent of the scores fall.* Thus, 50 percent of the scores fall at or below the 50th percentile, which is the median.

The 25th percentile is sometimes referred to as the *first quartile*, Q_1, the 50th percentile as the *second quartile*, Q_2, and the 75th percentile as the *third quartile*, Q_3. The first quartile is that value at or below which 25 percent of the scores fall. The third quartile is that value at or below which 75 percent of the scores fall.

> The median is more technically defined in terms of probability (the subject matter of Chapter 6). The layman uses the term probability to express the expectancy that an outcome will occur. Such expectancies can be expressed as probability numbers varying between 0 and 1. The higher the number is, the greater the expectancy that an outcome will occur.
>
> *The median is some value for which the probability of obtaining scores of equal or lower value is .5.* This can also be expressed:
>
> $$\text{Prob. }(X \leq \text{Mdn.}) = .5$$
>
> The probability of obtaining a score equal to or lower than the median is .5.
>
> Suppose that the median of a distribution of scores is 10. It is, therefore, known that the probability of obtaining a score of equal or lower value is .5. One way of interpreting probability numbers is as a proportion of future outcomes. From this perspective, if additional scores are obtained, 50 percent of these scores are expected to be equal to or lower than 10.

Computation

To compute the median from a set of unordered scores, the initial step is ordering or ranking the scores from highest to lowest. For example, the set of scores, 9, 9, 3, 10, 5, 1, 6, and 2, would be ordered 10, 9, 9, 6, 5, 3, 2, and 1. Computation of the median can then proceed according to the characteristics of the distribution being considered. We will consider two different cases.

CASE 1. Distributions with nonadjacent scores on either side of the middle interval.

Consider the following distribution: 8, 7, 5, and 4. Notice that the two middle scores, 7 and 5, are nonadjacent; i.e., have at least one place between them. According to convention, the median of such a distribution is the arithmetic mean of the two middle scores. For this example the median is 6, $(7 + 5) / 2 = 6$.

The convention of taking the arithmetic mean of the nonadjacent middle scores applies *regardless of the frequencies of the middle scores or of any scores in the distribution*. Thus, the convention applies for all of the following distributions: 8, 7, 7, 5, 5, and 4; 8, 7, 5, and 5; 9, 8, 8, 7, 7, 5, 1, 1, 1, and 1; and 9, 8, 4, and 3. For each of these distributions the median is the arithmetic mean of the two middle scores, or 6. Notice further that each of these distributions has an even number of scores. Middle intervals evenly dividing a distribution only occur with an even number of scores.

> What is the basis for this convention of taking the arithmetic mean of the two middle scores? Consider the example: 8, 7, 5, and 4. It is obvious that any number from 5 up to 7 has 50 percent of the scores at or below its value. In view of this fact, why pick the arithmetic mean of 7 and 5, 6? The convention is compelling from a probabilistic point of view. If one wishes to select a median so that 50 percent of the additional or future measurements will fall at or below its value, the arithmetic mean of the two middle scores is reasonable. The arithmetic mean of the two middle scores evenly divides the distance between them.

CASE 2. All distributions not covered by case 1.

Case 2 includes every possibility not included in case 1, for example, distributions with adjacent scores on either side of the middle interval such as the set: 8, 7, 7, 6, 6, and 6. Scores 7 and 6 are on either side of the middle interval. Also, distributions may occur with scores at the center of the middle interval; for example: 9, 7, 7, and 6. The 7 score is at the center of the middle interval.

The conventional procedure for calculating the median for case 2 involves interpolation into the middle interval. This procedure can be described in five steps:

TABLE 3.9. *Calculation of the Median for the Stroebe et al. High-Similarity Condition.*

	X	f(X)	CUMULATIVE FREQUENCY	
	7	12	60	*Median*
Mdn.	6	33	48	(60)(1/2) = 30
	5	4	15	Mdn. = 15/33 + 5.5 = 5.95
	4	7	11	
	3	4	4	

1. Construct a cumulative frequency column adjacent to the frequency column. This procedure is illustrated in Table 3.9 for the liking scores from the Stroebe *et al.* high-similarity condition. Thus, 4, the frequency for the lowest X scores, is taken as the bottom entry in the cumulative frequency column. The frequency for the next highest X score, 7, is added to 4 to give 11, the next entry in the cumulative frequency column. This general procedure is continued through the highest X score, which always has an associated cumulative frequency equal to n, the total frequency.

2. Find the number of scores at or below the median value by multiplying $\frac{1}{2}$ times n, the total frequency. For example in Table 3.9, take $\frac{1}{2}$ of 60, which equals 30.

3. Considering each value in the X column as the midpoint of an interval of width $i = 1$, locate the interval into which the median falls. This is accomplished by locating the lowest valued interval with an associated cumulative frequency greater than or equal to $\frac{1}{2}n$. In Table 3.9, the interval containing the median has been labelled "Mdn." This interval extends from 5.5 to 6.5. It is known that the median is equal to the lower real limit of this interval, plus some amount.

4. Determine the distance of the median into the interval. Subtract from $\frac{1}{2}n$ the cumulative frequency associated with the interval adjacent to and lower in value than the one containing the median. In Table 3.9, this is 15 scores into the interval. The interval, however, contains 33 scores. Therefore, the median is 15/33rds of the distance into the interval. The ratio 15/33rds is the distance ratio.

5. Add this distance ratio to the lower real limit for the interval containing the median. The resultant sum is the median. Since the lower real limit for the interval is 5.5, the median equals 5.5 plus 15/33, or 5.95.

Table 3.10 contains the calculation of the median for the low-similarity condition, 3.17. The medians of 5.94 and 3.17 for the high- and low-similarity conditions, respectively, are reasonably similar to the arithmetic means for the same scores, 5.70 and 3.27 (see Tables 3.1 and 3.2).

Essentially the same procedure can be used to calculate the various quartile and percentile values. The only difference occurs in the second step. For example, to calculate the first quartile, the total frequency, n, is multiplied by

TABLE 3.10. *Calculation of the Median for the Stroebe et al. Low-Similarity Condition.*

	X	$f(X)$	CUMULATIVE FREQUENCY	
	7	1	60	Median
	6	3	59	$(60)(1/2) = 30$
	5	6	56	Mdn. $= 14/21 + 2.5 = 3.17$
	4	13	50	
Mdn.	3	21	37	
	2	10	16	
	1	6	6	

¼ rather than by ½. To calculate the third quartile, the total frequency (n) is multiplied by ¾. The procedure for calculating both the first and third quartiles is illustrated with scores from the high-similarity condition in Table 3.11.

The student should note that application of the interpolative procedure to case 1 distributions will give an answer different from the arithmetic mean of the nonadjacent middle scores. Consider the distribution: 8, 7, 5, and 4. Interpolation gives an incorrect median, 5.5; while the true median is 6, the arithmetic mean of 7 and 5.

In every instance the median is located at a value at or below which 50 percent of the scores are expected to fall. In the case 1 situation above this dictates a median of 6—not 5.5. However, for all distributions not containing nonadjacent scores on either side of the middle interval, the case 2 (or interpolative) procedure will result in a value at or below which 50 percent of the scores are expected to fall. The difference between case 1 and case 2 procedures is thus justified on the basis of the probabilistic conception of the median.

TABLE 3.11. *Calculation of the First and Third Quartiles for the Stroebe et al. High-Similarity Condition.*

	X	$f(X)$	CUMULATIVE FREQUENCY	
	7	12	60	First Quartile
Q_3	6	33	48	$(60)(1/4) = 15$
Q_1	5	4	15	$Q_1 = 4/4 + 4.5 = 5.50$
	4	7	11	Third Quartile
	3	4	4	$(60)(3/4) = 45$
				$Q_3 = 30/33 + 5.5 = 6.41$

An adaptation of the case 1 procedure, for the situation in which non-adjacent scores bound the interval of interest, could be used for the various percentiles other than the fiftieth. The added complexity, however, does not appear to merit the slight improvement over the case 2 procedure. Thus, always use the case 2 procedure to calculate percentiles other than the median.

Computation for Distributions of Non-integers

All of the above discussion has focused upon distributions of integers or whole numbers. What does one do if the distribution contains numbers carried out to the first decimal place, 10.5 for example, or to the second decimal place, e.g., 10.55? The above described procedures will work equally well for such numbers. However, if the student prefers to work with whole numbers, a slight alteration in procedure makes this possible. For distributions containing numbers carried to the first decimal place, multiply all of the numbers by 10, compute the median of the transformed scores, and divide that median by 10. For distributions containing numbers carried to the second decimal place, multiply all of the numbers by 100, compute the median of the transformed scores, and divide that median by 100.

MODE

A third measure of central tendency is the mode. *The mode is the value on a scale of measurement that occurs most frequently in a set of scores.* To obtain the mode simply locate the value that occurs most frequently. For the Stroebe *et al.*

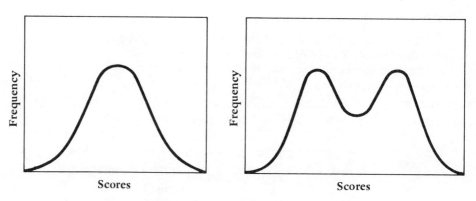

FIGURE 3–2. *Unimodal and bimodal distribution.*

high- and low-similarity conditions, the modes are 6.0 and 3.0, respectively (see Tables 3.9 and 3.10).

Usually a set of scores has only one most frequently occurring value, and thus, only one mode. The frequency distributions derived from such sets are referred to as *unimodal*. When a set of scores has two most frequently occurring values and thus more than one mode, its distribution is referred to as *bimodal*. Figure 3–2 illustrates unimodal and bimodal distributions.

COMPARISON OF THE MEAN, MEDIAN, AND MODE

Although the arithmetic mean, median, and mode are all commonly used indices of central tendency, some very real differences exist among them. These differences provide the researcher with information concerning which index is most appropriate in a given situation.

Ease of Computation

It is quite apparent that the mode is easier to compute than either the median or arithmetic mean. It is also true that with practice the median is easier to compute than the arithmetic mean. This is particularly true if the scores have already been ranked. Thus, ease of computation may be a consideration in choosing among the three indices of central tendency. Use of calculators and computers, however, does make this consideration less important.

Skewness

When a distribution of scores is unimodal and symmetrical, all three indices of central tendency coincide. This situation is illustrated in Fig. 3–3. With a skewed distribution, however, the three indices do not coincide. The mode is always at the peak of the curve and the arithmetic mean is located furthest toward the tail of the distribution. Such a distribution is illustrated in Fig. 3–4. Notice that the median is closer to the mean than to the mode. In general, for moderately skewed distributions, the median will be at a value approximately $\frac{2}{3}$ the distance from the mode to the mean. This rule of thumb works fairly well for the Stroebe et al. low-similarity condition (graphed in Fig. 2–4), which shows moderately positive skew. For this condition the mode is 3.0, the median 3.17, and the mean 3.27. The scores for the high-similarity condition (graphed in Fig. 2–3), show a markedly negative skew. For this more highly skewed distribution, the median is closer to the mode than to the mean (mode = 6.0, mdn. = 5.95, and

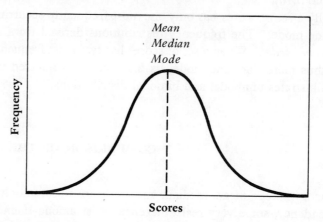

FIGURE 3-3. *A symmetrical unimodal distribution in which the mean, median, and mode all coincide.*

$\bar{X} = 5.70$). In this instance the degree of skewness is too great for the rule to be applied. For both conditions, however, the mode is at the peak of the curve and the arithmetic mean is furthest toward the tail of the distribution.

When a distribution is markedly skewed, either the median or the mode is considered a better measure of central tendency than the arithmetic mean because the arithmetic mean is so far toward the tail of the distribution. This type of situation occurs, for example, when the income level of a given community is reported. Because a few individuals can have very high incomes, the mean

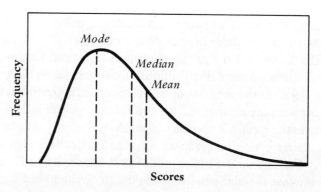

FIGURE 3-4. *A skewed distribution in which the mean, median, and mode do not coincide.*

income is likely to be greater than that obtained by 70 to 80 percent of the persons in the community. In such a situation it is advisable to report all three measures of central tendency.

In his book, *How to Lie with Statistics*, Darrel Huff (1954) makes much of the incorrect impression that can be conveyed through "the well-chosen average." He relates a hypothetical instance in which a real estate salesman convinces a prospective buyer of the snob appeal of a given neighborhood by telling him that the average income is $15,000. Some time later, after the sale is completed, the buyer encounters the real estate salesman circulating a petition to lower the tax rate—a petition justified on the basis of the neighborhood having an average income of only $3,500. As Huff points out, the real estate salesman was technically correct in both instances. It is just that the first average was the arithmetic mean, and the second, the median.

Types of Information

The three measures of central tendency differ in the types of information they provide. The mode provides the most frequent score; the median gives the value above and below which an equal number of scores fall; the arithmetic mean offers the balance or equilibrium point in the distribution of scores. In choosing among these measures of central tendency, the researcher can be guided by the type of information desired.

Algebraic Manipulations

Due to its mathematical properties, the arithmetic mean is more amenable to algebraic manipulation than either the median or the mode. For example, by calculating the weighted mean of a set of means, the mean of the total set of scores can be obtained. This type of manipulation is not possible with either the median or the mode. Amenability to algebraic manipulation is one of the reasons why researchers prefer the arithmetic mean as a measure of central tendency.

Stability and Accuracy

One of the most important advantages of the arithmetic mean is that it usually fluctuates less from sample to sample than either the median or mode. That is, if a large number of different samples are drawn from a given population and all three measures of central tendency are calculated for each sample, the arithmetic mean will usually vary less from sample to sample than either the median or the mode. This fact becomes important when one is attempting to infer the central tendency of an entire population from knowledge of the central tendency of a sample drawn at random from that population. Because of the greater stability of sample means across samples, it is likely that the sample mean is closer to the population mean than the sample median and mode are to the population median and mode.

GEOMETRIC MEAN

The geometric mean is a measure of central tendency, just as are the more commonly used arithmetic mean, median, and mode. The geometric mean is introduced for two reasons: First, it illustrates that measures of central tendency in addition to the common ones are possible. Second, the geometric mean is utilized in the discussion of correlation (see Chapter 5).

The geometric mean is the nth root of the product of the n scores. Symbolically, this can be expressed as:

$$G = \sqrt[n]{X_1 \cdot X_2 \cdot X_3 \cdots X_n} \qquad (3\text{-}9)$$

For example, the geometric mean of scores 3 and 12 is the square root of 36, or 6. The geometric mean of scores 2, 4, and 8 is the cube root of 64, or 4.

When n is 3 or more, computation of the geometric mean by formula 3–9 is somewhat difficult. In such a situation the simplest procedure is to convert the X scores to logarithms, compute the arithmetic mean of the logarithms, and obtain the antilogarithm of this arithmetic mean.

The geometric mean, unlike the arithmetic mean, weights the lower scores somewhat more heavily than the higher scores. Note that the geometric mean of 3 and 12 is 6, while the arithmetic mean is 7.5. This indicates that with a positively skewed distribution the geometric mean will not be pulled as far out into the tail of the distribution as the arithmetic mean. In such circumstances the geometric mean, like the median, may be a more appropriate measure of central tendency than the arithmetic mean. This is particularly true when the distribution of scores is sufficiently skewed so that the distribution of logarithms of these scores is symmetrical.

TABLING AND GRAPHING CENTRAL TENDENCY

Once an index or measure of central tendency has been calculated for the scores in the various conditions of an experiment, what is the best way to present the results? If there are only two conditions the simplest procedure is to present each index directly in the text of a report—although a table or graph also may be used.

When numerous experimental conditions are present, however, use of a table or graph is definitely preferable.

As an example, consider again the Stroebe *et al.* experiment. Half of the subjects in the high- and low-similarity conditions were males, and half were females. (Recall that the similarity information and pictures were always of an opposite sex person—male subjects receiving information about a female, and vice versa.) Suppose that the liking responses of males and females in the high- and low-similarity conditions are to be examined separately. Under these circumstances there are four experimental conditions: low-similarity-males, low-similarity-females, high-similarity-males, and high-similarity-females. These four conditions are generated by combining the two properties (psychologists call these "independent variables"), attitude similarity, and sex of subject. The means in these four conditions are low-similarity-males, 3.6; low-similarity-females, 3.0; high-similarity-males, 5.4; and high-similarity-females, 6.0.

The above four means are tabulated in Table 3.12 and graphed in Fig. 10–10. Examination of Table 3.12 reveals two interesting observations: First, for both male subjects and female subjects greater liking occurs with high- than with low-similarity. Second, the difference between low- and high-similarity is greater for females than for males. For females the means vary from 3.0 to 6.0, a difference of 3.0. For males the comparable means range from 3.6 to 5.4, a difference of 1.8.

Examination of Fig. 10–10 reveals the same two observations. The fact that both lines (one for males and one for females) rise from lower left to upper right indicates that as similarity increased from low to high liking increased correspondingly. This, then, is the first observation made above. Beyond this, however, it is apparent that the two lines are not parallel, and specifically, that the line for females is steeper than that for males. The difference between low- and high-similarity is greater for females than for males. This, then, is the second observation made above. Such a departure from parallelness is technically referred to as an *interaction*. Interactions will be discussed later in considerable detail (see Chapter 10).

Tables and graphs enable more ready recognition of effects such as interactions. But how does one decide between using a table or a graph? No firm

TABLE 3.12. *Liking as a Function of Similarity and Sex of Subject in the Stroebe et al. Experiment.*

		SEX OF SUBJECT	
		MALE	FEMALE
SIMILARITY	LOW	3.6	3.0
	HIGH	5.4	6.0

rule dictates stronger preference for one than the other. Tables have the advantage of presenting the exact values for the measures of central tendency (means in the above example). On the other hand, graphs frequently offer a clearer picture of the results.

In Chapter 2, five conventions were stated regarding the construction of graphs for frequency distributions. All but the second of these conventions also apply to graphs for measures of central tendency. This second convention, which dictates that the values of a variable be represented along the horizontal axis and the frequencies of these values along the vertical axis, is obviously not appropriate in the present context. In the case of central tendency, convention dictates that the property being measured by the experimenter (psychologists say the "dependent variable") is represented along the vertical axis. In the Stroebe *et al.* experiment the measured property was liking. This leaves the property or properties that the experimenter directly determines or manipulates. If there is only one such property it is represented along the horizontal axis. If two such properties (similarity and sex, for example) are present, one property is represented along the horizontal axis and the other by different types of lines.

Two general rules of thumb relate to the decision regarding which of the two properties is represented along the horizontal axis and which by different types of lines. First, in the case where the properties produce effects that obviously differ in magnitude, the property having the greater effect (similarity in the Stroebe *et al.* experiment) is represented along the horizontal axis. Second, in the case where the properties (or their symbolized variable) have different numbers of values (or levels), the property whose symbolization has the larger number of values is represented along the horizontal axis. The reason for this is simply that with too many different types of lines (one for each value) the graph may appear chaotic. Obviously, many instances occur in which the two rules are inconsistent; for example, the property (or variable) with the smaller number of values also can have the larger effect. In such situations the experimenter should use his or her own best judgment regarding which property is represented along the horizontal axis—or perhaps should use a table instead of a graph.

Tables are constructed by placing the measure of central tendency in the body of the table and the two manipulated properties across columns and rows. It does not matter which property is across columns or rows. However, for only one property, convention dictates that it be placed across columns, i.e., at the top.

GRAPHS THAT CREATE FALSE IMPRESSIONS

Before finally leaving the topic of graphs, it is worthwhile to discuss the manner in which they must be used to create false impressions. In his book, *How to Lie*

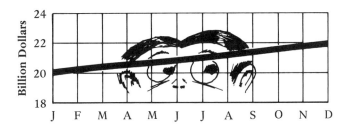

FIGURE 3–5. *A gee-whiz graph from* How to Lie with Statistics *by D. Huff, 1954, p. 62.*

with Statistics, Darrel Huff provides several interesting illustrations of what he calls the "gee-whiz graph." Two of his illustrations relate national income, in billions of dollars, to the month of the year. The first of these is reproduced in Fig. 3–5, and the second, in Fig. 3–6. The basic data indicate a 10 percent growth

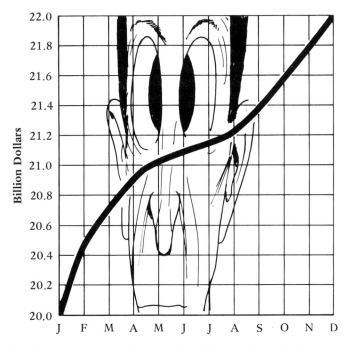

FIGURE 3–6. *A gee-whiz graph from* How to Lie with Statistics, *by D. Huff, 1954, p. 63.*

Graphs that Create False Impressions

over the 12 month period. (Whether the growth is an average of several measures or a single measure is, for present purposes, irrelevant). In Fig. 3–5, this modest 10 percent growth is made to appear much larger by failing to represent national income that is less than 18 billion down the vertical axis. This "oversight" could be justified as an attempt to save paper. However, not alerting the reader by inserting a break on the vertical axis between 18 and 0 as is called for by one of the conventions listed in Chapter 2 cannot be justified. In fact, the reason for the convention is to serve as a guard against such false impressions.

In Fig. 3–6, the false impression of a large growth rate is magnified even further. In this graph the vertical axis represents only the 20 to 22 billion range, and this range has been physically increased in size. Although the graph is technically accurate, it, in fact, creates a false impression.

HISTORICAL PERSPECTIVE ON THE INDICES OF CENTRAL TENDENCY

Mode

The various indices of central tendency have diverse historical roots. The concept of the most frequently occurring instance or score involves such a low level of abstraction that pinpointing its first formulation is difficult. Quite likely, many different individuals conceived of the concept without attaching much significance to it. It is known that the term "mode" comes from the Latin "modus," which means measure, size, or limit of quantity. In the sixteenth century, the Latin term changed; given a feminine ending, "e," it became used in the sense of "fashion." In the seventeenth century, the term "mode" was taken into the English language with the same meaning. This meaning of the term is, of course, very close to its meaning in statistics.

In 1902, Karl Pearson, one of the more important figures in the history of statistics, wrote in *Biometrika* of the "...now established use of the word 'mode'" (April 1, p. 305). Two years earlier in the second edition of his *Grammar of Science*, Pearson had defined the mode as: "A numerical value for which such frequency is greatest" (p. 382).

The Pythagoreans and Their Ten Means

As far as is known, the concept of the mean first was formulated by a group of pre-Socratic philosophers, the Pythagoreans. Many students will recall having previously encountered the Pythagorean theorem regarding right triangles. The Pythagorean school of thought was begun by Pythagoreas, who was born in the sixth century B.C. in a Greek settlement on Samos, an island in the Aegean Sea. The Pythagoreans were impressed with numbers, to which they attached mystical significance. According to the Pythagorean metaphysical view, numbers constituted the essence of substance or reality. Any object in space could be expressed as a certain geometric form that, in turn, could be reduced to numbers. The Pythagoreans believed that the earth and other heavenly bodies were perfect spheres moving in perfect circles, all of which could be reduced to and explained in terms of numbers.

The Pythagoreans discovered the numerical relationships on which the classical system of harmony is based. The sounds of two plucked strings harmonize only when their lengths have special ratios. Thus, the ratio of 2 to 1 creates the musical interval called an octave; the ratio of 3 to 2, the interval called a fifth, and so forth. The Pythagoreans constructed musical scales, each a succession of notes separated by musical intervals that permitted harmonizing by the chords composed of these notes. To the Pythagoreans pleasing sounds were harmonious because they were based on pleasing numbers and the ratios of such numbers. All natural phenomena, since they could be reduced to numbers, were thus harmonious. The heavenly bodies gave off sounds proportional to their speed, and this speed increased with the diameters of their circular paths. Futhermore, because the speeds and distances of the heavenly bodies existed in certain fixed ratios, they were in harmony with each other—thus giving rise to the "music of the spheres."

The theory of means was developed very early in the Pythagorean school. A fragment "On Music" defines three means: arithmetic, geometric, and subcontrary (later called the harmonic). Because of the Pythagorean fixation on ratios, however, these means were conceived of in terms of only two numbers. Thus, the arithmetic mean of two numbers was that value exceeding the smaller number by the same amount as the larger number exceeded the arithmetic mean. For example, the arithmetic mean of the numbers 2 and 8 is 5, and 5 exceeds 2 by the same amount that 8 exceeds 5. The geometric mean was conceived of as that value for which the ratio of the smaller number to the geometric mean was the same as the ratio of the geometric mean to the larger number. For example the geometric mean of the numbers 2 and 8 is 4 [$\sqrt{(2)(8)} = 4$], and the ratio of 2 to 4 is the same as the ratio of 4 to 8.

The harmonic mean, which has not been previously described, was conceived of as that value for which the ratio of the difference between the larger number and the harmonic mean to the larger number was the same as the ratio of the difference between the harmonic mean and the smaller number to the smaller number. For example, the harmonic mean of the numbers 2 and 6 is 3. This is true since the ratio of the difference between 6 and 3 to 6, 3/6, is the same as the ratio of the difference between 3 and 2 to 2, 1/2.

Although the harmonic mean is rarely used today, it can be calculated most simply for any n by taking the *reciprocal of the arithmetic mean of the reciprocals of the n scores*. The procedure is illustrated in Table 3.13. The theory of means was extended in the Pythagorean school by adding more and more means, each based on a different ratio. Eventually, ten means were developed (cf. Heath, 1930, pp. 51–53). Apparently, however, the Pythagoreans never applied their means to more than 2 numbers. Their fascination with numerical ratios simultaneously led them to some important discoveries, yet prevented them from proceeding any further with these discoveries.

Fechner and Galton

Relative to the mean, the median is a recent development. Gauss, an early statistician, conceived of the idea in 1816 but placed little emphasis upon it. The early names usually associated with the median are Gustav Fechner (1801–1887) and Sir Francis Galton (1822–1911), both of whom are important figures in the history of psychology.

Gustav Fechner had a scholarly and scientific career that was long and productive. He is chiefly remembered in psychology for his development of psychophysics. Like an intriguingly large number of other early figures in the history of psychology (e.g., Wundt, Pavlov, James, Thorndike, and Jung), he was a minister's son. After receiving his early training in medicine at the University of Leipzig, Fechner's interest shifted to physics. He remained at Leipzig, at first without official appointment, to translate more than a dozen books on physics and chemistry from French into German. He also published over forty articles. One of these articles on the quantitative measurement of direct current established his reputation as a physicist. All of this work resulted in a nervous breakdown, a condition complicated by an eye injury that was caused by gazing at the sun through colored glasses in the course of his research on after-images.

During the period of his invalidism and after his recovery, Fechner's religious convictions deepened. He was particularly troubled by the scientific materialism of his day, and developed a philosophical position asserting that the entire universe could be viewed as either mental and conscious, or nonconscious and inert. This view was advanced through a series of seven books. Fechner's development of psychophysics was an explicit attempt to give his philosophical position concrete, empirical form. In 1860, he published *Elemente der Psychophysik*, a treatment of the "exact science of the functional relations or dependency relations between body and mind." It was Fechner's belief that his philosophical

TABLE 3.13. *Calculation of the Harmonic Mean of 2 and 6.*

$$\Sigma = \frac{\frac{1}{2}}{\frac{1}{6}} = \frac{4}{6} \qquad \bar{X} = \left(\frac{1}{2}\right)\left(\frac{4}{6}\right) = \frac{4}{12}$$

harmonic mean = $1 / 4/12 = 3$

position would be buttressed by empirical demonstration of functional relationships between the variation in various physical stimulus dimensions (weight, brightness, distance, etc.) and the subjective experiences or sensations of such.

The next topic that caught Fechner's interest was aesthetics. During the next decade or so of his life, until 1876, he devoted his attention to pioneer studies in experimental aesthetics—a field that he founded. This period of his life was terminated when he was 75. Even at this advanced age, however, this remarkable person did not end his scientific and scholarly career. In 1878, he published a work which systematically discussed the median, which he called *der Centralwerth*, or C, and its characteristics. He appeared to have no practical intent, but merely a theoretical interest in finding other measures of central tendency. Fechner gave complete instructions for computing the median, including the case in which some of the frequencies are greater than one. When the median lies between two X scores, he believed that any point between the scores could be considered the median and the choice was, therefore, arbitrary.

In this same paper Fechner also discussed the mode, which he called *der dichteste Werth*, or D. He pointed out that if C and the arithmetic mean do not coincide the distribution is not symmetrical, and, in this instance, C will lie between D and the mean.

Francis Galton was one of the last of a now extinct breed—the gentleman scientist. When his father died leaving him, at age 22, a fortune, he quit his medical studies at London and devoted his life to travel, exploration, and scholarship. He never held an academic or professional post; but due to his wealth, his boundless curiosity, and his considerable intelligence (his IQ has been estimated at approximately 200), he succeeded in making important early contributions to both psychology and statistics. As will be explained in Chapter 5, Galton developed the concepts of regression and correlation.

Galton, a cousin of Charles Darwin, was deeply impressed with evolution and heredity. He did important early studies demonstrating that men of genius tended to cluster in families. In 1883, he coined the term "eugenics." He was deeply convinced that genetic selection for talent and healthiness was absolutely necessary in any society that wished to promote its quality and status. In order to carry out such a eugenic program, however, individual differences must be measured. He strongly believed that virtually anything is quantifiable. Galton invented the mental test and is considered the founder of the psychology of individual differences. He developed a wide array of tests and questionnaires. His tests were for such things as pitch discrimination, visual judgment of extension, olfactory discrimination, color discrimination, and so on. Galton maintained that tests of sensory discrimination are indicative of judgment, and thus, of intelligence. His questionnaires for determining types and vividness of mental imagery revealed remarkable individual differences. In the course of such studies Galton discovered synesthesia. Galton also established fingerprinting as an almost infallible means of human identification. His taxonomy of prints is basically that used today.

In order to facilitate the collection of data, Galton established, in 1884, the Anthropometric Laboratory at the International Health Exhibition. The laboratory was later transferred to the South Kensington Museum in London where it was maintained for six years. Persons were admitted to the Laboratory for threepence in order to be measured by an array of tests. Eventually, data from 9,337 persons were collected. Galton's aim was to collect the type of data that could be used in a eugenic program, but the collected data yielded no important generalizations regarding individual differences. Galton did conclude that women were inferior to men in all of their capacities. However, such a generalization, even if true, would be of dubious value in any eugenic program.

Galton's early concern with the rank order of talent is what initially led him to develop the median, as well as the various quartiles, and the concept of percentiles. Such concepts were developed in various writings during the 1870's up to 1880. He was apparently unaware of Fechner's discussion of the median in 1878. Galton's preference for the median was based on his belief that rank order is more important than absolute position. To quote Galton:

> A knowledge of the distribution of any quality enables us to ascertain the Rank that each man holds among his fellows, in respect to that quality. This is a valuable piece of knowledge in this struggling and competitive world, where success is to the foremost, and failure to the hindmost, irrespective of absolute efficiency. A

blurred vision would be above all price to an individual in a nation of blind men, though it would hardly enable him to earn his bread elsewhere. When the distribution of any faculty has been ascertained, we can tell from the measurement, say of our child, how he ranks among other children in respect to that faculty, whether it be a physical gift, or one of health, or of intellect, or of morals. As the years go by, we may learn by the same means whether he is making his way towards the front, whether he just holds his place, or whether he is falling back towards the rear. Similarly, as regards the position of our class, or of our nation, among other classes and other nations (1889, pp. 36–37).

We can agree with Galton that in many situations, as for example athletic contests, rank-order information is the most salient and important type. On the other hand, we should not be misled by our "competitive world" into thinking that knowledge concerning the magnitudes of intervals is not also important.

From our contemporary perspective, some of Galton's ideas appear quaint and oversimplified. He was, however, a pioneer and a creative genius. His emphasis on quantification and commitment to statistics is as relevant in this day as it was in his. No one has made a more eloquent statement on these matters than Galton:

Some people hate the very name of statistics, but I find them full of beauty and interest. Whenever they are not brutalized, but delicately handled by the higher methods, and are warily interpreted, their power of dealing with complicated phenomena is extraordinary. They are the only tools by which an opening can be cut through the formidable thicket of difficulties that bars the path of those who pursue the Science of Man (1889, pp. 62–63).

This is the kind of insight that Achenwall never had.

EXERCISES

1. Define the following terms:

(a) arithmetic mean
(b) median
(c) third quartile
(d) geometric mean
(e) mode

2. What are four characteristics of the arithmetic mean?

3. How do the arithmetic mean, median, and mode differ in the types of information that they provide?

4. State verbally the meaning of the following symbols:

(a) n (b) $\sum$ (c) x (d) X (e) a (f) $f(X)$ (g) $\bar{X}$ (h) n_k (i) Q_1 (j) G

5. Using formula 3–1, calculate the arithmetic mean of the twenty achievement scores in Exercise 8 of Chapter 2.

6. Arrange the above scores in an ungrouped frequency distribution and calculate their arithmetic mean using formula 3–8.

7. Demonstrate that subtracting 10 from each of the above scores results in reduction of the arithmetic mean by 10.

8. What are the median and mode of the above twenty achievement scores?

9. What are the first and third quartiles for these twenty achievement scores?

10. The following scores were obtained by a group of students on an algebra test. Calculate the arithmetic mean of these scores using formula 3–1.

100	73	51	47	40
99	70	50	46	40
98	65	49	46	39
80	60	48	46	38
79	55	47	45	25
75	52	47	45	20

11. Arrange the above scores in an ungrouped frequency distribution and calculate their arithmetic mean using formula 3–8.

12. Demonstrate that dividing each of the above scores by 10 results in division of the arithmetic mean by 10.

13. For the scores given in Exercise 10, calculate:
(a) the median
(b) the first quartile
(c) the third quartile
(d) the modes

14. One group of ten scores has a mean of 15 and another group of twenty scores has a mean of 25. What is the mean of all the scores?

15. Calculate the medians of the following scores:

(a) 10, 8, 7, 6
(b) 10, 8, 7, 6, 5
(c) 10, 8, 8, 7, 6
(d) 10, 8, 8, 7, 7, 6
(e) 10, 8, 8, 7, 6, 6
(f) 10, 8, 8, 8, 7
(g) 11, 10, 8, 8, 7, 6
(h) 7, 7, 7, 7
(i) 11, 10, 9, 7, 6, 5

16. Calculate the geometric mean of the following scores:

8, 2

17. Calculate the first and third quartiles for the scores in the Stroebe *et al.* low-similarity condition (see Table 3.10).

18. First, use the case 2 procedure to calculate the median of the scores 8, 7, 5, and 4; and then, the case 1 procedure. What basis is there for preferring the case 1 procedure?

19. Explain in your own words why formulas 3–1 and 3–8 give the same results.

20. Why is the concept of central tendency more meaningful for symmetrical, unimodal distributions than for skewed or bimodal distributions?

21. Calculate the arithmetic mean, median, and mode for the imagery vividness scores of the students in your statistics class.

Prologue to Chapter 4

Chapter 4 continues the discussion of the description of univariate frequency distributions through a consideration of variability. Variability relates to the extent to which scores in a distribution are spread out or vary in magnitude. Variability, along with central tendency, is one of the salient ways in which univariate frequency distributions differ. As was the case with central tendency, there are several different measures or indices of variability. The various measures of variability tend to be somewhat more complex than the measures of central tendency. However, the increase in complexity is slight. If you have mastered the contents of Chapter 3, this chapter should pose no insurmountable difficulties.

4

Variability

If scores are spread out widely, variability is high; if scores are not spread out widely, variability is low. Consider, for example, the following test scores from a class of fifteen high school trigonometry students: 50, 51, 59, 60, 63, 71, 74, 77, 86, 89, 94, 98, 98, 98, and 99. At another high school, also with a trigonometry class of fifteen students, scores on the same test were: 70, 70, 73, 73, 73, 75, 78, 79, 80, 81, 81, 83, 84, 85, and 85. Scores at the second high school are obviously more homogeneous, or less variable, than scores at the first high school.

Chapter 2 pointed out that descriptive statistics includes the description of univariate frequency distributions. Chapter 3 was concerned with describing the central tendency of univariate frequency distributions. This chapter is concerned with describing univariate frequency distributions in terms of variability. Just as there are several indices or measures of central tendency, there are several indices or measures of variability. The most important of these measures are the *range, variance, standard deviation,* and *semi-interquartile range*.

RANGE

The most easily comprehensible measure of variability is the range. *The range of a distribution is the difference between the highest score and the lowest score.* Symbolically, this can be expressed as:

$$R = H - L \qquad (4\text{-}1)$$

where R is the range, H is the highest score, and L is the lowest score. For example, the range of the set of scores, 100, 59, 50, 48, 47, 46, and 10, is simply 100 minus 10, or 90.

Visual comparison of the frequency polygon in Fig. 2–3 for the Stroebe *et al.* high-similarity condition with the frequency polygon in Fig. 2–4 for the Stroebe *et al.* low-similarity condition seems to indicate a greater degree of variability in the latter. This perceived difference in variability is at least partially due to the fact that the range in the low-similarity condition is greater. In the low-similarity condition, $R = 6(7 - 1 = 6)$; in the high-similarity condition, $R = 4(7 - 3 = 4)$.

In *How to Lie with Statistics*, Darrel Huff points out that knowledge as to the range may prevent someone from being misled by consideration of only the central tendency. Huff gives the example of someone who chooses a camp site on the basis of an annual mean temperature. The inland desert of California and San Nicolos Island off the south coast are both considered since they have an annual mean temperature of 61 degrees. However, the temperature at San Nicolos Island ranges from 47 to 87 degrees, while the temperature in the desert ranges from 15 to 104 degrees. As Huff observes "...you can freeze or roast if you ignore the range" (1954, p. 52).

Advantages and Disadvantages

The range has two principal advantages as a measure of variability: ease of comprehension and ease of computation. Very little statistical sophistication is required to comprehend that the range measures the variability of a set of scores. And computation of the range is, of course, exceedingly simple.

The range also has two main disadvantages: First, it is based on the two extreme scores; these may bear little relation to the intervening scores. If one or both of the extreme scores deviate markedly from the remainder of the scores in the distribution, the range then will give a distorted picture of the variability of the total distribution. This is the case for the above mentioned set of scores (100, 59, 50, 48, 47, 46, 10) with $R = 90$. Second, the range exhibits less sampling stability than any of the other measures of variability. If repeated samples are drawn from a given population and the range is calculated for each of these samples, considerable variation is likely in the ranges from sample to sample. This characteristic becomes important when one is attempting to infer the variability of a population from that of a sample. The range of a sample tends to be a relatively poor estimate of the variability in the population because it varies so much from sample to sample.

VARIANCE AND STANDARD DEVIATION

Since one of the disadvantages of the range relates to the fact that it is based on only two scores, it seems reasonable to develop a measure of variability that is based on all of the scores. The first step is to note that by subtracting each of the scores in a set from the mean; that is, by converting all the Xs to xs, we obtain information as to the deviation or departure of the X scores from the mean. The greater the deviation of the scores from the mean is, the greater is the variability. Then, it might be thought that the sum of the deviations from the mean, $\sum x$, would provide an overall index of variability. However, the $\sum x$ equals zero, as was demonstrated in Chapter 3. Table 4.1 (columns 2 and 5) illustrates this fact.

Two solutions are possible to this problem. One involves summing the absolute values of the x scores (the x scores being treated as if they were all positive), and the other involves summing the squares of the x scores. Both of these procedures eliminate negative numbers and provide that a sum other than zero may be obtained. Since using the absolute values limits the mathematical treatment of the scores, the preferred solution is summing the squares of the x scores. The expression $\sum x^2$, or $\sum (X - \bar{X})^2$ is referred to as a *sum of squares*. Thus, the second step is to square each of the x scores and sum the squares; the calculation is illustrated in Table 4.1 (columns 3 and 6).

A sum of squares reflects the variability of a set of scores, but it also reflects something else, n. In general, as n increases, the sum of squares also increases. Each additional score that is not exactly equal to the mean increases the sum of squares. Thus, sums of squares cannot be used to compare the variabilities of distributions with differing ns. The solution to this problem takes us to the third and last step in obtaining a measure of variability based on all the scores. This step is to divide the sum of squares by n. Such division "adjusts" for the fact that certain sums of squares may be large simply because of a large n, and some may be small simply because of a small n. We thus arrive at an index or measure of variability, a mean of squared deviations, that can be meaningfully applied to the distributions with differing ns. This index is referred to as the *variance* and is defined as:

$$s^2 = \frac{\sum (X - \bar{X})^2}{n} = \frac{\sum x^2}{n} \tag{4-2}$$

where s^2 symbolizes the variance.

Closely associated with the variance is the *standard deviation*. This is obtained by taking the square root of the variance. Symbolically, the standard deviation is defined as:

$$s = \sqrt{\frac{\sum (X - \bar{X})^2}{n}} = \sqrt{\frac{\sum x^2}{n}} \tag{4-3}$$

Although the variance is more amenable to subsequent mathematical treatment, the standard deviation is the more frequently reported measure of variability.

TABLE 4.1. *Calculation of the Variance and Standard Deviation for the Stroebe et al. High-Similarity Condition.*

X	$(X - \bar{X})$ or x	x^2	X	$(X - \bar{X})$ or x	x^2
7	1.30	1.69	6	0.30	0.09
7	1.30	1.69	6	0.30	0.09
7	1.30	1.69	6	0.30	0.09
7	1.30	1.69	6	0.30	0.09
7	1.30	1.69	6	0.30	0.09
7	1.30	1.69	6	0.30	0.09
7	1.30	1.69	6	0.30	0.09
7	1.30	1.69	6	0.30	0.09
7	1.30	1.69	6	0.30	0.09
7	1.30	1.69	6	0.30	0.09
7	1.30	1.69	6	0.30	0.09
7	1.30	1.69	6	0.30	0.09
6	0.30	0.09	6	0.30	0.09
6	0.30	0.09	6	0.30	0.09
6	0.30	0.09	6	0.30	0.09
6	0.30	0.09	5	−0.70	0.49
6	0.30	0.09	5	−0.70	0.49
6	0.30	0.09	5	−0.70	0.49
6	0.30	0.09	5	−0.70	0.49
6	0.30	0.09	4	−1.70	2.89
6	0.30	0.09	4	−1.70	2.89
6	0.30	0.09	4	−1.70	2.89
6	0.30	0.09	4	−1.70	2.89
6	0.30	0.09	4	−1.70	2.89
6	0.30	0.09	4	−1.70	2.89
6	0.30	0.09	4	−1.70	2.89
6	0.30	0.09	3	−2.70	7.29
6	0.30	0.09	3	−2.70	7.29
6	0.30	0.09	3	−2.70	7.29
6	0.30	0.09	3	−2.70	7.29
			$\Sigma = 342$	$\Sigma = 0.00$	$\Sigma = 74.60$

$$\bar{X} = \frac{\Sigma X}{n} = \frac{342}{60} = 5.70$$

$$s^2 = \frac{\Sigma x^2}{n} = \frac{74.60}{60} = 1.24$$

$$s = \sqrt{\frac{\Sigma x^2}{n}} = \sqrt{1.24} = 1.11$$

The standard deviation has the advantage of being expressed in terms of the same measurement units as the original scores, while the variance is expressed in terms of the square of these units.

Table 4.1 contains the calculation of the variance and standard deviation for the Stroebe *et al.* high-similarity condition; Table 4.2 contains the calculation of the variance and standard deviation for the Stroebe *et al.* low-similarity condition. Notice in Table 4.2, $\sum x = -0.20$ rather than zero. The discrepancy results from rounding error in calculation of the mean.

In the high-similarity condition, $s^2 = 1.24$, and $s = 1.11$; in the low-similarity condition, $s^2 = 1.83$, and $s = 1.35$. In agreement with the visual impression of Figs. 2–3 and 2–4, the low-similarity condition has greater variability.

Estimating a Population Variance and Standard Deviation

If an investigator simply wishes to know the variance and standard deviation of a set of scores, then, formulas 4–2 and 4–3 give the correct procedure. On the other hand, if an investigator wishes to infer the variance of a population from the variability within a sample subset, a slightly different formula is required. (The terms "sample" and "population" will be defined more precisely in Chapter 6. For now a population can be regarded simply as a large set from which sample subsets are drawn.) *The estimate of the variance of a population is:*

$$\hat{s}^2 = \frac{\sum (X - \bar{X})^2}{n - 1} = \frac{\sum x^2}{n - 1} \tag{4-4}$$

In formula 4–4, $\hat{s}^2$ ("s hat squared") stands for the estimate of a population variance, n for the number of cases in the sample, and x for $X - \bar{X}$. The statistic $\hat{s}^2$ is calculated from a sample for the purpose of inferring the variance of a population.

By taking the square root of $\hat{s}^2$, we obtain $\hat{s}$, *an estimate of the population standard deviation.* Symbolically, the estimate of a population standard deviation is expressed as:

$$\hat{s} = \sqrt{\frac{\sum (X - \bar{X})^2}{n - 1}} = \sqrt{\frac{\sum x^2}{n - 1}} \tag{4-5}$$

Table 4.3 illustrates the calculation of these estimates using the scores from the Stroebe *et al.* high-similarity condition (see Table 4.1).

What is the reason for the use of $\hat{s}$ and $\hat{s}^2$, as opposed to s and s^2 as estimates of the population standard deviation and variance? The statistics $\hat{s}$ and $\hat{s}^2$ are used because, on the average, they are closer to the population standard deviation

TABLE 4.2. *Calculation of the Variance and Standard Deviation for the Stroebe et al. Low-Similarity Condition.*

X	$(X - \bar{X})$ or x	x^2	X	$(X - \bar{X})$ or x	x^2
7	3.73	13.91	3	−0.27	0.07
6	2.73	7.45	3	−0.27	0.07
6	2.73	7.45	3	−0.27	0.07
6	2.73	7.45	3	−0.27	0.07
5	1.73	2.99	3	−0.37	0.07
5	1.73	2.99	3	−0.27	0.07
5	1.73	2.99	3	−0.27	0.07
5	1.73	2.99	3	−0.27	0.07
5	1.73	2.99	3	−0.27	0.07
5	1.73	2.99	3	−0.27	0.07
4	0.73	0.53	3	−0.27	0.07
4	0.73	0.53	3	−0.27	0.07
4	0.73	0.53	3	−0.27	0.07
4	0.73	0.53	3	−0.27	0.07
4	0.73	0.53	2	−1.27	1.61
4	0.73	0.53	2	−1.27	1.61
4	0.73	0.53	2	−1.27	1.61
4	0.73	0.53	2	−1.27	1.61
4	0.73	0.53	2	−1.27	1.61
4	0.73	0.53	2	−1.27	1.61
4	0.73	0.53	2	−1.27	1.61
4	0.73	0.53	2	−1.27	1.61
4	0.73	0.53	2	−1.27	1.61
3	−0.27	0.07	1	−2.27	5.15
3	−0.27	0.07	1	−2.27	5.15
3	−0.27	0.07	1	−2.27	5.15
3	−0.27	0.07	1	−2.27	5.15
3	−0.27	0.07	1	−2.27	5.15
3	−0.27	0.07	1	−2.27	5.15
			$\Sigma = 196$	$\Sigma = -0.20$	$\Sigma = 109.56$

$$\bar{X} = \frac{\Sigma X}{n} = \frac{196}{60} = 3.27$$

$$s^2 = \frac{\Sigma x^2}{n} = \frac{109.56}{60} = 1.83$$

$$s = \sqrt{\frac{\Sigma x^2}{n}} = \sqrt{1.83} = 1.35$$

TABLE 4.3. *Calculation of the Estimate of the Population Variance and the Estimate of the Population Standard Deviation for the Stroebe et al. High-Similarity Condition.*

$\sum x^2 = 74.60, \quad n = 60$

$$\hat{s}^2 = \frac{\sum x^2}{n-1} = \frac{74.60}{60-1} = \frac{74.60}{59} = 1.26$$

$$\hat{s} = \sqrt{\frac{\sum x^2}{n-1}} = \sqrt{1.26} = 1.12$$

and variance than s and s^2. This can be illustrated for $\hat{s}$ and s by the following procedure. First, a population of 100 scores is constructed and its standard deviation is computed. Second, each of the scores is written on a card and the entire deck of 100 cards is thoroughly shuffled. Third, a sample of 10 cards is drawn from the deck, and s and $\hat{s}$ computed for the ten scores. Fourth, the ten cards are returned to the deck, the deck is shuffled, another sample of 10 cards is drawn, and s and $\hat{s}$ are computed for the new sample. Fifth, step four is repeated a large number of times, thus generating many values of s and $\hat{s}$. Sixth and finally, the mean of all the sample ss is computed, and the mean of all the sample $\hat{s}$s is computed. If enough samples are drawn, such a procedure will reveal that the mean of the $\hat{s}$s is closer to the population standard deviation than is the mean of the ss. In general, the ss underestimate the population standard deviation. It is for that reason that $\hat{s}$, rather than s, is used as an estimate of the population standard deviation.

It is further evident that with larger and larger ns the -1 correction makes less and less difference and s and $\hat{s}$ become more and more similar in value. Stated differently, as n increases, s becomes a more accurate estimate of the population standard deviation.

Degrees of Freedom

The derivation of formulas 4–4 and 4–5 involves calculus, and thus, will not be presented here. It is important to know, however, that the $n-1$ term in these formulas is referred to as *degrees of freedom*. What is meant by degrees of freedom? Every sum of a sample set of quantities or sum of a sample set of squared quantities has associated with it a number called degrees of freedom. This number equals the number of quantities that are completely free to vary. For example, if a sample set is obtained from a larger set so that each member of the larger set has an equal chance of being selected, the mean of the sample has associated with it *degrees of freedom* equal to the number of scores in the sample set. This is true because the mean is derived directly from the sum of the sample set of scores:

$$\bar{X} = \frac{\sum X}{n}$$

and each score in the sum is completely free to vary; that is, it could be any score in the population.

In obtaining a sum of squared deviation scores, however, each quantity or score in the sample set is restricted prior to squaring by the subtraction of the sample mean. Subtraction of the sample mean renders the sum of such deviations equal to zero; that is, $\sum x = 0$. Thus, while the n original quantities are free to vary (free to be any value in the larger set), the n deviation scores are limited to a set that sums to zero. If we conceive of a set of deviation scores from which n scores are sampled, then, $n - 1$ of the deviation scores can be any value in the set; and one score is determined by the restriction that $\sum x = 0$.

This can be illustrated simply by a sample set of five deviation scores. Given only the restriction that the five deviation scores must sum to zero, we can guess at the values of the five scores. The first four guesses are completely free. We have no information that in any way would determine them. We might guess 4, 16, 28, and -30. Having guessed four scores, however, the restriction that $\sum x = 0$ determines the final choice. Since:

$$x_1 + x_2 + x_3 + x_4 + x_5 = 0$$

then,

$$4 + 16 + 28 - 30 + x_5 = 0$$

and,

$$x_5 = 30 - 4 - 16 - 28$$
$$= -18$$

Since a sample variance and standard deviation are derived directly from a sum of squared deviation scores, the sample variance and standard deviation have associated degrees of freedom equal to that associated with the sum of deviation scores, $n - 1$.

Advantages and Disadvantages

The variance and standard deviation have three principal advantages as measures of variability. First, they are based on all of the scores in a given distribution;

and thus, generally reflect the variability in the total distribution. Second, the variance and standard deviation have mathematical properties that make them very amenable to subsequent algebraic manipulations. And third, the statistics used to estimate a population variance and standard deviation exhibit considerable stability from one sample to the next.

One possible disadvantage of the standard deviation and variance as measures of variability is that they are greatly influenced by extreme scores. The operation of squaring score deviations from the mean magnifies the larger deviations, giving them greater weight in determining the size of the resultant indices.

Computational Formulas

Formulas 4–2 and 4–3 for the variance and standard deviation are the definitional formulas. While the variance and standard deviation may be calculated by these formulas, it is easier, particularly when using a desk calculator, to use alternative formulas. These so-called computational formulas, which are algebraically equivalent to the definitional ones, are:

$$s^2 = \frac{\sum X^2}{n} - \bar{X}^2 \qquad (4\text{--}6)$$

$$s = \sqrt{\frac{\sum X^2}{n} - \bar{X}^2} \qquad (4\text{--}7)$$

The first step in demonstrating the algebraic equivalence of the definitional and computational formulas is to prove that the sum of squares ($\sum x^2$) in the numerator of the definitional formulas 4–2 and 4–3 is equal to:

$$\sum X^2 - \frac{(\sum X)^2}{n}$$

By definition, $\sum x^2 = \sum (X - \bar{X})^2$. Expanding the term $(X - \bar{X})^2$, $\sum x^2 = \sum (X^2 - 2X\bar{X} + \bar{X}^2)$. Using summation rules 1, 2, and 3, $\sum x^2 = \sum X^2 - 2\bar{X} \sum X + n\bar{X}^2$. Substituting $\sum X / n$ for $\bar{X}$:

$$\sum x^2 = \sum X^2 - 2\frac{(\sum X)(\sum X)}{n} + n\frac{(\sum X)^2}{n^2}$$

$$= \sum X^2 - \frac{2(\sum X)^2}{n} + \frac{(\sum X)^2}{n}$$

$$= \sum X^2 - \frac{(\sum X)^2}{n}$$

Now, substituting for the sum of squares in the definitional formula of the variance as follows:

$$s^2 = \frac{\sum x^2}{n}$$

$$= \frac{\sum X^2 - [(\sum X)^2 / n]}{n}$$

$$= \frac{\sum X^2}{n} - \frac{(\sum X)^2}{n^2}$$

$$= \frac{\sum X^2}{n} - \bar{X}^2$$

This is, of course, the computational formula for the variance. By taking the square root of both sides of the equation, the computational formula for the standard deviation is obtained:

$$s = \sqrt{\frac{\sum X^2}{n} - \bar{X}^2}$$

Table 4.4 illustrates the use of the computational formulas for the variance and standard deviation. The obtained values for the variance and standard deviation, 1.24 and 1.11, are the same as those which were previously obtained when the definitional formulas were applied to the same scores (see Table 4.1).

The computational formulas for the estimate of a population variance and the estimate of a population standard deviation are:

$$\hat{s}^2 = \frac{n \sum X^2 - (\sum X)^2}{n(n-1)} \tag{4-8}$$

$$\hat{s} = \sqrt{\frac{n \sum X^2 - (\sum X)^2}{n(n-1)}} \tag{4-9}$$

Since it has already been proven that:

$$\sum x^2 = \sum X^2 - \frac{(\sum X)^2}{n}$$

the complete proof of formulas 4–8 and 4–9 is fairly simple. By definition:

$$\hat{s}^2 = \frac{\sum x^2}{n-1}$$

Substituting for $\sum x^2$:

$$\hat{s}^2 = \frac{\sum X^2 - [(\sum X)^2 / n]}{n-1}$$

$$= \frac{n \sum X^2 - (\sum X)^2}{n(n-1)}$$

TABLE 4.4. *Calculation of the Variance and Standard Deviation for the Stroebe et al. High-Similarity Condition Using the Computational Formulas.*

X	X^2	X	X^2
7	49	6	36
7	49	6	36
7	49	6	36
7	49	6	36
7	49	6	36
7	49	6	36
7	49	6	36
7	49	6	36
7	49	6	36
7	49	6	36
7	49	6	36
7	49	6	36
6	36	6	36
6	36	6	36
6	36	6	36
6	36	5	25
6	36	5	25
6	36	5	25
6	36	5	25
6	36	4	16
6	36	4	16
6	36	4	16
6	36	4	16
6	36	4	16
6	36	4	16
6	36	4	16
6	36	3	9
6	36	3	9
6	36	3	9
6	36	3	9
		$\Sigma = 342$	$\Sigma = 2024$

$$\bar{X} = \frac{\Sigma X}{n} = \frac{342}{60} = 5.70$$

$$s^2 = \frac{\Sigma X^2}{n} - \bar{X}^2 = \frac{2024}{60} - (5.70)^2 = 1.24$$

$$s = \sqrt{\frac{\Sigma X^2}{n} - \bar{X}^2} = \sqrt{1.24} = 1.11$$

This is the computational formula for variance. By taking the square root of both sides of this equation, the computational formula for the standard deviation is obtained:

$$\hat{s} = \sqrt{\frac{n \sum X^2 - (\sum X)^2}{n(n-1)}}$$

Table 4.5 illustrates the calculation of the estimate of a population variance and the estimate of a population standard deviation using the computational formulas. The scores are from the Stroebe *et al.* high-similarity condition (Table 4.4). The same values are obtained as when the definitional formulas were used (Table 4.3).

TABLE 4.5. *Calculation of the Estimate of the Variance and the Estimate of the Standard Deviation for the Stroebe et al. High-Similarity Condition Using the Computational Formulas.*

$\sum X = 342, \quad \sum X^2 = 2024, \quad n = 60$

$$\hat{s}^2 = \frac{n \sum X^2 - (\sum X)^2}{n(n-1)} = \frac{(60)(2024) - (342)^2}{60(60-1)} = 1.26$$

$$\hat{s} = \sqrt{\frac{n \sum X^2 - (\sum X)^2}{n(n-1)}} = \sqrt{1.26} = 1.12$$

Characteristics

The variance and standard deviation have two characteristics with which the student should be familiar.

1. *Adding a constant to each score in a set of scores or subtracting a constant from each score in a set of scores does not change either the standard deviation or the variance of the set of scores.* This is illustrated in Fig. 4–1. Here, a constant is added to each score in the set so as to shift the entire distribution of scores further to the right on the horizontal axis. The result of this addition is to increase the arithmetic mean by a value equal to the constant, but not to change the variation of the scores around the arithmetic mean. The variability of the scores has not been changed by adding a constant. Conversely, if the constant that was added is subsequently subtracted so as to move the distribution back to its original position on the horizontal axis, the variability is also not changed.
2. *The multiplication of each score in a set of scores by a constant results in multiplication of the standard deviation by that constant, and in multiplication of the variance by the square of that constant. The division of each score in a set of scores by a constant results in division of the standard deviation by the constant, and in division of the variance by the square of that constant.*

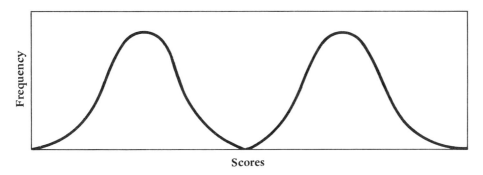

FIGURE 4-1. *Two frequency distributions differing by a constant value but having the same variance and standard deviation.*

The division rule of this second characteristic and the subtraction rule of the first characteristic can both be used, under certain circumstances, to provide coding techniques for calculating the standard deviation and variance. When the scores in a given set are very large, the labor involved in calculating the standard deviation and variance can be greatly reduced by subtracting a constant from each score or dividing each score by a constant. After subtracting a constant and calculating the standard deviation and variance, no correction is necessary. The resultant standard deviation and variance are equal to those for the original scores. After dividing by a constant, however, it is necessary to multiply the standard deviation by that constant and to multiply the variance by the square of that constant in order to obtain the standard deviation and variance of the original scores.

> In order to prove the multiplication and division rules of the second characteristic, it is first necessary to put the formula for the standard deviation into a form in which each of the X scores can be multiplied or divided by a constant.
> This can be accomplished by the following changes. By definition:
>
> $$s = \sqrt{\frac{\sum x^2}{n}}$$
>
> $$= \sqrt{\frac{\sum (X - \bar{X})^2}{n}}$$
>
> Substituting $\sum X / n$ for $\bar{X}$:
>
> $$s = \sqrt{\frac{\sum [X - (\sum X / n)]^2}{n}}$$

Variance and Standard Deviation

Now it is possible to multiply the X scores by a constant, proving that the result is the multiplication of the standard deviation by that constant. Multiplying each score by a constant, a, and computing the standard deviation of the resultant aX scores gives:

$$s_{aX} = \sqrt{\frac{\sum[aX - (\sum aX/n)]^2}{n}}$$

Substituting $a \sum X$ for $\sum aX$, according to summation rule 3:

$$s_{aX} = \sqrt{\frac{\sum[aX - (a\sum X/n)]^2}{n}}$$

Substituting $\bar{X}$ for $\sum X/n$:

$$s_{aX} = \sqrt{\frac{\sum(aX - a\bar{X})^2}{n}}$$

Expanding the $(aX - a\bar{X})^2$ term:

$$s_{aX} = \sqrt{\frac{\sum(a^2X^2 - 2a^2X\bar{X} + a^2\bar{X}^2)}{n}}$$

$$= \sqrt{\frac{a^2\sum(X^2 - 2X\bar{X} + \bar{X}^2)}{n}}$$

Reducing $(X^2 - 2X\bar{X} + \bar{X}^2)$ to $(X - \bar{X})^2$, and removing a^2 from the radical:

$$s_{aX} = a\sqrt{\frac{\sum(X - \bar{X})^2}{n}}$$

$$= as$$

The proof for division is analogous to the one for multiplication:

$$s_{X/a} = \sqrt{\frac{\sum\left(\frac{1}{a}X - \frac{(1/a)\sum X}{n}\right)^2}{n}}$$

$$= \sqrt{\frac{\sum\left(\frac{1}{a}X - \frac{1}{a}\bar{X}\right)^2}{n}}$$

$$= \sqrt{\frac{\sum\left(\frac{1}{a^2}X^2 - \frac{2}{a^2}X\bar{X} + \frac{1}{a^2}\bar{X}^2\right)}{n}}$$

$$= \sqrt{\frac{\frac{1}{a^2}\sum(X^2 - 2X\bar{X} + \bar{X}^2)}{n}}$$

$$= \sqrt{\frac{\sum(X - \bar{X})^2/a^2}{n}}$$

$$= \frac{1}{a}\sqrt{\frac{\sum(X - \bar{X})^2}{n}}$$

$$= \frac{s}{a}$$

The student can gain some practice in manipulating these symbols by demonstrating the effect on the variance of the X scores by multiplying and dividing by a constant.

The student's typical reaction after calculating his or her first standard deviation is: "So what?" A number has been obtained that seems almost totally devoid of meaning. Part of this problem can be handled by recalling that the calculated number is an index of variability. The larger the number, the greater the variability of scores. When two distributions have different standard deviations, the one with the greater standard deviation is the one with the greater variability.

But why should one wish to have information regarding the variability of a set of scores? There are two general reasons: First, information regarding variability may be interesting in its own right. Suppose, for example, that a teacher has difficulty "pitching" the level of instruction. If the level is too high, some students are confused; and, if the level is too low, some students are bored. In such a situation information regarding the variability of class aptitude would be of considerable interest. On the basis of this information, the teacher might construct relatively homogeneous (low variability) groups within the class. A second reason why information regarding variability is of interest is that variability within experimental conditions reflects something called *error variance*. Consider again the Stroebe *et al.* experiment. The variability within the high-similarity condition and within the low-similarity condition results from unknown, unspecified, or uncontrolled causes; i.e., is due to error. Thus, the variance of the scores within conditions is referred to as error variance. As later chapters will make clear, error variance is of crucial importance in statistical inference; i.e., in the determination of whether or not a manipulated property, such as similarity, has an effect that is not due to chance. In general, it is the case that the less the variability *within* experimental conditions, the more confidence an investigator may have that the difference *between* experimental conditions is not due to chance. For this reason, then, it is of interest to have information regarding the variabilities within the Stoebe *et al.* high- and low-similarity conditions. Such information is not particularly interesting in its own right, but nonetheless, is of crucial importance in judging the import of the similarity manipulation.

Variance and Standard Deviation of an Ungrouped Frequency Distribution

In order to calculate the variance and standard deviation of a frequency distribution, it is necessary to modify the basic formulas for variance and standard deviation, taking into account the frequencies associated with different score values. The formula for the variance of a frequency distribution is:

$$s^2 = \sum_{1}^{k}(X - \bar{X})^2 f(X) / n \qquad (4\text{-}10)$$

where $f(X)$ symbolizes the frequency of occurrence of a particular X value, and k symbolizes the number of different score values.

Notice that the variance of a frequency distribution (formula 4–10) is equivalent to the variance of scores (formula 4–2), which the frequency distribution represents. The numerator in formula 4–10, which calls for the summation of the products of the squared difference between each score value and the mean "times" the frequency of the score value, is equivalent to the numerator of formula 4–2, which calls for the summation of squared differences between each score in the set and the mean.

The computational formula for the variance of a frequency distribution (formula 4–11) is, likewise, equivalent to the computational formula for the variance of a set of scores:

Computational
$$s^2 = \left[\sum_{1}^{k} X^2 f(X) / n\right] - \left[\sum_{1}^{k} X f(X) / n\right]^2 \qquad (4\text{-}11)$$

The standard deviation of a frequency distribution is the square root of the variance; thus, the definitional and computational formulas for the standard deviation are:

Definitional
$$s = \sqrt{\sum_{1}^{k}(X - \bar{X})^2 f(X) / n} \qquad (4\text{-}12)$$

and

Computational
$$s = \sqrt{\left[\sum_{1}^{k} X^2 f(X) / n\right] - \left[\sum_{1}^{k} X f(X) / n\right]^2} \qquad (4\text{-}13)$$

Computation of the variance and standard deviation of a frequency distribution is illustrated in Table 4.6 for the Stroebe *et al.* high-similarity condition. The values in the $Xf(X)$ column are obtained by multiplying the values in the X column by the values in the $f(X)$ column. The values in the $X^2 f(X)$ column are obtained by multiplying the values in the $Xf(X)$ column by the values in the X column. Notice that the obtained values of s^2 and s (1.24 and 1.11) are the same as when the scores are not grouped in a frequency distribution (Table 4.4).

Review of Formulas

Table 4.7 reviews the definitional and computational formulas for the variance, standard deviation, estimate of the variance (of a population), and estimate of the

TABLE 4.6. *Calculation of the Variance and Standard Deviation of an Ungrouped Frequency Distribution for the Stroebe et al. High-Similarity Condition.*

X	f(X)	Xf(X)	X²f(X)
7	12	84	588
6	33	198	1188
5	4	20	100
4	7	28	112
3	4	12	36
	$\Sigma = 60$	$\Sigma = 342$	$\Sigma = 2024$

$$s^2 = \left[\sum_{1}^{k} X^2 f(X)/n\right] - \left[\sum_{1}^{k} Xf(X)/n\right]^2$$

$$s^2 = \frac{2024}{60} - \left(\frac{342}{60}\right)^2$$

$$s^2 = 1.24$$

$$s = \sqrt{1.24}$$

$$s = 1.11$$

standard deviation (of a population). Notice that the definitional and computational formulas for ungrouped frequency distributions are not given for the estimate of the variance and the estimate of the standard deviation. These formulas are omitted because they are rarely used.

SEMI-INTERQUARTILE RANGE

A final measure of variability is the semi-interquartile range. If the difference between the numerical values representing the third and first quartiles ($Q_3 - Q_1$) is referred to as the interquartile range, then, the *semi-interquartile range is one-half the interquartile range.* Symbolically this is expressed as:

$$Q = \frac{Q_3 - Q_1}{2} \qquad (4\text{--}14)$$

where Q stands for the semi-interquartile range. Because 25 percent of the scores in a distribution lie above Q_3, and another 25 percent lie below Q_1; 50 percent of the scores lie between Q_3 and Q_1. The size of Q is affected by the closeness of scores to the median. If scores tend to bunch up around the

TABLE 4.7. Review of Formulas.

	s^2 VARIANCE	s STANDARD DEVIATION	$\hat{s}^2$ ESTIMATE OF THE VARIANCE	$\hat{s}$ ESTIMATE OF THE STANDARD DEVIATION
Definitional Formulas	$\dfrac{\sum x^2}{n}$	$\sqrt{\dfrac{\sum x^2}{n}}$	$\dfrac{\sum x^2}{n-1}$	$\sqrt{\dfrac{\sum x^2}{n-1}}$
Definitional Formulas for Ungrouped Frequency Distribution	$\sum_{1}^{k}(X-\bar{X})^2 f(X)/n$	$\sqrt{\sum_{1}^{k}(X-\bar{X})^2 f(X)/n}$		
Computational Formulas	$\dfrac{\sum X^2}{n}-\bar{X}^2$	$\sqrt{\dfrac{\sum X^2}{n}-\bar{X}^2}$	$\dfrac{n\sum X^2-(\sum X)^2}{n(n-1)}$	$\sqrt{\dfrac{n\sum X^2-(\sum X)^2}{n(n-1)}}$
Computational Formulas for Ungrouped Frequency Distribution	$\left[\sum_{1}^{k}X^2 f(X)/n\right]-\left[\sum_{1}^{k}Xf(X)/n\right]^2$	$\sqrt{\left[\sum_{1}^{k}X^2 f(X)/n\right]-\left[\sum_{1}^{k}Xf(X)/n\right]^2}$		

Variability

median, Q then will be small, because the difference between Q_3 and Q_1 will be small. If scores deviate considerably from the median, Q then will be large, because the difference between Q_3 and Q_1 will be large.

> The semi-interquartile range can be regarded as the arithmetic mean of the distances between the third quartile and the median ($Q_3 - $ Mdn), and the median and first quartile (Mdn $- Q_1$). The following formulas illustrate the equivalence of this mean and formula 4–14:
>
> $$Q = \frac{(Q_3 - \text{Mdn}) + (\text{Mdn} - Q_1)}{2}$$
>
> $$= \frac{Q_3 - (\text{Mdn} - \text{Mdn}) - Q_1}{2}$$
>
> $$= \frac{Q_3 - Q_1}{2}$$

The first and third quartiles for the Stroebe *et al.* high-similarity condition are 5.50 and 6.41, respectively (see Table 3.11); thus:

$$Q = \frac{Q_3 - Q_1}{2}$$

$$= \frac{6.41 - 5.50}{2}$$

$$= 0.46$$

Although the first and third quartiles for the Stroebe *et al.* low-similarity condition have not been previously calculated, these values are 2.40 and 4.12. Given these values:

$$Q = \frac{Q_3 - Q_1}{2}$$

$$= \frac{4.12 - 2.40}{2}$$

$$= 0.86$$

The low-similarity condition has a greater semi-interquartile range (0.86) than does the high-similarity condition (0.46). Recall that the low-similarity condition also has a greater range, variance, and standard deviation than the high-similarity condition, and that such results are consistent with the visual comparison of the frequency polygons (Figs. 2–3 and 2–4).

Advantages and Disadvantages

The semi-interquartile range has two principal advantages. First, it ranks next to the range in ease of comprehension and calculation. And second, the semi-interquartile range is the preferred measure of variability in situations in which the median is the most appropriate measure of central tendency. Both the median and the semi-interquartile range are not as influenced by extreme scores as are the mean, standard deviation, variance, and range.

The semi-interquartile range has two principal disadvantages. First, it is not as amenable to meaningful algebraic manipulations as is the variance. Second, the semi-interquartile range does not take into account the extent of deviation of each individual score from the central tendency.

HISTORICAL PERSPECTIVE OF THE INDICES OF VARIABILITY

Range and Semi-Interquartile Range

The range, like the mode, involves such a low level of abstraction that its origin is unknown. Many individuals used the concept without making much of it. Galton, for example, frequently referred to the range of talent.

We do know that Galton invented the semi-interquartile range. As was pointed out in Chapter 3, Galton was primarily oriented to the rank-order characteristics of data. Thus, the semi-interquartile range was a natural accompaniment to the median. His discussions of data typically involved references to both M (median) and Q (semi-interquartile range). He was deeply impressed with variability and showed some impatience with statisticians who were concerned only with central tendency:

> It is difficult to understand why statisticians commonly limit their inquiries to Averages, and do not revel in more comprehensive views. Their souls seem as dull to the charm of variety as that of the native of one of our flat English counties, whose retrospect of Switzerland was that if its mountains could be thrown into its lakes two nuisances would be got rid of at once (1889, p. 62).

The student can doubtless sympathize with the views of those statisticians with whom Galton was impatient. Somehow or other, it seems inherently more interesting to describe frequency distributions in terms of their central tendency than in terms of their variability. As was indicated earlier, however, concern with central tendency in the context of statistical inference demands a concern with variability. To the extent that Stroebe *et al.* wished to generalize the central tendency difference between the high- and low-similarity conditions, they had to be concerned with the variabilities of the scores within these conditions. The less the variabilities were within conditions, the more confidence they could have that the central tendency differences between conditions were not due to chance. Thus, it is understandable that early concern with variability, and in particular with the standard deviation, arose in the context of concern with probability.

Standard Deviation

The standard deviation, as defined by formula 4–3, was incorporated by numerous early writers into formulas relating to probability and the normal distribution (cf. Walker, 1929). It was Karl Pearson, however, who first explicitly attended to the standard deviation as such and who coined the term "standard deviation." The date was 1894. Karl Pearson, a follower of Francis Galton, is also responsible for the mathematical development of the Pearson product–moment correlation coefficient, which will be discussed in Chapter 5.

EXERCISES

1. Define the following terms:
(a) range
(b) sum of squares
(c) variance
(d) standard deviation
(e) estimate of the variance
(f) semi-interquartile range

2. What are two characteristics of the variance and standard deviation?

3. Using the definitional formulas, calculate the variance, standard deviation, estimate of the variance, and estimate of the standard deviation for the following scores: 10, 9, 8, 6, 3, 2, and 1.

4. Using the computational formulas, calculate the variance, standard deviation, estimate of the variance, and estimate of the standard deviation for the scores in Exercise 3.

5. What is the range of scores in Exercise 3?

6. Using the definitional formulas, calculate the variance, standard deviation, estimate of the variance, and estimate of the standard deviation for the following set of scores: 1, 1, 2, 2, 2, 2, 3, and 3.

7. Perform the same calculations using the computational formulas.

8. What is the range of the scores in Exercise 6?

9. What is the semi-interquartile range of the scores in Exercise 6?

10. Using the computational formulas, calculate the variance, standard deviation, estimate of the variance, and estimate of the standard deviation for the Stroebe et al. low-similarity condition (see Table 4.2).

11. After arranging the above low-similarity scores in an ungrouped frequency distribution (as in Table 3.10), calculate the variance and standard deviation.

12. What is the rationale for the operations in the definitional formula for s^2 (formula 4–2)?

13. What are two reasons why it may be desirable to have information regarding variability?

14. Explain why formula 4–10 for the variance of an ungrouped frequency distribution is equivalent to formula 4–2 for the variance of a set of scores.

15. Explain the reason for the difference between the variance and the estimate of the population variance.

16. Explain the concept "degrees of freedom."

17. The mean is to the standard deviation as the median is to what? Why?

18. Explain why the recognition of the characteristics of the variance and standard deviation can sometimes simplify their calculation.

19. What are the advantages and disadvantages of the variance and standard deviation as measures of variability?

20. What are the advantages and disadvantages of the semi-interquartile range as a measure of variability?

Prologue to Chapter 5

Chapter 5 contains a discussion of correlation and related issues. What is correlation? Correlation is the co-relation between variables (or the properties represented by variables). There is, for example, a correlation between temperature and the height of a column of mercury. The warmer the temperature is, the greater the height of a column of mercury. Other examples are the co-relation between intelligence and school grades, and the co-relation between population density of city tracts and the crime rate. In general, as intelligence increases, so do grades; and as the population density of city tracts increases, so does the crime rate for these tracts.

The concept of correlation is an extremely important one in psychology, and indeed, in science generally. Unfortunately, however, the concept, and its mathematical development, are somewhat more complex than much of the previously discussed material. Thus, the mastery of the contents of Chapter 5 will require a somewhat greater expenditure of time and effort than was required for any of the preceding chapters. Furthermore, Chapter 5 is the longest single chapter in the book. Therefore, it is probably not advisable to read the entire chapter at one time. A good stopping place is just before the section titled COEFFICIENT OF DETERMINATION, or perhaps, just before the PREDICTION section.

5

Correlation

The topic of correlation, like the topics of central tendency and variability, belongs to the domain of descriptive statistics. Central tendency and variability relate to the description of univariate frequency distributions; correlation relates to the description of bivariate frequency distributions. Correlation literally means co-relation, and bivariate literally means two-variables. Thus, the present concern is with the description of the co-relation (or more simply, with the relation) between two variables. A bivariate frequency distribution was previously illustrated in Table 2.5. Table 2.5 gives the frequency with which pairs of history exam and English exam scores were obtained by a class of sixteen students. Careful examination of Table 2.5 reveals an interesting tendency for the students who received high grades in history to receive high grades in English, and for the students who received low grades in history to receive low grades in English. This is technically expressed by stating that a positive–linear correlation exists between history grades and English grades.

Before directly discussing correlation and the description of bivariate frequency distributions, it will be helpful to consider two background concepts. These are *standard scores* and the *function rule for a straight line*. Understanding these concepts is an essential prerequisite to the understanding of correlation—a fact that will be better appreciated when the topic of correlation is directly discussed.

BACKGROUND CONCEPTS

Standard Scores

Standard scores, also referred to as z scores, are a type of transformed raw scores. The transformation is accomplished by subtracting the raw score from the mean and dividing by the standard deviation:

$$z = \frac{X - \bar{X}}{s} \qquad (5\text{-}1)$$

Since $x = X - \bar{X}$, the transformation to z scores may also be written:

$$z = \frac{x}{s} \tag{5-2}$$

By subtracting an X score from the mean, or by converting X to x, the raw score is changed to a deviation score. Division of the deviation score by the standard deviation expresses the deviation score in multiples of the standard deviation, or in standard deviation units. It is analogous to the division of 60 inches by 12 inches in order to express 60 inches in a new unit, feet. A standard or z score thus can be defined as *a deviation score expressed in standard deviation units*.

Standard scores have two characteristics with which the student should be familiar.

1. *The mean of a distribution of z scores, $\bar{z}$, is always equal to zero.* If each of the X scores in a distribution of X scores is transformed to a z score, the resulting distribution of z scores will have a mean of zero. Expressed mathematically:

$$\bar{z} = 0 \tag{5-3}$$

The equality in formula 5–3 is illustrated in Table 5.1 for the liking scores from the Stroebe *et al.* high-similarity condition. The transformation of the X (or liking) scores to z scores is done using values for $\bar{X}$ and s that were calculated previously in Table 4.1. In the general context of bivariate frequency distributions where there are both X scores and Y scores, it is customary to indicate with subscripts whether an s is the standard deviation of the X scores, or the standard deviation of the Y scores. Since the liking scores have been designated as X scores, the standard deviation of these X scores is symbolized s_x. The standard deviation of Y scores is symbolized s_y.

The mean of the distribution of z scores in Table 5.1 is calculated by summing the z scores and dividing by the number of z scores; i.e., by calculating:

$$\frac{\sum z}{n}$$

As Table 5.1 indicates, this value is zero. Calculation of $\bar{z}$ will always produce an answer within rounding error of zero. In Table 5.1, the rounding errors happened to balance out so that an answer exactly equal to zero was obtained.

TABLE 5.1. *Mean and Standard Deviation of the z Transformed Liking Scores in the Stroebe et al. High-Similarity Condition.*

X	$X - \bar{X}$	$z = \dfrac{X - \bar{X}}{s_x}$	z^2	X	$X - \bar{X}$	$z = \dfrac{X - \bar{X}}{s_x}$	z^2
7	1.30	1.17	1.37	6	0.30	0.27	0.07
7	1.30	1.17	1.37	6	0.30	0.27	0.07
7	1.30	1.17	1.37	6	0.30	0.27	0.07
7	1.30	1.17	1.37	6	0.30	0.27	0.07
7	1.30	1.17	1.37	6	0.30	0.27	0.07
7	1.30	1.17	1.37	6	0.30	0.27	0.07
7	1.30	1.17	1.37	6	0.30	0.27	0.07
7	1.30	1.17	1.37	6	0.30	0.27	0.07
7	1.30	1.17	1.37	6	0.30	0.27	0.07
7	1.30	1.17	1.37	6	0.30	0.27	0.07
7	1.30	1.17	1.37	6	0.30	0.27	0.07
7	1.30	1.17	1.37	6	0.30	0.27	0.07
6	0.30	0.27	0.07	6	0.30	0.27	0.07
6	0.30	0.27	0.07	6	0.30	0.27	0.07
6	0.30	0.27	0.07	6	0.30	0.27	0.07
6	0.30	0.27	0.07	5	−0.70	−0.63	0.40
6	0.30	0.27	0.07	5	−0.70	−0.63	0.40
6	0.30	0.27	0.07	5	−0.70	−0.63	0.40
6	0.30	0.27	0.07	5	−0.70	−0.63	0.40
6	0.30	0.27	0.07	4	−1.70	−1.53	2.34
6	0.30	0.27	0.07	4	−1.70	−1.53	2.34
6	0.30	0.27	0.07	4	−1.70	−1.53	2.34
6	0.30	0.27	0.07	4	−1.70	−1.53	2.34
6	0.30	0.27	0.07	4	−1.70	−1.53	2.34
6	0.30	0.27	0.07	4	−1.70	−1.53	2.34
6	0.30	0.27	0.07	4	−1.70	−1.53	2.34
6	0.30	0.27	0.07	3	−2.70	−2.43	5.90
6	0.30	0.27	0.07	3	−2.70	−2.43	5.90
6	0.30	0.27	0.07	3	−2.70	−2.43	5.90
6	0.30	0.27	0.07	3	−2.70	−2.43	5.90
					$\sum = 0.00$	$\sum = 0.00$	$\sum = 60.33$

$$\bar{X} = 5.70$$

$$s_x = 1.11$$

$$\bar{z} = \frac{\sum z}{n} = \frac{0.00}{60} = 0$$

$$s_z = \sqrt{\frac{\sum z^2}{n} - \bar{z}^2} = \sqrt{\frac{60.33}{60} - 0^2} = 1.00$$

$$s_z^2 = 1.00$$

Keeping in mind the fact that $\sum x = 0$, the proof that $\bar{z} = 0$ is straightforward. Using formulas 3–1, 5–2, and summation rule 3, respectively:

$$\bar{z} = \frac{\sum z}{n}$$

$$= \frac{\sum (x/s)}{n}$$

$$= \frac{(1/s) \sum x}{n}$$

$$= \left(\frac{1}{s}\right)\left(\frac{\sum x}{n}\right)$$

$$= \left(\frac{1}{s}\right)\left(\frac{0}{n}\right)$$

$$= 0$$

2. *The variance and standard deviation of a distribution of z scores are always equal to one.* If each of the X scores in a distribution of X scores is transformed to a z score, the resulting distribution of z scores will have a variance and standard deviation that are equal to one:

$$s_z = 1 \qquad (5\text{--}4)$$

The equality in formula 5–4 is illustrated in Table 5.1. The standard deviation is calculated by altering the computational formula for s_x:

$$s_x = \sqrt{\frac{\sum X^2}{n} - \bar{X}^2}$$

so that zs are substituted for Xs:

$$s_z = \sqrt{\frac{\sum z^2}{n} - \bar{z}^2}$$

In Table 5.1, the distribution of z scores is found to have s_z and s_z^2 within rounding error of 1.00.

The proof of formula 5–4 assumes formula 4–3 (definitional formula for s), formula 5–3, formula 5–2, summation rule 3, and formula 4–3, respectively:

$$s_z = \sqrt{\frac{\sum (z - \bar{z})^2}{n}}$$

$$= \sqrt{\frac{\sum (z - 0)^2}{n}}$$

$$= \sqrt{\frac{\sum z^2}{n}}$$

$$= \sqrt{\frac{\sum (x^2 / s^2)}{n}}$$

$$= \sqrt{\frac{(1/s^2) \sum x^2}{n}}$$

$$= \sqrt{\frac{1}{s^2} \left(\frac{\sum x^2}{n} \right)}$$

$$= \sqrt{\frac{1}{s^2} \frac{s^2}{1}}$$

$$= \sqrt{1}$$

$$= 1$$

Function Rule for a Straight Line

The method for graphing functional relations should be a familiar one for most students. The first step is to *calculate* Y, given values of X. Table 5.2 illustrates the procedure for the function rule $Y = 1 + 2X$. Limiting consideration to positive X values (0, 1, 2, etc.), the corresponding Y values are calculated and placed in the table. The second step is to *plot* the corresponding values of X and Y, as illustrated in Fig. 5–1. Figure 5–1 is a graph of the function whose rule is $Y = 1 + 2X$. Notice that the function plots as a straight line. All functions of the form:

$$Y = a + bX \tag{5-5}$$

will plot as straight lines. The expression $Y = a + bX$ is the general equation for a straight line function, where Y and X are variables and a and b are constants. The constant a is referred to as the *intercept constant* because it determines the place at which the line intercepts the Y axis. The intercept value for the equation

TABLE 5.2. *Calculation of Several Corresponding X and Y Values for the Function Rule, $Y = 1 + 2X$.*

Y = 1 + 2X	
Y	X
1	0
3	1
5	2
7	3
9	4

Background Concepts

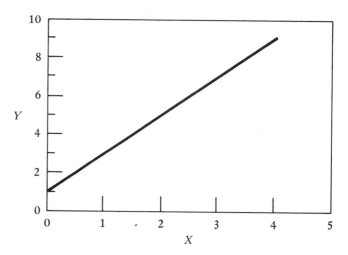

FIGURE 5-1. *Graph of the function rule,* $Y = 1 + 2X$.

$Y = 1 + 2X$ is 1, and the graph of this equation in Fig. 5-1 intersects the Y axis at 1. The constant b is referred to as the *slope constant*. The slope constant determines the slope or angle of inclination of the straight line with reference to the X axis.

The various straight line graphs in Fig. 5-2 illustrate some of the different values that can be assumed by b, the slope constant. In graphs (1) through (6), the line progressively rotates in a clockwise direction. The value of the slope constant can be obtained from any straight line by drawing a right triangle so that the hypotenuse is the line itself. The procedure is illustrated in Fig. 5-2. The side of the triangle paralleling the X or horizontal axis has been labeled H. The side paralleling the Y or vertical axis has been labeled V. The ratio of the length of V to the length of H gives the value of the slope constant. As long as the V and H lines form a right triangle, their exact location on the line forming the hypotenuse is unimportant. The V/H ratio will remain constant regardless of where the V and H lines are located. If the V line is increased in length, the H line will have to increase in length in order to form a right triangle. Thus, for any straight line the V/H ratio, or b, is a constant. In graph (1), V is 3 and H is 1, so V/H or b is 3. In graph (2), V and H are both 2, so b is 1. In graph (3), the line has rotated to a perfectly horizontal position so that V is now 0. In such a situation, V/H is, of course, 0; and the line has 0 slope. In graph (4), the right-hand side of the line slopes down from left to right. Such lines have a negative slope constant. With an H of 2 and a V of -2, the slope is -1. In

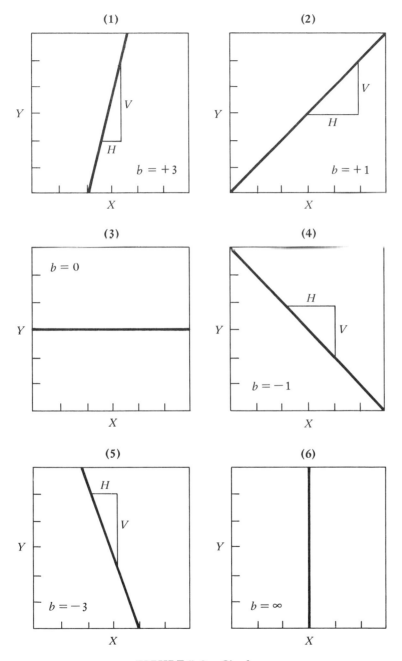

FIGURE 5-2. *Six slopes.*

graph (5), the line has rotated further so that V is now -3. With an H of 1, the slope is, of course, -3. Finally, in graph (6), the line is in a vertical orientation. The value of V here is infinite, and the slope of the line is thus infinite.

THE CONCEPT OF CORRELATION AND SLOPE

Correlation literally means co-relation—a relation between variables. Variables may, of course, be related in many ways. Table 2.5 illustrates one general type of relation. Examination of the bivariate frequency distribution in this table indicates that students who made high grades in history tended to make high grades in English, and that students who made low grades in history tended to make low grades in English. Table 2.5 illustrates what is technically referred to as a positive-linear correlation. In general, *two variables correlate linearly and positively when, as the value of one variable increases, the paired values of the other variable increase correspondingly*. A further example of a positive-linear correlation is the co-relation between height and weight. Although there are exceptions, tall people tend to weigh more than short people. Generally, as height increases, weight increases.

A second general type of co-relation between variables is a negative-linear correlation. For example, there is a negative-linear correlation between number of cigarettes smoked per day and life expectancy. Although there are exceptions, people who smoke a large number of cigarettes per day tend to have a lower life expectancy. Generally, as number of cigarettes increases, life expectancy decreases. *Two variables correlate linearly and negatively when, as the value of one variable increases, the paired values of the other variable decreases*. A negative correlation thus occurs when an *increase* in one variable is associated with a *decrease* in the other variable. A positive correlation, on the other hand, occurs when an *increase* in one variable is associated with an *increase* in the other variable.

Although there are several measures of correlation, the most important of these measures is the *Pearson product–moment correlation coefficient*. This correlation coefficient will be treated in the present chapter; and other less commonly used measures of correlation will be discussed in Chapter 12.

The Pearson product–moment correlation is an index of correlation that varies between $+1$ and -1. The sign of the coefficient indicates whether the correlation is positive or negative, as defined above. A coefficient of $+1$ indicates a perfect-positive-linear correlation; and a coefficient of -1 indicates a perfect-negative-linear correlation. A coefficient of zero indicates the total absence of any linear relation between two variables.

Graphing Bivariate Relations

In Chapter 2, a bivariate frequency distribution was employed to describe the relation between two variables. Recall that a bivariate frequency distribution describes the frequency of occurrence of measurement classes defined with reference to two different variables. That is, a bivariate frequency distribution specifies the frequency of occurrence of each of the pairs of values defining a relation between two variables. Since graphing a bivariate frequency distribution requires three dimensions (X and Y axes for the two variables and a third axis for frequency), such a distribution was represented by a table of cells (see Table 2.5) representing the pairs and including the frequency of occurrence of each given pair.

In the present chapter, bivariate relations will be represented by a plotting of points representing the occurring pairs of values. Such a plot of points with reference to coordinate axes (X and Y axes) is called a *scatterplot*. While such a plot does not directly inform one as to the frequency of occurrence of each pair, it provides a convenient form for illustrating the degree to which a relation is described by a straight line function.

Figure 5-3 is an example of a scatterplot relating two properties, height and weight. Each point in the distribution represents a pair of scores. Any

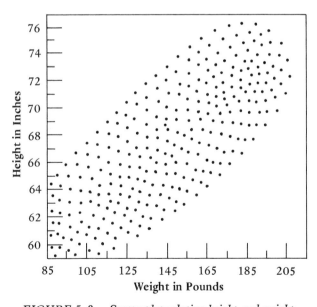

FIGURE 5-3. *Scatterplot relating height and weight.*

given point relates a particular height with a particular weight. The frequency with which pairs of scores occur is represented by the density of points in a certain area of the scatterplot. If the measurement scales are exact enough and the number of observations are not too large, two or more identical pairs are not likely to occur; however, in the event of such a circumstance one point or dot is simply placed on top of the other.

In Fig. 5–3, it is apparent that the points are clustered in an area from the lower left-hand corner to the upper right-hand corner. This is a reflection of the fact that small values on the height dimension are associated with small values on the weight dimension, and that large values on the height dimension are associated with large values on the weight dimension; i.e., there is a positive-linear correlation between height and weight. While there are both short, fat people and tall, skinny people, in general, tall people weigh more than short people.

Figure 5–3 clarifies an important point concerning correlation. In order to correlate two variables, the variables must be associated in some fashion. One could not correlate just any group of heights with just any group of weights. The correlation between heights and weights in Fig. 5–3 can be discussed because each instance of height is associated with one and only one instance of weight. The association results, of course, from the fact that any particular pair of height and weight scores is obtained from the same individual.

Figure 5–4 is a scatterplot relating income tax and number of dependents for a small group of people. Examination of the plot makes it apparent that the points tend to scatter from the upper left-hand corner to the lower right-hand

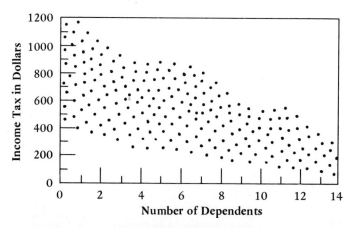

FIGURE 5–4. Scatterplot relating income tax and number of dependents.

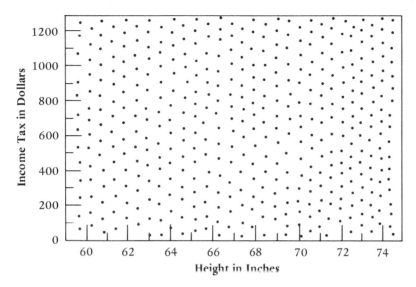

FIGURE 5-5. *Scatterplot relating income tax and height.*

corner. This is a reflection of the fact that high values on the income tax dimension are associated with low values on the number of dependents dimension, and that low values on the income tax dimension are associated with high values on the number of dependents dimension; i.e., there is a negative-linear correlation between income tax and number of dependents. For this particular group of people, the greater the number of dependents, the less is the income tax.

Figure 5-5 is a scatterplot relating income tax and height for a particular group of people. Inspection of this distribution reveals that the points are scattered more or less randomly over the entire plot. In this situation there is no tendency for values on the income tax dimension to be associated with values on the height dimension; i.e., there is a zero or minimal correlation between income tax and height for this group of people.

Slope as an Index of Correlation

The above three scatterplots make it apparent that the orientation of points in any scatterplot is related to the correlation between X and Y variables. When the points scatter from the lower left-hand corner to the upper right-hand corner, the correlation is positive; when the points scatter from the upper left-hand corner to the lower right-hand corner, the correlation is negative. This raises the

The Concept of Correlation and Slope

interesting possibility of drawing a straight line that fits or describes the points and then taking the slope of this line as an index of the correlation between X and Y. Once the line has been drawn, the slope can, of course, be obtained by taking the V to H ratio. Slope as measured by V/H would then serve as an index of correlation.

Such an approach has several difficulties, which will be considered subsequently. Nonetheless, the Pearson product–moment correlation coefficient is a *refined measure of slope*. In the following sections an attempt will be made to explain why and how slope is refined so as to obtain an index of the correlation between X and Y.

The Two Regression Lines

A straight line describing the relation between two variables is called a *regression line*. In attempting to fit a regression line to the points in a scatterplot, the distances between the points and the line are measured in either a vertical, up and down, direction or in a horizontal, left to right, direction. This, of course, follows from the nature of the scatterplot, in which distances are measures either along the Y axis or along the X axis. Measurement of distances in these two different directions results in *two different regression lines*. One of these lines is obtained when the distances between the points and the line is measured in a vertical direction along the Y axis. The other line is obtained when the distances between the points and the line are measured in a horizontal direction along the X axis. These two regression lines, referred to as the regression of Y on X and the regression of X on Y, will be considered in turn.

REGRESSION OF Y ON X. For the regression of Y on X, the distances between the points and the line are measured in the vertical (up and down) direction. Figure 5–6 illustrates these distances. Each point in the scatterplot has a value corresponding to its alignment with the Y axis. The distances or deviations between the points and the regression line are positive if the points are above the line, and negative if the points are below the line.

A line is desired that describes or fits these points when the departure or deviation of the points from the line is measured in a vertical direction. How is such a line obtained? An initial strategy might be to search for a line such that the sum of the distances, or deviations between the points and the line, would be as small as possible. Unfortunately, any number of lines will be found to generate a sum of deviations equal to zero. For example, a line drawn exactly horizontal through the mean of Y, $\bar{Y}$, will generate a sum of deviations equal to zero. The sum of deviations from any mean is always zero. The problem arises due to the fact that the deviations above the line are positive and the deviations below the

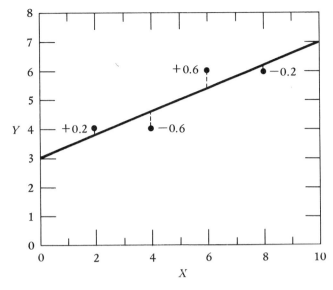

FIGURE 5-6. *Vertical deviations from the regression of Y on X.*

line negative. Since the negative deviations may balance out the positive deviations, it is quite possible for numerous simple sums to be zero or near zero.

One possible solution is to obtain a line that minimizes the sum of absolute deviations. Another solution is to obtain a line that minimizes the sum of squared deviations. Squaring the negative deviations, of course, eliminates the minus signs. The theory of correlation and regression relies upon this latter solution. With the aid of calculus, the straight line, $Y = a + bX$, can be proven to be the line that minimizes the sum of squared vertical deviations when:

$$b = \frac{n \sum XY - (\sum X)(\sum Y)}{n \sum X^2 - (\sum X)^2} \tag{5-6}$$

And:

$$a = \bar{Y} - b\bar{X} \tag{5-7}$$

In formula 5-7, $\bar{Y}$ symbolizes the mean of the Y scores just as $\bar{X}$ symbolizes the mean of the X scores. When the equation for the straight line has its slope constant (b) and intercept constant (a) as determined by formulas 5-6 and 5-7, respectively, then the equation becomes the formula for the regression of Y on X. The regression of Y on X is sometimes referred to as a *line of least-squares-best-fit*; it is a line that minimizes the sum of squared deviations in terms of Y values.

Since the slope and intercept constants for the regression of X and Y will be introduced in a subsequent section, it is helpful to distinguish with subscripts the slope and intercept constants of the two regression lines. For the regression of Y on X, we will write b_y and a_y; for the regression of X on Y, b_x and a_x.

If the Xs and Ys in formula 5–6 are expressed in deviation score form, b_y is given by:

Ignore this Formula

$$b_y = \frac{\sum xy}{\sum x^2} \qquad (5\text{–}8)$$

Formulas 5–6 and 5–8 are algebraically equivalent.

In order to prove that:

$$\frac{n \sum XY - (\sum X)(\sum Y)}{n \sum X^2 - (\sum X)^2} = \frac{\sum xy}{\sum x^2}$$

we will work first with the numerator, $\sum xy$, and then with the denominator, $\sum x^2$:

$$\sum xy = \sum [(X - \bar{X})(Y - \bar{Y})]$$
$$= \sum (XY - \bar{X}Y - X\bar{Y} + \bar{X}\bar{Y})$$
$$= \sum XY - \bar{X} \sum Y - \bar{Y} \sum X + n\bar{X}\bar{Y}$$
$$= \sum XY - \frac{(\sum X)(\sum Y)}{n} - \frac{(\sum Y)(\sum X)}{n} + \frac{\cancel{n}(\sum X)(\sum Y)}{\cancel{n} \cdot n}$$
$$= \sum XY - \frac{(\sum X)(\sum Y)}{n}$$

$$\sum x^2 = \sum (X - \bar{X})^2$$
$$= \sum (X^2 - 2X\bar{X} + \bar{X}^2)$$
$$= \sum X^2 - 2\bar{X} \sum X + n\bar{X}^2$$
$$= \sum X^2 - \frac{2(\sum X)(\sum X)}{n} + \frac{\cancel{n}(\sum X)^2}{n^2}$$
$$= \sum X^2 - \frac{2(\sum X)^2}{n} + \frac{(\sum X)^2}{n}$$
$$= \sum X^2 - \frac{(\sum X)^2}{n}$$

Substituting these values in the formula $\sum xy / \sum x^2$:

$$\frac{\sum xy}{\sum x^2} = \frac{\sum XY - \frac{(\sum X)(\sum Y)}{n}}{\sum X^2 - \frac{(\sum X)^2}{n}}$$
$$= \frac{n \left[\sum XY - \frac{(\sum X)(\sum Y)}{n}\right]}{n \left[\sum X^2 - \frac{(\sum X)^2}{n}\right]}$$
$$= \frac{n \sum XY - (\sum X)(\sum Y)}{n \sum X^2 - (\sum X)^2}$$

Thus:

$$\frac{n \sum XY - (\sum X)(\sum Y)}{n \sum X^2 - (\sum X)^2} = \frac{\sum xy}{\sum x^2}$$

The paired X and Y scores in Table 5.3 can be used to illustrate the calculation of b_y by formula 5–6, and a_y by formula 5–7. As Table 5.3 indicates, $b_y = +.4$ and $a_y = +3$.

Going further, these values for b_y and a_y can be substituted into the straight-line function for the regression of Y on X, $Y = a_y + b_y X$, so as to give the formula for the regression of Y on X for the above scores, $Y = 3 + .4X$. As Table 5.3 indicates, different values for X can be substituted into the formula so as to obtain corresponding Y values. Thus, when $X = 0$, $Y = 3$; when $X = 5$, $Y = 5$, etc. These corresponding X and Y points can then be used to plot the regression of Y on X. This has been done in Fig. 5–6.

TABLE 5.3. *Calculation of the Regression of Y on X, and X on Y.*

X	Y	XY	X²	Y²
8	6	48	64	36
6	6	36	36	36
4	4	16	16	16
2	4	8	4	16
20	20	108	120	104

$b_y = \dfrac{n \sum XY - (\sum X)(\sum Y)}{n \sum X^2 - (\sum X)^2}$

$= \dfrac{(4)(108) - (20)(20)}{(4)(120) - (20)^2}$

$= +.4$

$a_y = \bar{Y} - b_y \bar{X}$

$= 5 - (.4)(5)$

$= +3$

$Y = a_y + b_y X$

$= 3 + .4X$

Y	X
3	0
5	5
7	10

$b_x = \dfrac{n \sum XY - (\sum X)(\sum Y)}{n \sum Y^2 - (\sum Y)^2}$

$= \dfrac{(4)(108) - (20)(20)}{(4)(104) - (20)^2}$

$= +2$

$a_x = \bar{X} - b_x \bar{Y}$

$= 5 - (2)(5)$

$= -5$

$X = a_x + b_x Y$

$= -5 + 2Y$

X	Y
1	3
5	5
9	7

The Concept of Correlation and Slope

The line in Fig. 5–6 is the regression of Y on X; it is the line that minimizes the sum of squared deviations in terms of Y values. Figure 5–6 indicates the deviations of the points from the line: $+0.2$, -0.6, $+0.6$, and -0.2. The sum of the squares of these deviations is .80 ($.04 + .36 + .36 + .04 = .80$). It is impossible to draw a line through the points in Fig. 5–6 such that the sum of the squared deviations in terms of Y values will be less than .80. It is for this reason that such a line is sometimes referred to as a line of least-squares best fit.

The slope of the regression line is, of course, dependent on the position of the points in the scatterplot. When the X and Y scores have equal standard deviations, the regression of Y on X has an area of rotation that is shown in Fig. 5–7. The regression line rotates counterclockwise between a line running at a 45 degree angle from upper left to lower right ($b_y = -1$) to a line running exactly horizontally through the mean of Y ($b_y = 0$), to a line running at a 45 degree angle from lower left to upper right ($b_y = +1$).

REGRESSION OF X ON Y. For the regression of X on Y, the distances between the points and the line are measured in the horizontal (left to right) direction. Figure 5–8 illustrates these distances for the data in Table 5.3. The regression of X on Y is a line that minimizes the sum of the squared horizontal distances or deviations.

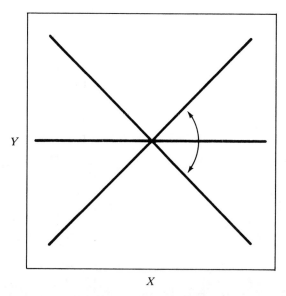

FIGURE 5–7. *Area of rotation for the regression of Y on X when the Y and X scores have equal standard deviations.*

Since the present concern is the horizontal or X axis, we need a straight line equation for X:

$$X = a_x + b_x Y \qquad (5\text{-}9)$$

In equation 5-9, a_x and b_x are the intercept and slope constants. The intercept constant determines the point at which the line intercepts the X axis (rather than the Y axis as is the case for the regression of Y on X). The slope constant determines the slope of the line as measured by the H (horizontal) to V (vertical) ratio (rather than the V to H ratio as is the case for the regression of Y on X). With the aid of calculus, it can be shown that the straight line, $X = a_x + b_x Y$, will be the line that minimizes the sum of squared horizontal deviations, when:

$$b_x = \frac{n \sum XY - (\sum X)(\sum Y)}{n \sum Y^2 - (\sum Y)^2} \qquad (5\text{-}10)$$

And:

$$a_x = \bar{X} - b_x \bar{Y} \qquad (5\text{-}11)$$

When formula 5-9 has its slope and intercept constants determined by formulas 5-10 and 5-11, respectively, it then becomes the formula for the regression of X on Y. The regression of X on Y is a *line of least-squares-best-fit*—it is a line that minimizes the sum of squared deviations in terms of X values.

If the Xs and Ys in formula 5-10 are expressed in deviation form, we can write:

$$b_x = \frac{\sum xy}{\sum y^2} \qquad (5\text{-}12)$$

Formulas 5-10 and 5-12 are algebraically equivalent.

Table 5.3 illustrates the use of formula 5-10 for b_x and formula 5-11 for a_x, with the same data that were previously used for the regression of Y on X. Table 5.3 also contains the calculation of the corresponding X and Y values for the regression of X on Y. The regression of X on Y is plotted in Fig. 5-8.

The horizontal deviations from the line are indicated in Fig. 5-8: $+1.0$, -1.0, $+1.0$, and -1.0. The sum of the squares of these deviations is 4.0 ($1.0 + 1.0 + 1.0 + 1.0 = 4.0$). It is impossible to draw a line through the points in Fig. 5-8 such that the sum of squared deviations in terms of X values is less than 4.0. Thus, the regression of X on Y is also a line of least-squares best fit.

Figure 5-9 presents the regression of Y on X from Fig. 5-6, and the regression of X on Y from Fig. 5-8. The two regression lines obviously do not coincide. One line minimizes the sum of squared vertical deviations, and the other line minimizes the sum of squared horizontal deviations.

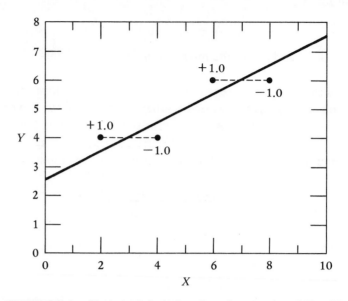

FIGURE 5-8. *Horizontal deviations from the regression of X on Y.*

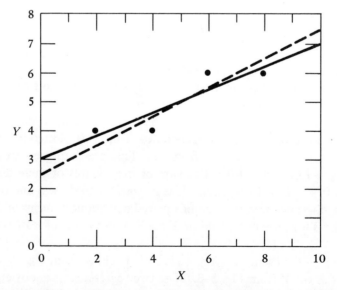

FIGURE 5-9. *The regression of Y on X (solid line) and the regression of X on Y (dashed line).*

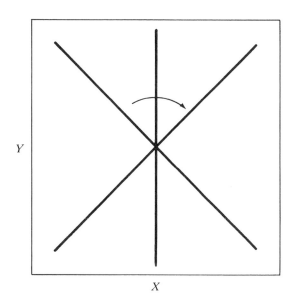

FIGURE 5-10. *Area of rotation for the regression of X on Y when the X and Y scores have equal standard deviations.*

When the X and Y scores have equal standard deviations, the regression of X on Y has the area of rotation that is shown in Fig. 5-10. The regression line rotates clockwise between a line running at a 45 degree angle from upper left to lower right ($b_x = -1$) to a line running exactly vertically through the mean of X ($b_x = \infty$), to a line running at a 45 degree angle from the lower left to upper right ($b_x = +1$).

Difficulties with Slope as an Index of Correlation

Two difficulties are encountered when slope is used as an index of correlation. First, since there are two regression lines, there are two slopes. How does one obtain a single index or measure of correlation from two slope constants? Second, slope is influenced both by the correlation between X and Y and by the variabilities of the X and Y scores. Each of these difficulties will be discussed in turn.

Two Slopes. If our intent is to use slope as an index of correlation, we are confronted with the fact that there are two slopes, one for each regression line. Which one shall we use? What is done is to take the geometric mean of the two

The Concept of Correlation and Slope

slope constants. The geometric mean of the slopes of the two regression lines is, in fact, the Pearson product–moment correlation coefficient.

$$r = \sqrt{(b_y)(b_x)} \qquad (5\text{-}13)$$

This equation is acceptable, as stated, if it is understood that when both of the slopes are positive, the positive square root is taken, and when both of the slopes are negative, the negative square root is taken. (It is impossible to have a situation in which the slopes are of differing signs.)

Substituting formula 5–8 for b_y and formula 5–12 for b_x into formula 5–13 gives a second commonly presented formula for r:

$$r = \frac{\sum xy}{\sqrt{(\sum x^2)(\sum y^2)}} \qquad (5\text{-}14)$$

Since $b_y = \sum xy / \sum x^2$ and $b_x = \sum xy / \sum y^2$, we can substitute in formula 5–13 as follows:

$$r = \sqrt{(b_y)(b_x)}$$

$$= \sqrt{\frac{\sum xy}{\sum x^2} \frac{\sum xy}{\sum y^2}}$$

$$= \frac{\sqrt{(\sum xy)(\sum xy)}}{\sqrt{(\sum x^2)(\sum y^2)}}$$

$$= \frac{\sum xy}{\sqrt{(\sum x^2)(\sum y^2)}}$$

The reason for taking the geometric mean of the slope constants as an index of correlation is not at all obvious. In attempting to solve the second difficulty, relating to the use of slope as an index of correlation, it turns out that certain modifications are made and these result in r being the geometric mean of the two slope constants.

SLOPE AS AFFECTED BY VARIANCE. The second difficulty with using slope as an index of correlation is that slope is affected by the variance of the X and Y variables as well as by the correlation between them. Consider the two sets of correlated scores in Table 5.4. The correlation between the X and Y scores is $+1$ for both set (1) and set (2). In set (1), every increment of 1 on the X variable is matched by an increment of 1 on the Y variable. In set (2), every increment of 1 on the X variable is matched by an increment of 2 on the Y variable. In each case a linear functional relationship exists between X and Y. It is the constant ratio of increments on the two variables that produces the perfect positive correlation for both sets (1) and (2). When these relationships are plotted in Fig. 5–11, however, it becomes very obvious that the two slopes are different. The

TABLE 5.4. *Two Sets of Perfectly Correlated X and Y Scores.*

(1)		(2)	
X	Y	X	Y
8	7	8	11
7	6	7	9
6	5	6	7
5	4	5	5
4	3	4	3
3	2	3	1

differing slopes are a result of the fact that the variability of the Y scores in set (2) is somewhat greater than the variability of the Y scores in set (1).

While slope is affected by the correlation between X and Y, it is also influenced by the variabilities of the X and Y scores. In other words, slope is

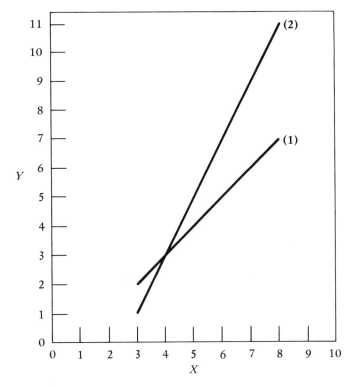

FIGURE 5-11. *Graphic presentation of relationships (1) and (2) in Table 5.4.*

an impure index of correlation. What can be done about this problem? Asked more precisely, how can we hold constant the variabilities of the X and Y scores? The solution to this problem taken by the Pearson product–moment correlation coefficient is to transform the X and Y scores to z scores. The standard deviation and variance of z scores are always equal to one. Thus, regardless of the variabilities of the X and Y scores, the transformation to z scores will produce a standard situation in which s_{z_x} and s_{z_y} equal one.

The transformation of X and Y to z scores results in a very interesting consequence—the numerical equality of the two slope constants. After transforming X and Y to z scores, the regression of Y on X and the regression of X on Y acquire numerically identical slopes. This new slope equals the geometric mean of the original slopes before transformation of the X and Y scores to z scores. *The Pearson product–moment correlation coefficient is simply a measure of slope after the X and Y scores have been transformed to z scores.*

The most commonly used computational formula for the Pearson product–moment correlation coefficient is:

$$r = \frac{n \sum XY - (\sum X)(\sum Y)}{\sqrt{[n \sum X^2 - (\sum X)^2][n \sum Y^2 - (\sum Y)^2]}} \quad (5\text{–}15)$$

Although it is by no means apparent, formula 5–15 simply involves the transformation of the X and Y scores to z scores and the calculation of the slope of the regression lines relating these transformed scores.

In deriving formula 5–15, we will begin with formula 5–8 for b_y:

$$b_y = \frac{\sum xy}{\sum x^2}$$

$$= \frac{\sum xy}{n(\sum x^2 / n)}$$

$$= \frac{\sum xy}{n s_x^2}$$

$$= \frac{\sum (X - \bar{X})(Y - \bar{Y})}{n s_x^2}$$

Formula 5–12 for b_x may be manipulated in a similar fashion:

$$b_x = \frac{\sum xy}{\sum y^2}$$

$$= \frac{\sum xy}{n(\sum y^2 / n)}$$

$$= \frac{\sum xy}{n s_y^2}$$

$$= \frac{\sum (X - \bar{X})(Y - \bar{Y})}{n s_y^2}$$

After transforming X and Y to z scores, we may write:
$$b_{z_y} = \frac{\sum(z_x - \bar{z}_x)(z_y - \bar{z}_y)}{ns_{z_y}^2}$$
and:
$$b_{z_x} = \frac{\sum(z_x - \bar{z}_x)(z_y - \bar{z}_y)}{ns_{z_x}^2}$$

where $\bar{z}_x$ is the mean of the X scores after transformation to z scores, $\bar{z}_y$ is the mean of the Y scores after transformation to z scores, $s_{z_x}^2$ is the variance of the X scores after transformation to z scores, and $s_{z_y}^2$ is the variance of the Y scores after transformation to z scores. Since the variance of z scores is 1, the formulas for b_{z_y} and b_{z_x} become algebraically identical:

$$b_{z_y} = \frac{\sum(z_x - \bar{z}_x)(z_y - \bar{z}_y)}{n(1)}$$
$$= b_{z_x}$$

The identity of b_{z_y} and b_{z_x} means that after transformation of X and Y to z scores, the slopes for the two regression lines are numerically equal.

When X and Y have been transformed to z scores, b_{z_y} and b_{z_x} are by definition r. Thus:
$$r = b_{z_x} = b_{z_y}$$

Further, the mean of z is always zero and thus:
$$r = b_{z_x}$$
$$= b_{z_y}$$
$$= \frac{\sum(z_x - 0)(z_y - 0)}{n(1)}$$
$$= \frac{\sum z_x z_y}{n} \tag{5-16}$$

Formula 5-16 is sometimes given as the definitional formula for r. Notice that:

$$r = \frac{\sum z_x z_y}{n}$$
$$= \frac{\sum \frac{x}{s_x} \frac{y}{s_y}}{n}$$
$$= \frac{\frac{1}{s_x}\frac{1}{s_y}\sum xy}{n}$$
$$= \frac{\sum xy}{n s_x s_y}$$
$$= \frac{\sum xy}{n\sqrt{\left(\frac{\sum x^2}{n}\right)\left(\frac{\sum y^2}{n}\right)}}$$
$$= \frac{\sum xy}{\sqrt{(\sum x^2)(\sum y^2)}}$$

The Concept of Correlation and Slope

This latter expression is formula 5–14, the formula that was previously obtained by expressing r as the geometric mean of the two slope constants, b_x and b_y.

We have previously proved that:

$$\sum xy = \sum XY - \frac{(\sum X)(\sum Y)}{n}$$

and that:

$$\sum x^2 = \sum X^2 - \frac{(\sum X)^2}{n}$$

Thus:

$$r = \frac{\sum xy}{\sqrt{(\sum x^2)(\sum y^2)}}$$

$$= \frac{\sum XY - \frac{(\sum X)\sum Y}{n}}{\sqrt{\left[\sum X^2 - \frac{(\sum X)^2}{n}\right]\left[\sum Y^2 - \frac{(\sum Y)^2}{n}\right]}}$$

$$= \frac{n\left[\sum XY - \frac{(\sum X)(\sum Y)}{n}\right]}{n\sqrt{\left[\sum X^2 - \frac{(\sum X)^2}{n}\right]\left[\sum Y^2 - \frac{(\sum Y)^2}{n}\right]}}$$

$$= \frac{n\sum XY - (\sum X)(\sum Y)}{\sqrt{[n\sum X^2 - (\sum X)^2][n\sum Y^2 - (\sum Y)^2]}}$$

This, as you know, is the computational formula for r (formula 5–15). It is simply a formula for b or slope that involves the transformation of X and Y to z scores.

Calculation and Coding

The use of the computational formula (formula 5–15) to calculate the Pearson product–moment correlation coefficient is illustrated in Table 5.5. Before calculating his or her first correlation coefficient, the student should note two things about formula 5–15. First, n is *not* the total number of X and Y scores, but it is the number of X and Y pairs; i.e., the number of points in the scatterplot. Thus, in Table 5.5, n is 5, not 10. Second, there is a crucial difference between $\sum X^2$ and $(\sum X)^2$ and between $\sum Y^2$ and $(\sum Y)^2$. In the first instances, the scores are squared and then summed; and in the second, the scores are summed and then squared. For the data in Table 5.5, $\sum X^2 = 55$, and $(\sum X)^2 = (15)^2 = 225$.

Examination of the X and Y scores in Table 5.5 makes it apparent that every increment of 1 on the X variable is matched by a uniform increment of 2 on the Y variable; i.e., that the correlation between X and Y is $+1$. Whenever the increments on the X variable maintain a fixed ratio to the increments on the Y variable, the correlation is $+1$. Such a fixed ratio of increments guarantees that all of the points in the scatterplot will fall exactly on a straight line. As can be seen from Table 5.5, the application of formula 5–15, in fact, reveals a correlation of +1.

In some instances the calculation of r can be simplified through coding. The simplest coding procedure involves subtracting the lowest X score from all of the X scores, and the lowest Y score from all of the Y scores. Such subtraction will not change the correlation between X and Y. This fact can be intuitively grasped by examining the two graphs in Fig. 5–12. In graph (1), with Y scores ranging from 50 to 80 and X scores ranging from 40 to 70, the points all lie on a straight line and the correlation is thus +1. Graph (2) was constructed by subtracting the lowest Y score, 50, from all of the Y scores; and the lowest X score, 40, from all of the X scores. Notice, however, that all of the points still lie on a straight line with the same slope as in graph (1). Subtracting 50 from all of the Y scores and 40 from all of the X scores did not change the correlation between X and Y.

TABLE 5.5 Calculation of the Pearson Product Moment Correlation Coefficient.

X	Y	XY	X^2	Y^2
5	10	50	25	100
4	8	32	16	64
3	6	18	9	36
2	4	8	4	16
1	2	2	1	4
15	30	110	55	220

$$r = \frac{n \sum XY - (\sum X)(\sum Y)}{\sqrt{[n \sum X^2 - (\sum X)^2][n \sum Y^2 - (\sum Y)^2]}}$$

$$= \frac{5(110) - (15)(30)}{\sqrt{[5(55) - (15)^2][5(220) - (30)^2]}}$$

$$= \frac{550 - 450}{\sqrt{[275 - 225][1100 - 900]}}$$

$$= \frac{100}{\sqrt{[50][200]}}$$

$$= \frac{100}{\sqrt{10{,}000}}$$

$$= \frac{100}{100}$$

$$= +1$$

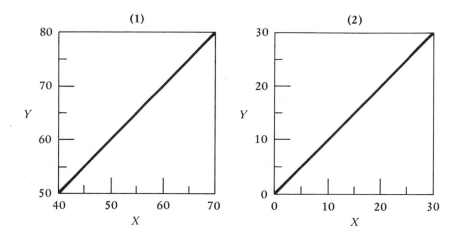

FIGURE 5–12. *A +1 slope (graph 1) is unchanged by subtraction of constants (graph 2).*

Within-Cell and Zero-Order Correlation

In Chapter 1, it was pointed out that Stroebe *et al.* measured a number of different aspects of interpersonal attraction. As Table 1.1 indicated, these included liking and preference as a date. Up to this point only the liking scores have been considered. When attention is also directed toward the dating scores, an obvious question concerns the correlation between dating and liking. Table 5.6 contains the calculation of the correlation within the high-similarity condition, and Table 5.7 the calculation of the correlation within the low-similarity condition. These correlations are +.70 and +.73, respectively.

The above Pearson product–moment correlations are technically referred to as *within-cell* correlations. As the name suggests, within-cell correlations are correlations calculated within the cells or conditions of an experimental design. In some instances it is of interest to calculate the Pearson product–moment correlation between all of the scores in all of the cells or conditions of an experimental design. Such a correlation is technically referred to as a *zero-order* correlation. Table 5.8 contains the calculation of the zero-order correlation for the Stroebe *et al.* experiment. Notice that for the calculations the difference between the high- and low-similarity conditions is ignored and an overall correlation is computed. Thus, ΣX, for example, equals the total sum of all the liking scores. This is most easily obtained by summing the totals for the high- and low-similarity conditions (342 + 196 = 538).

TABLE 5.6. Calculation of the Pearson Product–Moment Correlation between Liking and Dating within the Stroebe et al. High-Similarity Condition.

LIKING X	DATING Y	XY	X^2	Y^2	LIKING X	DATING Y	XY	X^2	Y^2
7	7	49	49	49	6	7	42	36	49
7	7	49	49	49	6	7	42	36	49
7	7	49	49	49	6	5	30	36	25
7	7	49	49	49	6	7	42	36	49
7	7	49	49	49	6	4	24	36	16
7	7	49	49	49	6	5	30	36	25
7	7	49	49	49	6	5	30	36	25
7	7	49	49	49	6	5	30	36	25
7	6	42	49	36	6	7	42	36	49
7	5	35	49	25	6	6	36	36	36
7	4	28	49	16	6	6	36	36	36
7	5	35	49	25	6	4	24	36	16
6	5	30	36	25	6	5	30	36	25
6	4	24	36	16	6	3	18	36	9
6	5	30	36	25	6	3	18	36	9
6	6	36	36	36	5	7	35	25	49
6	7	42	36	49	5	2	10	25	4
6	6	36	36	36	5	2	10	25	4
6	5	30	36	25	5	4	20	25	16
6	6	36	36	36	4	4	16	16	16
6	3	18	36	9	4	2	8	16	4
6	5	30	36	25	4	5	20	16	25
6	6	36	36	36	4	3	12	16	9
6	4	24	36	16	4	3	12	16	9
6	5	30	36	25	4	2	8	16	4
6	4	24	36	16	4	5	20	16	25
6	4	24	36	16	3	3	9	9	9
6	3	18	36	9	3	1	3	9	1
6	4	24	36	16	3	1	3	9	1
6	6	36	36	36	3	1	3	9	1

$\sum = 342 \quad \sum = 288 \quad \sum = 1723 \quad \sum = 2024 \quad \sum = 1566$

$$r = \frac{n \sum XY - (\sum X)(\sum Y)}{\sqrt{[n \sum X^2 - (\sum X)^2][n \sum Y^2 - (\sum Y)^2]}}$$

$$= \frac{60(1723) - (342)(288)}{\sqrt{[60(2024) - (342)^2][60(1566) - (288)^2]}}$$

$$= +.70$$

The Concept of Correlation and Slope

TABLE 5.7. *Calculation of the Pearson Product–Moment Correlation between Liking and Dating within the Stroebe et al. Low-Similarity Condition.*

LIKING X	DATING Y	XY	X^2	Y^2	LIKING X	DATING Y	XY	X^2	Y^2
7	7	49	49	49	3	2	6	9	4
6	7	42	36	49	3	3	9	9	9
6	5	30	36	25	3	3	9	9	9
6	5	30	36	25	3	3	9	9	9
5	4	20	25	16	3	2	6	9	4
5	7	35	25	49	3	4	12	9	16
5	4	20	25	16	3	3	9	9	9
5	4	20	25	16	3	1	3	9	1
5	3	15	25	9	3	3	9	9	9
5	5	25	25	25	3	2	6	9	4
4	5	20	16	25	3	3	9	9	9
4	5	20	16	25	3	1	3	9	1
4	4	16	16	16	3	2	6	9	4
4	3	12	16	9	3	2	6	9	4
4	3	12	16	9	2	1	2	4	1
4	1	4	16	1	2	1	2	4	1
4	1	4	16	1	2	1	2	4	1
4	3	12	16	9	2	1	2	4	1
4	3	12	16	9	2	1	2	4	1
4	4	16	16	16	2	1	2	4	1
4	1	4	16	1	2	1	2	4	1
4	2	8	16	4	2	1	2	4	1
4	2	8	16	4	2	2	4	4	4
3	2	6	9	4	2	3	6	4	9
3	6	18	9	36	1	1	1	1	1
3	3	9	9	9	1	1	1	1	1
3	3	9	9	9	1	1	1	1	1
3	2	6	9	4	1	2	2	1	4
3	3	9	9	9	1	2	2	1	4
3	3	9	9	9	1	1	1	1	1

$$\sum = 196 \quad \sum = 165 \quad \sum = 636 \quad \sum = 750 \quad \sum = 613$$

$$r = \frac{n \sum XY - (\sum X)(\sum Y)}{\sqrt{[n \sum X^2 - (\sum X)^2][n \sum Y^2 - (\sum Y)^2]}}$$

$$= \frac{(60)(636) - (196)(165)}{\sqrt{[60(750) - (196)^2][60(613) - (165)^2]}}$$

$$= +.73$$

TABLE 5.8. *Calculation of the Zero-Order Pearson Product-Moment Correlation between Liking and Dating for the Stroebe et al. Experiment.*

$\sum X = 342 + 196 = 538, \quad \sum Y = 288 + 165 = 453$
$\sum X^2 = 2024 + 750 = 2774, \quad \sum Y^2 = 1566 + 613 = 2179$
$\sum XY = 1723 + 636 = 2359, \quad n = 120$

$$r = \frac{n \sum XY - (\sum X)(\sum Y)}{\sqrt{[n \sum X^2 - (\sum X)^2][n \sum Y^2 - (\sum Y)^2]}}$$

$$= \frac{120(2359) - (538)(453)}{\sqrt{[120(2774) - (538)^2][120(2179) - (453)^2]}}$$

$$= +.80$$

Within-cell correlations indicate the correlation between two measured properties, holding the manipulated property at some value or level. Thus, holding similarity constant at a high value, the correlation between liking and dating in the Stroebe *et al.* data is +.70. Holding similarity constant at a low value, the correlation is +.73. These data indicate that for the Stroebe *et al.* subjects a fairly strong correlation exists between liking and dating, even when similarity is held constant.

Zero-order correlations indicate the correlation between two measured properties when the manipulated property is not held constant. For the Stroebe *et al.* data, the zero-order correlation is +.80. The correlation between liking and dating increased somewhat when similarity was not held constant.

COEFFICIENT OF DETERMINATION

Several possible interpretations or meanings can be given to a Pearson product-moment correlation coefficient. One of these interpretations is in terms of slope. This, of course, is the matter discussed above. Another possible interpretation is in terms of the proportion of variance in one variable accounted for by the other variable. This matter will be discussed briefly in this section.

Consider the total variance in the Y scores. Substituting Ys for Xs in the definitional formula for the variance (formula 4–2), gives:

$$s_y^2 = \frac{\sum (Y - \bar{Y})^2}{n}$$

In the context of one scatterplot, n is a constant. Thus, n can be eliminated from the above formula and the expression $\sum (Y - \bar{Y})^2$ used as an index of the variability of the Y scores.

From the perspective of a scatterplot, expression $\sum (Y - \bar{Y})^2$ is an index of the total up and down variability of the points from the mean of Y. As formula 5–17 indicates, this total variability of Y has two additive components:

$$\sum (Y - \bar{Y})^2 = \sum (Y - \tilde{Y})^2 + \sum (\tilde{Y} - \bar{Y})^2 \qquad (5\text{–}17)$$

Formula 5–17 contains two symbols, Y and $\tilde{Y}$ (pronounced Y tilde), which need to be distinguished. Symbol Y stands for the Y value for each point in the scatterplot. Symbol $\tilde{Y}$ stands for all the Y values making up the Y on X regression line. We have previously not distinguished between the values of Y for the regression of Y on X:

$$Y = a_y + b_y X$$

and the values of Y for the points in the scatterplot. In the present context, such a distinction is essential. Thus, the above function rule can be rewritten:

$$\tilde{Y} = a_y + b_y X$$

The $\tilde{Y}$ values are the values of Y making up the regression of Y on X.

Thus, the expression $\sum (Y - \tilde{Y})^2$ in formula 5–17 is an index of the vertical variability of the points from the regression of Y on X. The $Y - \tilde{Y}$ distance is indicated in Fig. 5–13. The expression $\sum (Y - \tilde{Y})^2$ can be regarded as that part of the total variability in Y that is not accounted for by the linear relationship between X and Y.

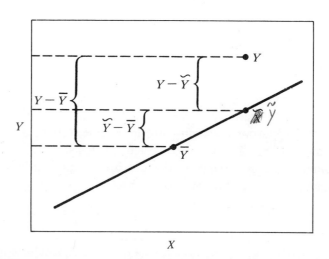

FIGURE 5–13. Illustration of the distances measured in formula 5–17.

The final expression in formula 5–17, $\sum(\tilde{Y} - \bar{Y})^2$, is an index of the vertical deviation of the regression of Y on X from the mean of Y. The $\bar{Y} - \tilde{Y}$ distance is also indicated in Fig. 5–13. Expression $\sum(\tilde{Y} - \bar{Y})^2$ can be regarded as that part of the total variability in Y that can be accounted for by the linear relationship between X and Y.

According to formula 5–17, the total variability in Y, $\sum(Y - \bar{Y})^2$, is composed of two additive components. These are, first, the variability in Y not accounted for by the linear relationship between X and Y, or an index of the vertical deviation of the points from the regression of Y on X, $\sum(Y - \tilde{Y})^2$. And, second, the variability in Y that is accounted for by the linear relationship between X and Y, or an index of the vertical deviation of the regression of Y on X from an exactly horizontal line through the mean of Y.

Suppose that we hold $\sum(Y - \bar{Y})^2$ constant and increase $\sum(Y - \tilde{Y})^2$. This implies that $\sum(\tilde{Y} - \bar{Y})^2$ will of necessity decrease. Expressed graphically, this means that if we hold the total variability of Y scores constant, an increase in the vertical deviations of scores from the regression line will result in a regression line that is closer to the mean of Y (has a more nearly horizontal slope). On the other hand, suppose that we hold $\sum(Y - \bar{Y})^2$ constant and decrease $\sum(Y - \tilde{Y})^2$. This implies that $\sum(\tilde{Y} - \bar{Y})^2$ will of necessity increase. Expressed graphically, this means that if we hold the total variability of Y scores constant, a decrease in the deviation of the Y scores from the regression line will result in a regression line that is further from the mean of Y (has a less horizontal slope).

The above relationships make it apparent that formula 5–17 expresses something very basic about slope, and hence, the correlation between X and Y. In fact, the ratio of $\sum(\tilde{Y} - \bar{Y})^2$ to $\sum(Y - \bar{Y})^2$, referred to as the *coefficient of determination*, is the square of the Pearson product–moment correlation coefficient:

$$r^2 = \frac{\sum(\tilde{Y} - \bar{Y})^2}{\sum(Y - \bar{Y})^2} \qquad (5\text{–}18)$$

Since $\sum(\tilde{Y} - \bar{Y})^2$ is that part of the total variability that is accounted for by the linear relationship between X and Y, and $\sum(Y - \bar{Y})^2$ is the total variability, the ratio of these two is the proportion of accountable variance. Thus, *the Pearson product–moment correlation coefficient can be interpreted as the square root of the proportion of accountable variance.*

As indicated above, the zero-order correlation between liking and dating in the Stroebe *et al.* experiment was +.80. Since $(.80)^2$ equals .64, liking accounts for 64 percent of the variance in dating (or conversely, dating accounts for 64 percent of the variance in liking). On the other hand, within the high-similarity condition the correlation was +.70. With this correlation the percent of accountable variance is 49, since $(.70)^2$ equals .49. Thus, from the perspective of accountable variance, a 15 percent difference occurs between the correlations of $+.70$ and $+.80$ $(.64 - .49 = .15)$. Holding similarity constant at a high level reduces the accountable variance by 15 percent.

PREDICTION

One of the most important uses of correlation-related concepts is in the making of predictions. This section offers a brief discussion. Predictions involve inferences that go beyond a simple description of frequency distributions. Thus, although intimately related to the topic of correlation, the topic of predictions is part of inferential, not descriptive, statistics.

Illustrative Problem

Suppose that someone is interested in predicting the college grade-point averages that will be made by the individuals in a group of incoming students. The only information available about these particular students is their high school grade-point averages. Accurate predictions about these students can be made if three conditions hold:

1. Information is possessed with regard to the high school and college grade-point averages of a previous group of students.
2. There is a sizeable correlation between the high school and college grade-point averages of the previous group of students.
3. The previous and present group of students can be assumed to be comparable.

The actual prediction procedure is very simple in essence. The initial step is to calculate and draw the linear regression line of college grades upon high school grades, as has been done in Fig. 5–14. Next, a vertical straight line is drawn from the value of a student's high school grade-point average to the point of intersection with the regression line. The value on the Y axis (college grades) corresponding to the point of intersection is the predicted value of college grade-point average for a given high school grade-point average. The procedure is illustrated in Fig. 5–14. Here we see that a high school grade-point average of slightly less than 2.8 predicts a college grade-point average of slightly less than 3.0.

In the previous discussion of correlation and slope, the existence of two regression lines was a "problem." Such is not the case, however, when the intent

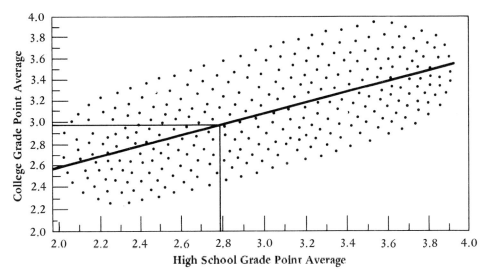

FIGURE 5-14. *The regression of college grade-point average on high school grade-point average.*

is making predictions. In this context it is important to distinguish carefully between the two lines. In making predictions from X to Y, the regression of Y on X is used; and in making predictions in the other direction, from Y to X, the regression of X on Y is used. The regression line can be considered a series of points corresponding to pairs of predictor and predicted values.

The accuracy of predictions depends upon the comparability of the past and present sets of subjects and upon the degree of correlation between X and Y. If the correlation between X and Y were zero, the regression line of Y on X would dictate that any particular X would always predict the mean of Y. For example, if the correlation between college grade-point averages and high school grade-point averages were zero, the regression line in Fig. 5-14 would be perfectly horizontal. Thus, any particular high school grade-point average would predict one and only one college grade-point average, the mean of all the college grade-point averages.

Regression Equations

The above graphic procedure of drawing a straight line from the predictor score to the regression line is somewhat cumbersome and, of course, not perfectly accurate. Predictions can be made more easily and accurately by the use of regression equations:

$$\tilde{Y} = \bar{Y} + r\frac{s_y}{s_x}(X - \bar{X}) \tag{5-19}$$

$$\tilde{X} = \bar{X} + r\frac{s_x}{s_y}(Y - \bar{Y}) \tag{5-20}$$

The $\tilde{Y}$ values are the predicted Y values, the Y values of points on the regression of Y on X; and the $\tilde{X}$ values are the predicted Xs, the X values of points on the regression of X on Y. Formula 5-19 is used for predicting Y from X, and formula 5-20 for predicting X from Y.

Formulas 5-19 and 5-20 are simply "fancy" equations for straight lines. Formula 5-19 is a variant of the formula for the regression of Y on X, and formula 5-20 is a variant of the formula for the regression of X on Y. In predicting Y from X, for example, the predictor X value is "plugged" into the formula for the regression of Y on X, and a predicted $\tilde{Y}$ is obtained. The procedure is an algebraic duplication of the graphic procedures employed above.

The two straight-line equations can be written as:

$$\tilde{Y} = a_y + b_y X \tag{5-21}$$

And:

$$\tilde{X} = a_x + b_x Y \tag{5-22}$$

Since the derivation of the regression equation, formula 5-20, from the straight-line equation, 5-22, is strictly parallel to the derivation of the regression equation, 5-19, from the straight-line equation, 5-21; only the latter will be presented.

As was previously indicated, formula 5-7 for a_y, obtained from calculus, is given by:

$$a_y = \bar{Y} - b_y \bar{X}$$

Substituting for a_y in formula 5-21:

$$\tilde{Y} = \bar{Y} - b_y \bar{X} + b_y X$$
$$= \bar{Y} + b_y(X - \bar{X}) \tag{5-23}$$

Formula 5-23 could be used in making predictions if formula 5-6 were used to calculate b_y. Since we typically do not compute slope coefficients, however, but do compute correlation coefficients and standard deviations, it is frequently easier to make use of the following identity:

$$b_y = r\frac{s_y}{s_x} \tag{5-24}$$

Substituting formula 5-24 in formula 5-23 gives:

$$\tilde{Y} = \bar{Y} + r\frac{s_y}{s_x}(X - \bar{X})$$

which is the $\tilde{Y}$ regression equation, formula 5-19.

Example Problems

The actual mechanics of using formula 5–19 to predict Y from X is illustrated in Table 5.9. Here an X of 5 is found to predict a $\hat{Y}$ of -2. In order to use formula 5–19, it is first necessary to calculate $\bar{Y}$, r, s_y, s_x, and $\bar{X}$. The only remaining symbol in formula 5–19 is X. The symbol X stands for the predictor value, which is 5 in Table 5.9. The predictor value is, of course, so chosen as to satisfy the particular needs of the investigator.

The data in Table 5.9 are unique in that the correlation between the X and Y variables is -1. Whenever the correlation between the X and Y variables is either $+1$ or -1, it is possible to make predictions without recourse to a

TABLE 5.9. *Prediction of $\hat{Y}$ for an X of 5.*

X	Y	XY	X^2	Y^2
4	0	0	16	0
3	2	6	9	4
2	4	8	4	16
1	6	6	1	36
0	8	0	0	64
10	20	20	30	120

$$r = \frac{n \sum XY - (\sum X)(\sum Y)}{\sqrt{[n \sum X^2 - (\sum X)^2][n \sum Y^2 - (\sum Y)^2]}}$$

$$= \frac{5(20) - (10)(20)}{\sqrt{[5(30) - (10)^2][5(120) - (20)^2]}}$$

$$= -1$$

$$\bar{X} = \frac{\sum X}{n} = \frac{10}{5} = 2$$

$$s_x = \sqrt{\frac{\sum X^2}{n} - \bar{X}^2} = \sqrt{\frac{30}{5} - (2)^2} = 1.41$$

$$s_y = \sqrt{\frac{\sum Y^2}{n} - \bar{Y}^2} = \sqrt{\frac{120}{5} - (4)^2} = 2.82$$

$$\bar{Y} = \frac{\sum Y}{n} = \frac{20}{5} = 4$$

$$\hat{Y} = \bar{Y} + r \frac{s_y}{s_x}(X - \bar{X})$$

$$= 4 + (-1)\left(\frac{2.82}{1.41}\right)(5 - 2)$$

$$= 4 - 2(3)$$

$$= 4 - 6$$

$$= -2$$

regression equation. In Table 5.9, the X scores are 4, 3, 2, 1, and 0. A predictor X score of 5 is one unit higher than any of the existing scores. The corresponding Y scores are 0, 2, 4, 6, and 8. Since an X score of 3 corresponds to a Y score of 2 and an X score of 4 corresponds to a Y score of 0, a predictor X score of 5 will by extrapolation correspond to a Y score of -2. This is, of course, the same answer that was obtained by application of the regression equation. The regression equation predicts the values lying on the regression line. When the correlation is either $+1$ or -1, these values can be calculated easily through simple examination of the correlated scores. If the correlation is not a perfect one, this cannot be done with any accuracy, however, and it becomes necessary to use a regression equation.

As an example of the latter situation, suppose that Stroebe *et al.* wished to predict the dating score of a subject for whom they only had the liking score (X). Suppose this liking score were 5 ("I feel that I would probably like this person to a slight degree"). On the assumption that this subject is comparable to the previously tested ones, the prediction procedure described above can be used. Table 5.10 summarizes the calculations. The values for r, $\sum X$, $\sum X^2$,

TABLE 5.10. *Prediction of Dating from Liking.*

$$r = +.80$$

$$\bar{X} = \frac{\sum X}{n} \qquad s_x = \sqrt{\frac{\sum X^2}{n} - \bar{X}^2} \qquad s_y = \sqrt{\frac{\sum Y^2}{n} - \bar{Y}^2}$$

$$= \frac{538}{120} \qquad = \sqrt{\frac{2774}{120} - (4.48)^2} \qquad = \sqrt{\frac{2179}{120} - (3.78)^2}$$

$$= 4.48 \qquad = 1.75 \qquad = 1.97$$

$$\bar{Y} = \frac{\sum Y}{n}$$

$$= \frac{453}{120}$$

$$= 3.78$$

$$\hat{Y} = \bar{Y} + r \frac{s_y}{s_x}(X - \bar{X})$$

$$= 3.78 + (.80)\frac{1.97}{1.75}(5 - 4.48)$$

$$= 3.78 + .90(.52)$$

$$= 3.78 + .47$$

$$= 4.25$$

$\sum Y$, $\sum Y^2$, and n are contained in Table 5.8. The predicted Y, or dating, score is 4.25. This is a score between 4 and 5, but closer to 4. Since Stroebe *et al.* did not allow subjects to make responses in between alternatives (see Table 1.1), the most reasonably predicted response is 4. Response 4 indicated that the subject "might or might not" consider the other person as a potential date.

EFFECT OF RANGE UPON CORRELATION

One of the primary difficulties with the correlation coefficient is that it may be affected by the range of scores on the X and Y variables. Unless the correlation coefficient is either -1, $+1$, or 0, a reduction in the range of either the X variable or the Y variable may reduce the size of the correlation. Why is this? The reason can be grasped by examining the scatterplot presented in Fig. 5–15. This scatterplot depicts the relationship between verbal aptitude and quantitative aptitude for a set of persons.

It is apparent from the orientation of the points that the two aptitude variables are related. Suppose, however, that we reduce the range of the two aptitude variables by looking at just those people who are fairly extreme on both dimensions. This gives us a new scatterplot, which is indicated by the dashed lines in the upper right-hand corner of Fig. 5–15. It is apparent that within this

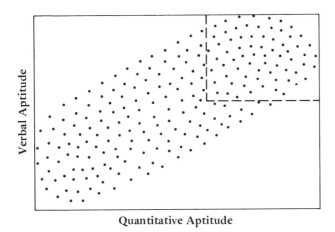

FIGURE 5–15. *Scatterplot relating verbal and quantitative aptitude.*

scatterplot the points are more or less randomly scattered and that the correlation between verbal and quantitative aptitude is close to 0.

Many students are aware of the fact that someone who does well in a course in calculus will not necessarily do well in a course in English literature. On the other hand, these students may also remember that in grammar school the pupils who did well in arithmetic also seemed to do well in reading. Expressed somewhat more exactly, in grammar school the correlation between verbal and quantitative aptitude seemed to be much higher than in college. If we make the reasonable assumption that the range of verbal and quantitative aptitude is greater in grammar school than in college, Fig. 5–14 provides a possible interpretation for this shifting impression. The correlation is lower in college due to the fact that students with lower verbal and quantitative aptitudes are not represented, thereby reducing the range of both aptitudes.

MULTIPLE AND PARTIAL CORRELATION

Before leaving the topic of correlation, we will discuss briefly multiple and partial correlation.

Multiple Correlation

The previous discussion of correlation has related entirely to the correlation between two variables. It is possible, however, to compute a multiple correlation between a set of two, three, four, five, six, or any number of variables and another single variable. Such multiple correlations are frequently useful in making predictions. Thus, one could, for example, predict college grade-point average on the basis of both high school grade-point average and aptitude scores. Assuming that the multiple correlation (symbolized R) is higher than the simple r between college grade-point average and either high school grade-point average or aptitude scores, the multiple regression approach will result in more accurate predictions. (It is mathematically impossible for R to be lower than r.)

A thorough discussion of multiple correlation is subject matter for a more advanced treatment of statistics. Our present concern is just to acquaint the student with the existence of the topic.

Partial Correlation

Sometimes it is of interest to find the correlation between two variables holding constant a third variable; i.e., one may wish to know the correlation between X and Y, holding Z constant. Such a problem can be solved easily with the aid of the partial correlation technique. Once the correlations between all possible pairs of the variables are computed, the partial r can be obtained with the following formula:

$$r_{xy \cdot z} = \frac{r_{xy} - r_{xz}r_{yz}}{\sqrt{(1 - r_{xz}^2)(1 - r_{yz}^2)}} \tag{5-25}$$

where $r_{xy \cdot z}$ is the correlation between X and Y holding Z constant; r_{xy} is the correlation between X and Y; r_{xz} is the correlation between X and Z; and r_{yz} is the correlation between Y and Z.

Partial correlation is a very important technique for various research problems in sociology and social psychology. One such problem relates to the effects of density and crowding. Research with non-human animals has demonstrated convincingly that with a fixed amount of space an increase in the number of animals above certain levels leads to social disorganization, increased aggressiveness, high infant mortality, and so forth. This occurs even with ample food and water, and sufficient space for mobility and exercise. Does such a relation between density and social disorganization also exist for humans? Superficially, the answer appears to be yes. Crime rates, for example, are higher in larger cities than in smaller cities; and within cities, crime rates are higher in the more dense inner city areas than in the suburbs. If cities are divided into geographical areas, it is possible to compute a correlation between the number of people per acre and some index of crime, such as the number of juvenile delinquents per 100 male juveniles. It comes as no surprise that such correlations are moderately high. Further research, however, has demonstrated that if other variables, such as socio-economic status, are held constant or are partialled out, the correlation between density and juvenile delinquency drops to near zero. Thus, it is at least possible that crime results, not from high density, but from the low socio-economic conditions that are associated with high density.

Partial correlation is conceptually related to within-cell correlation. In both cases the relationship between two variables is examined while a third variable is held constant. A within-cell correlation, however, holds the third variable constant at one particular value or level. Thus, the correlation of +.70 between liking and dating within the high-similarity condition or cell held similarity constant at a high value or level. Partial correlation holds a third variable constant across all the measured values of that variable. It is thus of more general interest.

CORRELATION, CAUSATION, AND EXPERIMENTATION

One very important methodological problem has to do with the relation between correlation and causation. Many investigators like to point out that "correlation does not imply causation." Consider, for example, the positive correlation between the number of saloons and the number of churches in U.S. cities. Cities with a large number of saloons have a large number of churches, and cities with few saloons have few churches. This correlation, however, does not imply that saloons cause churches or that churches cause saloons. Common sense indicates that a causal inference would be absurd. However, in other instances correlation does appear to be an indication of causation. One well-known example is the correlation between the positions of the moon and sun and the magnitudes of the tides. Another is the correlation between the number of cigarettes smoked and the incidence of lung cancer. In many instances, however, it is unclear whether or not correlation results from a direct causal linkage.

Consider two properties, A and B, which are positively correlated. Low values of A are associated with low values of B, and high values of A are associated with high values of B. Such a correlation could result from any of four possibilities: First, A causes B; second, B causes A; third, A and B are both caused by C, where C is any single property, or network of properties, connecting A and B; fourth, any combination of the above. Consider the correlation between attitude similarity and interpersonal attraction. It has been found among various groups of people that when attraction is high, similarity is high; and when attraction is low, similarity is low. First, such a correlation may result from attitude similarity causing attraction. Second, the correlation may result from attraction producing a conformity (similarity) in attitude. Third, the correlation may result from common group experiences producing both attitude similarity and attraction. Fourth, and finally, the correlation may result from any combination of the above causal possibilities. Thus, similarity may produce attraction, which, in turn, produces more similarity. In view of all these possibilities, extreme caution should be exercised in the causal interpretation of such a correlation.

The ambiguity surrounding the causal meaning of correlation is one reason for utilizing the experimental method. An experiment can be defined as *the measurement of observed outcomes resulting from the random assignment of subjects to conditions in which a manipulative procedure occurs.* By random assignment is meant that the experimenter makes a chance determination as to each subject's condition. For example, before each subject arrives at the laboratory, the experimenter may flip a coin to determine in which of two conditions a subject is to be placed. By a manipulative procedure is meant a procedure in which the experimenter determines the differences between or among conditions. Stated somewhat differently, the experimenter manipulates a property. For example, in the Stroebe *et al.* experiment, subjects in the high-similarity condition were given information indicating that the anonymous other person had beliefs and attitudes similar to that of each subject; and subjects in the low-similarity condition were given information indicating that the anonymous other person had beliefs and attitudes dissimilar to that of each subject. The experimenter *manipulated* the property attitude–similarity.

As Figs. 2–3 and 2–4 descriptively indicate, Stroebe *et al.* found a greater degree of reported liking or attraction in the high-similarity condition than in the low-similarity condition. Stroebe *et al.* interpreted such results as indicating that, in the preacquaintance situation, attitude similarity causes attraction. In the most general sense of the term correlation, Stroebe *et al.* obtained a correlation between similarity and attraction. High similarity was associated with high attraction, and low similarity with low attraction. Why then was the causal inference that similarity causes attraction justified? Consider the other possibilities listed above. It is not plausible to argue that attraction caused similarity in this situation. It is not plausible, first, because similarity was determined by the experimenter; and second, because the similarity manipulation occurred before the assessment of liking or attraction. What about the possibility that one or more third properties caused both similarity and liking? Again, it is known that similarity was caused by the experimenter. It is, however, still possible that some unidentified property, or set of properties, caused the higher liking scores in the high-similarity condition than in the low-similarity condition. At this point the importance of random assignment of subjects becomes apparent. In view of the random assignment, whatever liking differences that do exist must have arisen either as a result of chance or as a result of the difference between the high- and low-similarity conditions. As will be explained in subsequent chapters, the probability that experimentally obtained results do arise from chance alone can be calculated. If this calculated probability is sufficiently small, say .05, or 5 / 100, an investigator may be willing to assume that the obtained results did not arise from chance. This then leaves the difference between conditions as the most likely cause of the obtained results. To the extent that this difference between conditions reflects only the property of interest (e.g., attitude similarity), the causal inference from this property to the measured result (e.g., liking) is justified.

While the above statement is correct, there is a real problem concerning the extent to which the difference between conditions reflects properties other than the one that is of primary interest. Consider an experiment in which some subjects are given a drug and others are not. The drug condition and the non-drug condition thus differ in the intended way. These conditions, however, also differ in an unintended way. Subjects in the drug condition know that they have received a drug and subjects in the non-drug condition know that they have not received a drug. In view of this problem, it has become standard practice to give the non-drug subjects a non-drug placebo in order to keep all subjects ignorant as to whether they have, or have not, received the drug.

However, it is still possible that in the future someone may discover some other unintended difference between conditions. In view of this fact, it is possible to overemphasize the difference between experimental and non-experimental research. The difference is in the *degree* to which competing causal interpretations can be eliminated. The experimental method allows for a more complete elimination of competing causal interpretations than does a non-experimental method.

In view of the fact that the experimental method is the best method for identifying causal relationships, the student may wonder why psychologists would ever correlate non-experimentally manipulated properties. Stated another way, why do not psychologists exclusively use the experimental method? While, undoubtedly, many answers can be given to this question, one answer is that the laboratory setting in which experiments are conducted may differ in certain respects from the non-laboratory setting to which the investigator wishes to generalize. Thus, while Stroebe *et al.* had a general interest in preacquaintance, short-term, and long-term attraction, they only studied preacquaintance attraction. Also, the Stroebe *et al.* subjects knew they were participating in an experiment, and this knowledge may have had an effect on the obtained results. In view of considerations such as these, it is reassuring to find that in a non-laboratory setting similarity and attraction are correlated.

Another reason why experimental methods may not be used is that investigators may not know how to manipulate the properties of concern. Until the launching of the first sputnik, this was, of course, true for the entire field of astronomy. Still another reason for not using the experimental method is an ethical concern for the welfare of subjects. It would be unethical, for example, to assign subjects randomly either to a condition in which they were required to smoke a certain number of cigarettes per day or to another condition in which they were not allowed to smoke at all, and then to note the incidence of lung cancer over some time period.

Finally, it should be pointed out that just because non-experimental methods are used does not mean that in some instances a correlation may not be interpreted confidently as arising from a direct causal relation. These are situations in which further study has revealed a pattern of relationships that are all consistent with the causal inference. This appears to be the case, for example, with the inference that the position of the sun and moon causes the tides, and that smoking causes lung cancer.

HISTORICAL PERSPECTIVE ON CORRELATION

Thinkers before Galton

It is generally recognized that Francis Galton first formulated correlation as we now understand it. As Walker (1929, p. 92) points out, a number of individuals in the nineteenth century "hovered on the verge of the discovery of correlation," but none did. Most of these individuals were mathematicians, or physicists who were concerned with probability and the simultaneous occurrence of two errors. One notable non-mathematician of this period was Henry

P. Bowditch. In 1877, under the sponsorship of the Massachusetts Board of Health, he made an extensive, statistical study of children. Of particular interest are his data on height and weight. In attempting to relate these two variables, he developed a "measure of stoutness," the ratio of the annual increase in pounds weight to the annual increase in inches height for each age from 5 to 18. He was obviously groping for the concept of correlation. It is interesting how conceptions that are so obvious today were beyond the grasps of the most talented minds of the previous century—an observation that doubtlessly also will be made by residents of the 21st century.

Francis Galton

In the third volume of his opus, *The Life, Letters, and Labours of Francis Galton*, Karl Pearson writes a detailed description of Galton's struggle toward the concept of correlation. As Pearson points out, prior to 1889 men of science had only the concept of causation to express relationships. They had no way of dealing with non-causal relationships, as, for example, the relationship between head length and foot length.

In 1876, with Darwin's assistance, Galton began experimenting with sweet-pea seeds. He noted that mother plants with seeds of large diameter tended to have daughter plants with seeds of large diameter, and that mother plants with seeds of small diameter tended to have daughter plants with seeds of small diameter. Galton, however, went beyond this verbal statement and attempted to display the relationship graphically. Galton's notebook for this period contains what Pearson considers the first regression line ever drawn. It was not, however, a line of least-squares best fit. Galton's method was much simpler and more intuitive.

Galton's method involved the following steps. First, the diameter of the parent seeds was represented along the horizontal axis, and the diameter of the offspring seeds along the vertical axis. Second, for each diameter of a parent seed the mean diameter of the offspring seeds was calculated. This involved the calculation of the mean value for each column of the graph. Third, these means were plotted in the graph and a straight line was visually fit to the points. Since the means appeared to lie along a straight line, the visual fitting was intuitively compelling. Fourth, the slope of this line was noted, and in particular, the deviation of this line from a 45 degree slope (1.00).

(In Galton's later work, he invariably used medians rather than means; but otherwise, up until his discovery of correlation, the procedure was as described.)

Galton noted that his line was flatter than 45 degrees. Pearson tells us, after calculating the constants, that its slope was actually .33. These results suggested to Galton that there is a "reverting" towards "the average ancestral type," a phenomenon that he later called "regression." Mother plants with large diameter seeds tended to have daughter plants with large diameter seeds; however, the diameters of the offspring seeds were not as extreme or deviant as were the diameters of the parental seeds. Galton noted that the extent of reversion was indicated by the departure of his straight line from a 45 degree angle. Since the parental seeds were located along the horizontal axis, reversion involved a line that had lesser slope, or was flatter, than a line with a slope of 1.00.

Pearson points out that since Galton was dealing with heredity in the same sex the approximate equality of variabilities in the two generations "preserved him from any greater error." In this circumstance the two regression lines have the same numerical slope, and that slope is the correlation between the variables. In terms of Galton's procedure the straight line that is visually fit to the column medians would have approximately the same numerical slope as the straight line fit to the row medians.

In 1877, Galton presented his ideas in a lecture delivered to the Royal Institution. Here the symbol "r" was introduced as the coefficient of "reversion," the slope of the straight-line relationship. For the next eight years, Galton published nothing on the subject of reversion or regression. During this period, though, he did collect data on inheritance in man. In 1885 and 1886, he published a number of papers dealing with inheritance of stature. He constructed graphs relating a weighted average of the parents' height (the "midparents'" height) to children's height. These data again indicated the presence of reversion or regression. Tall parents had tall children, but the deviancy of the children was less apparent than the deviancy of the parents. At this time Galton had no clear idea of the distinction between slope and correlation. Thus, one can appreciate the seeming paradox he confronted when it turned out that the two straight lines had numerically differing slopes.

Two years later, in 1888, while Galton was taking a walk on the grounds of Naworth Castle, it occurred to him that the solution to the problem lay in the expression of the X variable, in units of its own variability, and the expression of the Y variable, in units of its own variability. In his *Memories of My Life* (1908), Galton describes this event as follows:

> As these lines are being written, the circumstances under which I first clearly grasped the important generalization that the laws of Heredity were solely concerned with deviations, expressed in statistical units, are vividly recalled to my memory. It was on the grounds of Naworth Castle, where an invitation had been given to ramble freely. A temporary shower drove me to seek refuge in a reddish recess in the rock by the side of the pathway. There the idea flashed across me, and I forgot everything else for a moment in my great delight (p. 300).

Pearson rightly suggests that this "recess" deserves a commemorative tablet. This flash of creative insight is one of those things that has occurred all too infrequently in the course of human thought. On December 20, 1888, Galton read a paper to the Royal Society entitled, "Co-relations and Their Measurement, Chiefly from Anthropometric Data." Pearson considers this date the "birthday of the conception of correlation."

It is important to realize, however, that Galton did not use means and standard deviations. Instead, he employed medians and semi-interquartile ranges, which were, of course, his own inventions. As previously indicated, Galton was committed to the importance of rank-order relationships. His procedure thus involved dividing the difference between column medians and the overall Y median by the semi-interquartile range for the Y scores. The details of this particular procedure, which are now somewhat of a historical curiosity, can be found in Pearson (1930, Vol. IIIA, p. 51). Despite the mathematical crudeness of Galton's procedure, however, it did yield correlation coefficients remarkably close to those obtained by the Pearson product–moment correlation coefficient. Pearson observes that this is the case because Galton dealt with variables, height and cubit for example, which were symmetrically distributed and thus had medians and means of approximately the same magnitudes.

Karl Pearson

Karl Pearson (1857–1936) obtained a degree in mathematics in 1879 at Cambridge. His initial interests were in applied mathematics, mechanics, and the theory of elasticity. He subsequently studied law at the Universities of Heidelberg and Berlin, and also attended lectures on physics, metaphysics, and Darwinism. He developed wide interests in German folklore, history of the Reformation, the German humanists, and Martin Luther. In 1884, he was appointed to the chair of applied mathematics and mechanics at University College, London. In this post Pearson taught mathematics to engineering students. Although the core of his interests at that time related to the physical sciences, the groundwork for other interests had already been laid by Henry Bradshaw, the Cambridge University Librarian. From Pearson's student days at Cambridge until Bradshaw's death in 1886, he encouraged and directed Pearson's scholarly interests.

In 1890, Pearson obtained a lectureship in geometry at Grisham College, London. The late 1880s and early 1890s marked a turning point in Pearson's thought. It was during this period that Pearson established a lasting friendship with W. F. R. Weldon, professor of zoology at University College, London. As E. S. Pearson, Karl Pearson's son and biographer describes it, both young professors "...lectured from 1 to 2, and the lunch table, between 12 and 1, was the scene of many a friendly battle, the time when problems were suggested, solutions brought, and even worked out on the back of the *menu* or by the aid of pellets of bread" (1938, p. 19). Weldon was committed to the idea of "...demonstrating Darwinian evolution by mathematical enquiries" and his "infectious enthusiasm" had a profound influence on Pearson. It was Weldon who introduced Pearson to Francis Galton. Both Weldon and Pearson fell under Galton's spell, particularly after reading his *Natural Inheritance*, published in 1889. According to Pearson,

> I interpreted...Galton to mean there was a category broader than causation, namely correlation, of which causation was only the limit, and this new conception of correlation brought psychology, anthropology, medicine, and sociology in large parts into the field of mathematical treatment. It was Galton who first freed me from the prejudice that sound mathematics could only be applied to natural phenomena under the category of causation. Here, for the first time was a possibility, I will not say a

certainty, of reaching knowledge—as valid as physical knowledge was then thought to be—in the field of living forms, and above all, in the field of human conduct (quoted by E. S. Pearson, 1938, p. 19).

In retrospect it is difficult for us to appreciate how revolutionary it was to conceive of applying mathematics to the life and social sciences. Pearson's biography is a case study in the directing of a great mind by social contacts, first with Bradshaw and then with Weldon and Galton.

Beginning in 1900, Pearson's lectures and writing reflected an increasing concern with statistical matters. During the following years Pearson developed a remarkable number of statistical concepts and procedures. He introduced the standard deviation and first used the term "normal curve" to refer to what others had called the "curve of error." He is responsible for the mathematical statement of the product-moment correlation coefficient as we now know it, and also the mathematical procedures for calculating multiple and partial correlation coefficients. He developed the correlation ratio (to be treated in Chapter 12) as a measure of non-linear correlation, the tetrachoric correlation as a measure of correlation when the X and Y variables are each divided into two subsets, and the biserial correlation coefficient when only one of the variables is divided into two subsets. One of Pearson's more notable achievements, however, was the development of chi square (the subject matter of Chapter 11).

The motivating force behind the development of these techniques was a desire to quantify and understand various hereditary relationships. Multiple correlation, for example, grew directly out of Galton's so-called Law of Ancestral Heredity. Galton's Law, expressed in statistical form, is simply a statement that the offspring's height, for example, is a linear function of the parents' height, the grandparents' height, the great-grandparents' height, and so forth. From this beginning, Pearson developed the mathematical procedure for calculating R, the multiple correlation coefficient.

One of the more intriguing aspects of Pearson's professional life had to do with his involvement in a continuing controversy with the Mendelians. In 1900, Mendel's work relating to the mechanism of individual inheritance was rediscovered. The simplicity and elegance of Mendel's idea attracted the attention of many biologists—the most notable of whom was William Bateson. From our current perspective it is obvious that there is nothing fundamentally incompatible between Mendel's theory concerning the mechanism of inheritance and the biometricians' attempt to quantify and relate inherited characteristics. However, what is obvious today was not obvious then—particularly to the Mendelians. The difference in emphasis was fueled by a clash of personalities, the unwillingness of the Mendelians to accept the importance of quantification, and the biometricians' unwillingness to accept the importance of Mendel's theory concerning the mechanism of inheritance.

The heat and emotion of this controversy is illustrated by an incident that E. S. Pearson relates in detail. Karl Pearson submitted a paper to the Royal Society; and William Bateson, who was a referee for the society, returned a critical report. Thereafter, Bateson's detailed criticisms were printed before Pearson's original paper was printed and Pearson had even been informed as to the fate of his paper. It is easy to imagine the outrage that Pearson must have felt. As a result of this incident, Weldon wrote a letter to Pearson containing the following passage:

The contention "that numbers mean nothing and do not exist in Nature" is a serious thing that will have to be fought. Most other people have got beyond it, but most biologists have not....

Do you think it would be too hopelessly expensive to start a journal of some kind? (E. S. Pearson, 1938, p. 39)

Thus, Pearson and Weldon began a journal, *Biometrika*, a journal that remains today as one of the leading technical periodicals in statistics.

In 1903, Pearson and Weldon published a paper, "On a Generalized Theory of Alternative Inheritance, with Special Reference to Mendel's Laws," which attempted to show, among other things, that Mendelian principles led to certain predictions concerning a population mating at random that could be tested statistically. The essential correctness of this perspective, however, was lost on the Mendelians; and Pearson, for his part, remained very skeptical of and aloof from the Mendelian perspective. Through this period Weldon and Pearson gained moral support from Galton, who maintained an elder statesmanlike role and did not personally participate in the controversy.

When Galton died in 1906, he left the remainder of his estate to endow a chair of eugenics at University College, London. Consistent with Galton's will, the chair was offered to Pearson. After some negotiations with the University to ensure funds for a laboratory, Pearson accepted the chair. The laboratory and chair were incorporated in a new Department of Applied Statistics. In 1915, the department started teaching undergraduates—a development that Pearson welcomed with some ambivalence. He was pleased that statistics had achieved sufficient stature to warrant an undergraduate degree, but he was concerned that the department might change from a "research" to a "teaching" orientation. In 1933, Pearson resigned his professorship. After long deliberation the Department of Applied Statistics was split into two independent units, a Department of Eugenics with which the Galton chair was associated and a new Department of Statistics. E. S. Pearson became the head of the new Department of Statistics; and R. A. Fisher, along with Pearson one of the most illustrious names in the history of statistics, became the second holder of the Galton chair.

During the years of his retirement, Pearson continued to work on *Biometrika*, and before his death in 1936 had almost seen the final proofs of the first half of Volume 28—a remarkable achievement. Pearson can be safely assumed to have established the discipline of mathematical statistics in Great Britain.

EXERCISES

1. Define the following terms:
(a) standard scores
(b) intercept constant
(c) slope constant
(d) bivariate frequency distribution
(e) negative correlation
(f) regression line
(g) regression of Y on X
(h) regression of X on Y
(i) Y tilde
(j) coefficient of determination
(k) multiple correlation
(l) partial correlation

2. What are two characteristics of z scores?

3. What are two difficulties with slope as an index of correlation?

4. How are the difficulties with slope as an index of correlation overcome?

5. Calculate the coefficient of determination for the following correlations: .90, .80, .70, .60, .50, and .40.

6. What is the correlation between X and Y, if the two regression lines have slopes of .50 and .50?

7. Consider the following paired observations:

X	Y	X (cont.)	Y (cont.)
7	1	8	5
3	2	1	6
5	3	9	7
5	4	6	8
2	5	4	9

(a) What is r?
(b) What is $\bar{X}$?
(c) What is $\bar{Y}$?
(d) What is s_x?
(e) What is s_y?
(f) What value of Y is predicted by an X of 5?

8. Transform the X and Y scores in Exercise 7 to z scores.
(a) What is the mean of the z_x scores?
(b) What is the mean of the z_y scores?
(c) What is the standard deviation of the z_x scores?
(d) What is the standard deviation of the z_y scores?

9. Consider the following paired observations:

X	Y
5	3
6	7
3	4
1	8
0	9

(a) What is r?
(b) What is $\bar{X}$?
(c) What is $\bar{Y}$?
(d) What is s_x?
(e) What is s_y?
(f) What value of Y is predicted by an X of 4?
(g) What value of Y is predicted by an X of 3?

10. Transform the X and Y scores in Exercise 9 to z scores.

(a) What is the mean of the z_x scores?
(b) What is the mean of the z_y scores?
(c) What is the standard deviation of the z_x scores?
(d) What is the standard deviation of the z_y scores?

11. Calculate $r_{xy \cdot z}$ for the following situations:

(a) $r_{xy} = .8, r_{xz} = .5, r_{yz} = .3$
(b) $r_{xy} = .8, r_{xz} = .5, r_{yz} = .2$
(c) $r_{xy} = .8, r_{xz} = .5, r_{yz} = 0$
(d) $r_{xy} = .8, r_{xz} = .5, r_{yz} = .5$
(e) $r_{xy} = .8, r_{xz} = 0, r_{yz} = .5$

12. Subtract a constant, 4, from all of the X scores in Exercise 7; and calculate the correlation between the $X - 4$ scores and Y scores.

13. In attempting to obtain a measure of correlation, what two difficulties are solved by transforming the X scores to z scores, and the Y scores to z scores?

14. Consider the equation, $\Sigma(Y - \bar{Y})^2 = \Sigma(Y - \hat{Y})^2 + \Sigma(\hat{Y} - \bar{Y})^2$.

(a) If we hold $\Sigma(Y - \bar{Y})^2$ constant and increase $\Sigma(Y - \hat{Y})^2$, how is the regression of Y on X affected?
(b) If we hold $\Sigma(Y - \bar{Y})^2$ constant and decrease $\Sigma(Y - \hat{Y})^2$, how is the regression of Y on X affected?

15. Calculate the correlation between the following paired observations. Think!

X	Y	X (cont.)	Y (cont.)
101	1003	110	1007
104	1004	110	1006
105	1004	110	1008
107	1005	110	1009
109	1004	110	1010

16. What is the effect of range upon correlation? Explain.

17. What is the difference between within-cell and zero-order correlation?

18. What are two possible mathematical interpretations, or meanings, of the correlation coefficient?

19. What are four possible causal interpretations of a correlation between two properties?

20. What is the relevance of partial correlation to causal inference?

21. Why is it possible to overemphasize the difference between experimental and non-experimental research with regard to causal inference?

Prologue to Chapter 6

Chapter 6 attempts to lay the groundwork for inferential statistics through a discussion of elementary probability. After giving a formal definition of probability, there is an extended discussion of two of the models for calculating probability, the classical model and the long-run relative frequency model. These two so-called objective models continue to occupy "center stage" in statistics. (There is no treatment of the so-called subjective model for calculating probability.) The remainder of the chapter continues with a discussion of sampling, the computation of probabilities, and probability distributions. Once the contents of this chapter are mastered, the following chapter dealing with sampling distributions should pose no problem. In general, you should appreciate the importance of learning any given chapter before going on to a subsequent one. The present chapter, and in particular the section on probability distributions, should not be read unless central tendency and variability (Chapters 3 and 4) are understood.

6

Probability

Earlier in the text (Chapter 1), inferential statistics was defined as that subfield of statistics concerned with drawing inferences about characteristics of a population set using knowledge about characteristics of a sample of observations. Probability concepts provide the foundation for statistical inference and must be understood before the inference process is discussed. This chapter defines and discusses the necessary probability concepts. Subsequent chapters will use these concepts as the basis for statistical inference.

The layman most often uses the word *probability* to express a subjective judgment about the likelihood or degree of certainty that a particular event will occur. People say such things as: "It will probably rain tomorrow." "It is unlikely that we will win the ball game." "It is quite probable that we will have an exam next week." It is possible to assign a number to the *outcome being predicted*. This number represents the *degree of confidence* that the outcome will occur. For example, someone might say that the likelihood that the instructor will give an exam next week is about .9. If 1 represents certainty, .9 would mean that he is almost certain that the instructor will give an exam. If, on the other hand, someone assigned the number .6, he would be judging the likelihood of an exam to be just slightly greater than the likelihood of no exam. A rating of 0 would indicate complete certainty that the outcome will not occur.

PROBABILITY AND PROBABILITY NUMBERS

The numbers used above as ratings of likelihood or certainty are called *probabilities* or *probability numbers*. Mathematicians approach probability from the standpoint of three axioms; these serve to provide the framework for a formal definition of probability numbers.

The Axioms of Probability

Consider an ordinary deck of 52 playing cards. Such a deck, of course, consists of 13 clubs, 13 hearts, 13 diamonds, and 13 spades. For each card that can be drawn from this deck we can assign a number. These numbers are called probability numbers if they satisfy three conditions. The conditions constitute the axioms of probability.

1. The values of the probability numbers assigned to the individual cards must be equal to or greater than zero and equal to or less than one. Symbolizing a probability number as $p(X)$, we can express this first axiom as $0 \leq p(X) \leq 1$.
2. The sum of the values of the probability numbers assigned to all 52 cards must equal one; i.e., $\sum p(X) = 1$.
3. If the set of 52 cards is divided into mutually exclusive and exhaustive subsets (e.g., clubs, hearts, diamonds, and spades), the values of the probability numbers associated with each subset must also sum to 1. Notice that the four suits in the deck are both *mutually exclusive* and *exhaustive*. The suits are mutually exclusive because no card belongs to more than one suit; and the suits are exhaustive because the deck has no card that is not a club or a heart or a diamond or a spade. Thus, if a probability number is assigned to each of the suits (subsets), these numbers must sum to one; i.e., $\sum p(\text{subsets}) = 1$.

When numbers can be assigned to each instance of a set of outcomes consistent with the above axioms, these numbers are referred to as probability numbers or probabilities.

This, then, is the formal definition of probability. But how do we determine the probability number for any particular outcome, for example, the ace of spades? How are probability numbers for particular outcomes from the set of possible outcomes determined? The formal definition does not tell us. Rather, the answer is given by various models that specify calculational procedures consistent with certain assumptions. We will consider two such models, the classical model and the long-run, relative frequency model. These follow after a preliminary discussion of measurement and procedures.

Probability, Measurement, and Procedures

Before considering the models for calculating probability numbers, it is worth noting that probability concepts generally are applicable to situations in which observed outcomes are measured. This is true, regardless, whether the measured outcomes are liking responses (as in the Stroebe *et al.* experiment), chemical reactions (as might occur in a chemistry laboratory), or, more simply, the number of cars passing through an intersection in a fixed amount of time.

The sequence of activities preceding the measurement of outcomes can be referred to as a *procedure*. Procedures can vary, from elaborate activities like those involved in the Stroebe *et al.* experiment all the way to the simple matter of rolling a die (singular form of dice).

> The term "procedure" is preferred to the term "experiment" because psychologists would not like to consider activities like rolling a die an experiment. In Chapter 5, an experiment was defined as *the measurement of observed outcomes resulting from the random assignment of subjects to conditions in which a manipulative procedure occurs.* This definition appears most nearly to capture the meaning of the term as it is used in psychology, and hence, it will be followed here. In other contexts where subjects are not randomly assigned, however, the term "experiment" is defined more broadly as *the measurement of observed outcomes resulting from the application of a manipulative procedure.* The mixing of chemicals and the measurement of the observed outcome (or result) would be an experiment. Also, the rolling of a die and the measurement of the observed outcome would be an experiment. In the present context, however, such occurrences can be referred to simply as *manipulative procedures*. Some statisticians, however, define the term "experiment" yet even more broadly as *the measurement of observed outcomes*. Thus, the measurement of the number of cars that pass through an intersection during a given time interval would be an experiment. For present purposes, such occurrences can be referred to as *measurement procedures*.
>
> Thus, all together, procedures are of three types: measurement procedures, manipulative procedures, and experimental procedures (or more simply experiments). Probability concepts are generally applicable to all procedures, regardless as to type. Therefore, much of the subsequent discussion will simply refer to procedures without specification of type. The common quality of all procedures is that observed outcomes or results are measured.

Classical Model of Probability

Consider a fair die with sides containing one, two, three, four, five, or six dots. According to the classical model of probability, the probability that a roll of the die will result in a two is 1/6th. This follows from the fact that of the 6 possibilities (all equally likely) only 1 contains two dots. The classical model of probability can be formally stated as follows. *If a procedure results in n equally likely and mutually exclusive outcomes, and if $f(X)$ of these outcomes have the value X, the probability of X is the ratio of $f(X)$ to n:*

$$p(X) = \frac{f(X)}{n} \qquad (6\text{-}1)$$

In formula 6-1, $p(X)$ symbolizes the probability of X, n the number of possible outcomes, and $f(X)$ the number of the possible outcomes that have the value X. Thus, in the above example of a die whose sides each contained between one and six dots, the probability of rolling a two is calculated as follows:

$$p(2) = \frac{f(2)}{6} = \frac{1}{6}$$

In order to use the classical model of probability, two assumptions must hold. First, the n outcomes must be equally likely. Second, the n outcomes must be mutually exclusive. Both of these assumptions hold for the above example. Rolling a fair, or perfectly balanced, die and accurately noting the results of each roll assures that the 6 outcomes are equally likely. Second, the 6 outcomes are mutually exclusive; i.e., all 6 outcomes are in non-overlapping sets.

Consider a second example. Suppose that we have a group of 100 people consisting of 60 males and 40 females. If the name of each individual is written on a card and the cards are thoroughly shuffled, what is the probability that the selection of one of the cards will result in a male? Note: the shuffling of the cards assures that each of the 100 outcomes is equally likely, and the fact that no person's name is on more than one card assures that the 100 outcomes are mutually exclusive. Substituting in formula 6–1:

$$p(\text{male}) = \frac{f(\text{male})}{100} = \frac{60}{100} = \frac{3}{5}$$

The probability of selecting a male is 3/5ths.

As mentioned above, use of the classical model can only be justified when two assumptions are made: that the n outcomes are mutually exclusive, and that the n outcomes are equally likely. Consider, first, the assumption regarding mutually exclusive outcomes. Suppose that we have a group of 100 people, 60 of whom are male and 60 of whom have blue eyes. If the name of every male is written on a card and the name of every person with blue eyes is written on a card, there would be 120 cards. The males with blue eyes have their names on two different cards. Because of the lack of mutual exclusiveness, to calculate the probability of drawing a male as the ratio of the number of males (60) to the number of cards (120) would be an error.

The second assumption involves equally likely outcomes. Perhaps the biggest problem with the classical model of probability is that it cannot be applied when outcomes are not equally likely to occur. Consider a six-number die that has been loaded by slightly overweighting the side with one dot. Now what is the probability that a roll of the die will result in a certain number of dots, for example, 6? A gambler may avoid the question by refusing to use the loaded die. However, there are many practical situations in which probability questions cannot be avoided easily. An insurance company, for example, may wish to know the probability that a person who is alive at age 20 will also be alive at age 40. It is obvious that the possibilities of being alive or dead at age 40 are not equally likely possibilities, and yet, the survival of insurance companies is dependent upon having answers to questions such as this.

Long-Run, Relative Frequency Model of Probability

Questions such as the above led to the development of the long-run, relative frequency model of probability. Consider again the loaded die. Since the six sides are not equally likely, the probability of rolling a six cannot be directly calculated as the ratio of the number of sixes to the total possible number of outcomes. Does this mean, however, that it is impossible to discover the probability of rolling a six with this loaded die? A moment's thought will suggest the obvious possibility of repeatedly rolling the loaded die and observing what happens; i.e., to use what philosophers call the empirical method. Table 6.1 presents the results of having rolled the die between 20 and 1,000 times. The first column of Table 6.1 indicates $f(6)$; i.e., the number of times a roll of the die resulted in a six. The second column indicates the number of times the die was rolled. Since each roll of the die, or repetition of the procedure, results in a score, the total number of repetitions is n. The third or final column of Table 6.1 indicates the relative frequency of six. Thus, after 20 rolls, or repetitions, a six occurred 3 times, so the relative frequency is 3/20 or .150.

The third column of Table 6.1 illustrates a very important point. As the number of repetitions increases, the relative frequency of six fluctuates less and less, and tends to converge to a constant value. This value appears to be in the neighborhood of .22.

The long-run, relative frequency model of probability can be stated as follows. *The probability of X is the ratio of the frequency of X, $f(X)$, to the number of repetitions of the procedure, n, after a very large number of repetitions:*

$$p(X) = \frac{f(X)}{n}$$

Thus, if after a very large number of repetitions the relative frequency of rolling a six stabilized at some value, that value would be considered the probability of six.

TABLE 6.1. *Relative Frequency of Rolling a Six with a Particular Loaded Die.*

NUMBER OF SIXES $f(6)$	NUMBER OF REPETITIONS n	RELATIVE FREQUENCY $f(6)/n$
3	20	.150
11	40	.275
20	80	.250
45	200	.225
110	500	.220
222	1000	.222

The student may be confused by the fact that the above formula for the long-run, relative frequency model appears to be just like formula 6–1 for the classical model. While the formulas are the same, the models are different. In the classical model, n represents the number of possible outcomes; and $f(X)$, the number of these outcomes that have the value X. In the long-run, relative frequency model, n represents the number of repetitions of the procedure, and $f(X)$, the number of these repetitions that resulted in X. Furthermore, it is assumed that the procedure has been repeated a large number of times; i.e., that n is a very large number. The symbol, n, is appropriate in both cases since it represents number of scores. It is just that the models focus upon different scores. In the case of the classical model, the scores are the possible outcomes; and in the case of the relative frequency model, the scores are the outcomes resulting from numerous repetitions of the procedure. Likewise, symbol $f(X)$ is appropriate in both cases since it represents the frequency with which a given score value occurs. For the classical model, $f(X)$ represents the frequency of possible outcomes with the value X; and for the relative frequency model, $f(X)$ represents the frequency of repetitions of the procedure that results in X.

As has already been indicated, the long-run, relative frequency model does not assume equally likely outcomes. Mutually exclusive outcomes, however, are assumed. Such mutual exclusiveness results from the decision to regard each repetition of the procedure as producing an outcome belonging to a distinct set.

The long-run, relative frequency model is, of course, closely related to the notion of repeatability or reoccurrence. This raises an interesting question concerning the use of the long-run, relative frequency model for outcomes that have not occurred or have not occurred repeatedly. Suppose a physician is asked whether or not his patient's broken arm will heal within a month's time. If the patient has never previously broken his arm, the relative frequency model appears to be of little help. This is not the case, however, if other similar outcomes are considered. The physician could consider the relative frequency with which other broken arms have healed within a month's time. In general, then, a probability statement can be made about a non-repeated outcome by referring to a set of other similar outcomes. While this sounds straightforward enough, determining what outcomes are similar and what outcomes are not similar can be a problem. In the case of the above example, the physician might determine similarity on the basis of the age of the patient, the health of the patient, the location of the break, and other similar factors. If the set of similar outcomes is narrowed down too much, however, the number of outcomes may become undesirably small. It is, therefore, evident that the long-run, relative frequency model, like the classical model, has shortcomings.

> A more precise interpretation of probability as a long-run, relative frequency follows. The probability number associated with a particular outcome or set of outcomes may

be interpreted as the limiting value of the proportion $f(X)/n$, as n approaches infinity. The numerator, $f(X)$, is the frequency of occurrence of a particular outcome or set of outcomes symbolized by X. Thus, $f(X)$ should be read, "frequency of X." The denominator n is the number of repetitions of the procedure. Stating the probability of X symbolically as $p(X)$, this interpretation can be written:

$$p(X) = \lim \frac{f(X)}{n}, \quad \text{as } n \to \infty \tag{6-2}$$

As n becomes so large as to approach infinity, the proportion, $f(X)/n$, becomes less variable and stabilizes at a constant value called its limiting value. This constant value or limiting value is the probability of X.

At any value of n less than infinity, there is a likelihood that $f(X)/n$ will differ by a very small amount or more from its limiting value; and this likelihood changes as the number of repetitions changes. In fact, the likelihood that $f(X)/n$ will differ from $\lim f(X)/n$ by any small amount or more approaches zero as n becomes larger and larger. This statement can be expressed symbolically as:

$$p\left[\left(\frac{f(X)}{n} - \lim \frac{f(X)}{n}\right) \geq d\right] \to 0 \quad \text{as } n \to \infty \tag{6-3}$$

where d is a very small difference. As the number of repetitions of an experiment increases, it becomes less and less probable that $f(X)/n$ differs from $\lim f(X)/n$.

SAMPLING AND PROBABILITY

Population and Sample

In earlier chapters the terms *population* and *sample* have been used. *A population is the set of all elements of concern.* In many contexts the elements are real objects, living or non-living. In psychology, of course, the objects are typically living. For example, a population might consist of all the full-time undergraduates enrolled at the University of North Carolina. In many fields, however, the objects are typically non-living, for example, the television picture tubes manufactured in 1974.

The above definition, however, is broad enough so that the elements need not be living or non-living objects but simply may be the set of outcomes observed in connection with a procedure. A procedure, of course, always guarantees that the outcome is assigned to a measurement category. A population might consist of the set of heights of all the full-time undergraduates enrolled at the University of North Carolina. Therefore, in appropriate contexts one may refer both to a total set of organisms as a population and to a total set of numerical values resulting from application of a procedure as a population.

One final complication is that all of the values in an outcome population need not actually exist; i.e., not all of the measurements need to have been taken.

For example, one can consider the population of outcomes (as represented by numerical values) that could result from the repetition of an experimental procedure an indefinite number of times. Such populations are only conceptual possibilities. For purposes of statistical inference, they are, however, mathematically useful.

A sample is a subset of a population. It is a subset of elements selected from a population. *The process of selecting elements from a population is called sampling.* When the population consists of actual objects (living or non-living), sampling consists of selecting a subset of these objects. When the population elements are observed outcomes (or numerical representations of observed outcomes), sampling consists of the application of a procedure. Procedures, of whatever type, always result in numerical values.

The type of sampling usually assumed is *simple random sampling.* *Simple random sampling consists of selecting elements from a population set in such a way that on each selection each element has an equal chance of being selected.* Simple random sampling is always assumed by the classical model of probability, and also by the long-run, relative frequency model.

> Since random sampling is defined as selecting elements from a population in such a way that on each selection each element has an equal chance of being selected, how can random sampling be used for the above example of a loaded die? Since the die is loaded, each side does not have an equal chance of occurring when the die is rolled. The problem can be handled by conceptualizing the population as a theoretical one consisting of a very large number of rolls of this particular die. Random sampling would then consist of selecting rolls of the die from the population in such a way that each roll has an equal chance of being selected. An example of non-random sampling from this population would be the selection of only those rolls in which the die turned over once after hitting the table. Violation of the assumption of random sampling can have effects varying all the way from trivial to extreme.

There are instances in which violation of random sampling can lead to rather biased conclusions. An understanding and appreciation of random sampling can serve as a guard against being misled by the seemingly precise statistical statements that are frequently encountered in the mass media. This is the main point of Darrell Huff's first chapter in *How to Lie with Statistics.*

Huff notes that *Time* magazine once commented that "The average Yaleman, Class of '24, makes $25,111 a year." According to Huff the likeliest source of error in this statement is the sampling procedure. Twenty-five years after graduation the addresses of all graduates undoubtedly were not known. Furthermore, not all those graduates whose addresses were known would respond to a questionnaire enquiring about their income. In some instances response rates to mail questionnaires may be as low as 5 to 10 percent—although it probably was higher in this instance. Therefore, it is highly unlikely that the above statement was based on a random sample—a sample in which each person had an equal probability of being included.

Non-random samples may lead to biased conclusions. In this instance Huff argues that the direction of the bias was to raise the level of reported income. This is true for two reasons. First, since the addresses of the wealthy (corporation directors, executives, etc.) are easier to locate than the addresses of the poor (tramps, unemployed, alcoholics, etc.), the original mailing list probably contained disproportionately more wealthy. Second, since the wealthy are more likely to be proud of their incomes than the poor, the wealthy were more likely to respond to the questionnaire.

Sampling with and without Replacement

Two types of simple random sampling can be distinguished: *sampling with replacement* and *sampling without replacement*. *When sampling with replacement, the particular element selected is returned to the population before making the next selection. When sampling without replacement, the element selected is not returned before making the next selection.* When sampling without replacement, the element selected at one sampling is not available for selection at the next sampling. Technically speaking, that element is no longer a member of the population being sampled.

When sampling with replacement, the probabilities associated with each subset of elements in the population remain constant from selection to selection. Since each element being selected is returned to the population prior to the next selection, the frequency of elements in any particular subset remains the same. For example, when sampling with replacement from the population of 100 students (60 male, 40 female), each time a male is selected he is returned to the population prior to the next selection; thus, 60 males always are available for selection and the probability of selecting a male remains a constant .6.

When sampling without replacement, the probabilities associated with each subset of elements change from selection to selection. After each selection, the population contains one less element, and the number of remaining elements in the subset to which the selected element belongs is reduced by one. If a student is selected without replacement from the population of 100 students, the population for the next selection then contains just 99 students. And, if the selection is a male, the number of males available for the next selection then is 59 rather than 60. Thus, the probability of selecting a male on the next sampling is 59/99, or .596, rather than .600. Further, while all 40 females remain available, the probability of selecting a female changes to 40/99, or .404, from .400.

It can be seen easily that when sampling without replacement, the probability associated with each different kind of element on each sampling *depends,* or is *conditional,* upon the outcomes of prior selections. In the above example, if the outcome of the first selection had been a female rather than a male, the probabilities associated with obtaining a male or female on the second selection then would have been respectively .606 and .394, rather than .596 and .404. The probabilities on the second sampling depend on the outcome of the first sampling.

When probabilities of outcomes (e.g., male or female) *depend on prior selections, the selection process is termed dependent sampling.* Sampling without replacement is a form of *dependent sampling.* Simple random sampling with replacement is a form of *independent sampling.* When sampling with replacement, the probabilities of outcomes at each sampling do not *depend* on prior selections—they are completely *independent* of prior selections. In fact, they are constant from selection to selection.

> Probabilities occurring in the process of sampling without replacement are examples of conditional probabilities. As stated, the probability that a member of a particular subset will occur on a given sampling depends, or is conditional upon, the outcomes of prior samplings. In the above example, the probability of sampling a male on the second selection was .596, if the first selection was a male, and .606, if the first selection was a female. These two probabilities are examples of conditional probabilities and can be written symbolically as:
>
> $$p(Male_2 \mid Female_1) = .606$$
>
> $$p(Male_2 \mid Male_1) = .596$$
>
> These forms are called conditional probability statements and should be read, respectively, as: the probability of sampling a male on the second selection, given the condition that a female has occurred on the first selection, is .606; the probability of sampling a male on the second selection, given the condition that a male has occurred on the first selection, is .596.

Inferential methods usually require that probabilities of elements selected be independent of prior selections and constant across selections. When simple random sampling is employed and the population is infinite, independence and constancy then are assured, whether sampling with or without replacement. The actual conduct of experiments, however, typically includes sampling without replacement from a finite population. The researcher usually samples from a finite set of subjects and does not want to replace a subject and allow him or her to be available for selection again. One legitimately can assume independence and constancy when sampling proceeds without replacement, as long as the population from which one is sampling is very large and the ratio of sample size to population size is very small. Under these conditions, the probabilities associated with different subsets of elements remain essentially constant and independent across samplings, even when sampling without replacement.

If the number of elements in a population is very large, then, the removal of an element without replacement will not alter appreciably the probabilities associated with the various kinds of elements. For example, suppose a population includes 10,000 persons, 1,000 having incomes between $15,000 and $20,000. The probability associated with selecting a person in the income range $15,000–$20,000 (assuming simple random sampling) is .10. If, in the process of sampling, several persons possessing incomes between $15,000 and $20,000 are removed from the population and not returned, the probability associated with this income range will change on succeeding samplings. However, the change will not be appreciable, providing the number of people removed is not too large. If twenty samplings are completed, and on five of these the persons selected possess incomes in the $15,000–$20,000 range, then, on the twenty-first sampling, the probability associated with this income range will be 995 / 9,980 = .0997, which is not appreciably different from .10. Thus, as long as the population is very large and the ratio of the sample size to the size of the population is small, the researcher can use statistical procedures that require random and independent sampling even though the sampling is done without replacement.

COMPUTING PROBABILITIES

We now turn to a consideration of the computation of probabilities. In order to do this systematically, however, it is first necessary to introduce some background concepts—and some rules.

Background Concepts

n FACTORIAL. *The factorial of any number is the joint product of that number times each whole number of lower value down to 1.* A factorial is symbolized !. For example, if n is 3, $n! = (3)(2)(1) = 6$. One further point is that zero factorial, $0!$, is defined as 1.

PERMUTATIONS. *A permutation is a serial order or sequence.* The number of permutations of a group of objects is the number of serial orders in which the objects may be arranged. For example, consider three objects labeled A, B, and C. These three objects may be arranged in six permutations:

$$\begin{array}{ccc} A, B, C & B, A, C & C, A, B \\ A, C, B & B, C, A & C, B, A \end{array}$$

Likewise, consider three rolls of a die that result in the outcomes 1, 5, and 6. These three outcomes may occur in six permutations:

$$\begin{array}{ccc} 1, 5, 6 & 5, 1, 6 & 6, 1, 5 \\ 1, 6, 5 & 5, 6, 1 & 6, 5, 1 \end{array}$$

In general, the number of permutations of n elements (whether real objects, observed outcomes, or whatever) is $n!$:

$$_nP_n = n! \qquad (6\text{-}4)$$

In formula 6-4, $_nP_n$ refers to the number of permutations of n elements where all n elements are different. For the above example:

$$_3P_3 = 3! = (3)(2)(1) = 6$$

Formula 6-4 applies only in the situation in which all n of the elements are different. What about the situation in which the n elements are of just two types? Many probability problems concern just such a situation. The <u>number of permutations of n elements when only two types of elements are symbolized, is $_nP_{r,n-r}$</u>. In general:

$$_nP_{r,n-r} = \frac{n!}{r!\,(n-r)!} \tag{6-5}$$

Where: n is the number of elements
   r is the number of elements of one kind
   $n - r$ is the number of elements of the other kind.

For example, suppose that a coin is tossed four times. One wishes to know the number of permutations in which a head occurs three times. Formula 6-5 is appropriate in this case since the outcomes are of just two kinds, head (H) and tail (T). Using formula 6-5, n equals 4 (tosses), r equals 3 (heads), and $n - r$ equals 1 (tail), thus:

$$\begin{aligned}_nP_{r,n-r} &= \frac{n!}{r!\,(n-r)!} \\ &= \frac{4!}{3!\,(4-3)!} \\ &= \frac{(4)(3)(2)(1)}{(3)(2)(1)} \\ &= 4\end{aligned}$$

There are four permutations in which a head (H) occurs three times: HHHT, HHTH, HTHH, THHH.

Notice in the above example that it is arbitrary whether one asks for the number of permutations in which a head occurs three times or the number of permutations in which a tail occurs once; i.e., it does not matter whether r equals 3 or 1. If r equals 1:

$$\begin{aligned}_nP_{r,n-r} &= \frac{n!}{r!\,(n-r)!} \\ &= \frac{4!}{1!\,(4-1)!} \\ &= \frac{(4)(3)(2)(1)}{(3)(2)(1)} \\ &= 4\end{aligned}$$

just as when r equals 3.

Some readers might note that the ratio, $n!/[r!(n-r)!]$, is identical to that used to compute the number of different combinations of r elements that can be obtained when making r selections from a set of n elements; i.e., $_nC_r = n!/[r!(n-r)!]$. The same ratio tells us both the number of distinct sequences of n elements where r are of one kind and $n - r$ are of another kind, and the number of distinct combinations of r elements that can be sampled from a set of n elements.

Some Rules Concerning Probabilities

Several rules regarding the manipulation of probabilities now will be described. These rules were developed for *a priori* probabilities, like those generated by the classical model, but we will assume that they also apply to empirically estimated probabilities.

ADDITION RULE FOR THE UNION OF OUTCOMES. In Appendix A, two sets are defined as mutually exclusive if they contain no elements in common; i.e., no element in one set is contained in the other set. If the elements in the sets are outcomes, we then have mutually exclusive outcome sets, or more simply, mutually exclusive outcomes. Mutually exclusive outcomes are very important in psychology because measurement involves the placement of observed outcomes into mutually exclusive sets or categories. No outcome is assigned to more than one measurement set or category.

The addition rule is applicable to the union of outcome sets. As indicated in Appendix A, the union of two sets A and B is a new set comprised of those elements belonging to A or to B, or to both A and B. When the sets are mutually exclusive, of course, no elements are found in both A and B, and the union of A and B simply refers to those elements in A or B. The simple addition rule for the union of mutually exclusive outcomes can be stated as follows: *The probability of the union of two or more mutually exclusive outcomes is equal to the sum of the probabilities of the separate outcomes.*

This rule is quite useful to the researcher. As noted previously, measurement scales consist of mutually exclusive sets or categories. Provided one can empirically estimate the probabilities associated with each measurement category, the probability associated with any grouping of categories can be computed. For example, in a running time experiment, the running time of rats was measured to the nearest .5 second. Thus, a running time in the range of 9.75–10.24 was classified as 10 seconds; and a time in the range of 10.25–10.74 seconds was classified as 10.5 seconds. The recorded values were 5.0, 5.5, 6.0, 6.5, 7.0, etc., with each value representing the midpoint of a range of possible values. The relative frequency with which each value occurred was used as the estimated probability of that value. Table 6.2 presents the recorded values and empirically estimated probabilities. By means of the information in Table 6.2 and the

TABLE 6.2. *Empirically Established Probabilities Associated with Possible Scores in Running Time Experiment.*

SCORES	PROBABILITIES	SCORES	PROBABILITIES
5.0	.01	10.0	.05
5.5	.02	10.5	.03
6.0	.04	11.0	.03
6.5	.04	11.5	.02
7.0	.05	12.0	.01
7.5	.07	12.5	.01
8.0	.08	13.0	.01
8.5	.13	13.5	.01
9.0	.17	14.0	.01
9.5	.15	14.5	.01
		15.0	.05

$$\sum_{1}^{k} = 1.00$$

addition rule, the probability that an animal will complete the run within any range of running time values can be computed. The probability that an animal will complete the run within 8 to 10 seconds is computed by summing the probabilities associated with 8.0, 8.5, 9.0, 9.5, and 10.0. This sum equals .58 (.08 + .13 + .17 + .15 + .05 = .58). The probability is .58 that any rat will complete the run in between 8 and 10 seconds.

Students are sometimes confused about the situation in which the simple addition rule is appropriate. A useful rule of thumb is to determine whether or not the problem can be stated so that the word "or" occurs between all the outcomes. This is obviously the case for the above problem. The problem is to determine the probability of completing the run in between 8 and 10 seconds. By way of restatement, the problem is to determine the probability of completing the run in 8.0 *or* 8.5 *or* 9.0 *or* 9.5 *or* 10.0 seconds. The little word "or" signals the union of outcomes and, thus, the possible appropriateness of the addition rule. The existence of "or," of course, does not guarantee appropriateness. The outcomes still have to be mutually exclusive.

Although it will not be used in subsequent parts of this book, a more general addition rule is offered that does not assume mutual exclusiveness. In the case of only two outcomes this rule can be stated as follows:

$$p(A \cup B) = p(A) + p(B) - p(A \cap B) \qquad (6\text{--}6)$$

The probability of the union for two outcomes A and B is equal to the probability of A "plus" the probability of B, "minus" the intersection of A and B. In the case in which A and B are mutually exclusive, $p(A \cap B)$ equals 0, and formula 6–6 reduces to the simple addition rule.

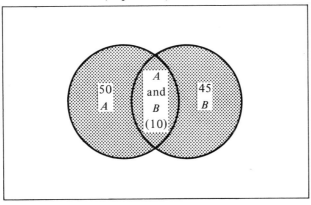

FIGURE 6-1. Venn diagram illustrating a population of 200 outcomes and two subsets, A and B.

The rationale for formula 6-6 can be grasped through a consideration of Fig. 6-1. In Fig. 6-1, the two outcome sets, A and B, are represented by a Venn diagram. The population consists of 200 outcomes; set A consists of 50 outcomes (all outcomes being within the circle labeled A, which includes those in $A \cap B$); set B consists of 45 outcomes (all outcomes being within the circle labeled B, which also includes those in $A \cap B$); and 10 outcomes belong to both A and B [i.e., $f(A \cap B) = 10$]. Assuming that all of the 200 outcomes in the population are equally likely to occur:

$$p(A) = \frac{f(A)}{f(\text{population})} = \frac{50}{200}$$

The probability of A equals the ratio of the number of outcomes in A to the total number of outcomes, likewise:

$$p(B) = \frac{f(B)}{f(\text{population})} = \frac{45}{200}$$

And finally:

$$p(A \cap B) = \frac{f(A \cap B)}{f(\text{population})} = \frac{10}{200}$$

In calculating the probability of the union of A and B, the intent is to determine the ratio of the number of outcomes in the shaded areas to the total population. However, by adding $p(A)$ and $p(B)$, the ratio of $f(A \cap B)$ to the total population [i.e., $p(A \cap B)$] is represented twice. It is for this reason that formula 6-6 calls for the subtraction of $p(A \cap B)$. Thus, applying formula 6-6 to the example in Fig. 6-1:

$$p(A \cup B) = \frac{50}{200} + \frac{45}{200} - \frac{10}{200} = \frac{85}{200}$$

The general addition rule can be extended to any number of sets, but the extension is not a simple one. For three sets A, B, and C:

$$p(A \cup B \cup C) = p(A) + p(B) + p(C) - p(A \cap B) - p(A \cap C) \\ - p(B \cap C) + p(A \cap B \cap C) \qquad (6\text{--}7)$$

Superficially, it may appear that the general addition rule cannot be used in conjunction with the classical model of probability, since that model assumes mutual exclusiveness. In fact, however, the general addition rule eliminates the lack of mutual exclusiveness (by subtracting out the area of overlap) so that the classical model can be applied.

MULTIPLICATION RULE FOR THE INTERSECTION OF OUTCOMES. The intersection of two sets A and B refers to the set containing all of the elements common to both A and B. When the elements are outcomes, we have the intersection of outcome sets, or more simply, the intersection of outcomes. *Two types of situations give rise to the intersection of outcomes.* First, the intersection of outcomes occurs when objects are measured with regard to more than one property. For example, a population of students could be measured with regard to the properties of sex and age. Table 6.3 illustrates four joint outcomes that could arise from the measurement of sex and age; where sex is measured using two classes, male and female, and age is measured using two classes, over 20 years and 20 years or younger. In this table, variable X represents either of the two classes measuring age, and variable Y represents either of the two classes measuring sex. The four pairs of measurements represent intersections of outcomes.

Second, the intersection of outcomes also occurs as a result of a series of samplings or repetitions of a procedure. Suppose a coin is tossed twice. The first toss can result in either a head or a tail, and the second toss can result in either a head or a tail. The resulting intersections of outcomes are given in

TABLE 6.3. *Interesection of Outcomes Resulting from the Measurement of Sex and Age.*

		\(X\)	
		X_1 (over 20)	X_2 (20 or under)
Y	Y_1 female	Y_1, X_1 (female and over 20)	Y_1, X_2 (female and 20 or under)
	Y_2 male	Y_2, X_1 (male and over 20)	Y_2, X_2 (male and 20 or under)

TABLE 6.4. Intersections of Outcomes Resulting from Two Tosses of a Coin.

		FIRST TOSS	
		Head	Tail
SECOND TOSS	Head	Head and head	Tail and head
	Tail	Head and tail	Tail and tail

Table 6.4. Each combination of one of the two outcomes for the first roll with one of the two outcomes for the second roll is an intersection of outcomes.

In the discussion of sampling earlier in this chapter, outcomes to sampling were defined as independent when probabilities of outcomes at each sampling are not dependent on prior selections. They are independent of (unaffected by) prior selections. This concept of independent outcomes can be broadened and related to the concept of the intersection of outcomes. The outcomes comprising *an intersection of outcomes are said to be independent if the probability of each outcome is in no way altered by, or dependent on, the occurrence of the other outcome(s).* In sampling from a population of persons, outcomes comprising the intersection of outcomes "male, female, male" are independent if the probability of male on the third selection is in no way altered by the occurrence of "male, female" on the first two selections; and the occurrence of female on the second selection is in no way altered by the occurrence of male on the first selection.

In measuring age and sex as in Table 6.3, the outcomes of male and over 20 are independent if the probability of selecting a male is in no way altered by the fact that the person is also over 20; and the probability of selecting a person over 20 is in no way altered by the fact that the person is also a male. If, for example, the probability of selecting a male, given that the person is certain to be over 20, is different than it would be if the person were certain to be under 20; then, the the outcomes of male and over 20 would not be independent—they would be dependent.

The simple multiplication rule is for the intersection of independent outcomes. That rule can be stated as follows: *The probability of the intersection of two or more independent outcomes is equal to the product of the probabilities of the separate outcomes.* For example, if someone samples randomly with replacement from a population of 100 students (60 male, 40 female), he or she can compute the probability of obtaining a particular series of outcomes (an intersection of outcomes), such as "male, female, male, male," after four samplings by multiplying the probabilities associated with each of these outcomes: $(.6)(.4)(.6)(.6) = .0864$. Or, consider the probability of obtaining the intersection of outcomes "head,

Computing Probabilities

tail, tail" in three independent tosses of a coin. This probability is $(.5)(.5)(.5) = .125$. Consider also the age and sex example in Table 6.3. If the probability of selecting a male is .6, and if the probability of selecting a person over 20 is .5, and further, if the outcomes of male and over 20 are independent; the probability of the intersection of outcomes "male over 20" is $(.6)(.5)$ or .30.

Just as the word "or" between each of the outcomes signals the possible appropriateness of the addition rule, the word "and" between each of the outcomes signals the possible appropriateness of the multiplication rule. Thus, if the problem is to determine the probability of obtaining the intersection of outcomes "head, tail, tail" in three tosses of a coin, the problem, by way of restatement, is to determine the probability of obtaining a head *and* a tail *and* a tail. Similarly, if the problem is to determine the probability of sampling a male who is over 20, by way of restatement the problem is to determine the probability of obtaining someone who is both male *and* over 20. The cue word "and" between the outcomes signals the intersection of outcomes and the possible appropriateness of the multiplication rule. Of course, in order to apply the simple multiplication rule, the outcomes must also be independent.

> A discussion of *conditional probability* should add some meaning to the concept of independence and the multiplication rule. *Given two sets of outcomes, A and B, the conditional probability, $p(A \mid B)$, is the probability of A, given the condition that B has occurred or is certain to occur.* Referring to Table 6.3, the conditional probability $p(Y_1 \mid X_1)$ is the probability of selecting a female given the condition that selecting a person over 20 is a "sure thing."
>
> When each outcome has an equal probability of occurring, the conditional probability of any outcome A (i.e., an outcome within the set A), given the condition that outcome B is certain, then is equal to the proportion of outcomes in B that also belong to A. In Fig. 6-1, the two outcome sets, A and B, are represented in a Venn diagram. Recall that the population consists of 200 outcomes. Set A consists of 50 outcomes—all outcomes within the circle labeled A, which includes those in $A \cap B$. Outcome B consists of 45 outcomes—all outcomes within the circle labeled B, which also includes those in $A \cap B$. And, 10 outcomes belong to both A and B, $f(A \cap B) = 10$. Given the condition that outcome B will occur, or has occurred, the probability of A occurring is the ratio of the number of outcomes that belong to B and A, $f(A \cap B)$, to the number that belong to B, $f(B)$:
>
> $$p(A \cap B) = \frac{f(A \cap B)}{f(B)} \qquad (6\text{-}8)$$
>
> If B is certain to occur, then, the only elements that can appear in sampling are the set of 45 elements belonging to B. Thus, the probability of A becomes the proportion of elements in set B that also belong to set A. For the particular population illustrated in Fig. 6-1, the proportion of A in B is given by:
>
> $$\frac{f(A \cap B)}{f(B)} = \frac{10}{45} = .22$$

The ratio $f(A \cap B)/f(B)$ is also equivalent to the fraction, the probability of $A \cap B$ divided by the probability of B, since:

$$\frac{f(A \cap B)}{f(B)} = \frac{\frac{f(A \cap B)}{f(\text{population})}}{\frac{f(B)}{f(\text{population})}}$$

$$= \frac{p(A \cap B)}{p(B)}$$

Thus, the conditional probability of $p(A \mid B)$ is equal to the probability of joint event $A \cap B$ divided by the probability of B:

$$p(A \mid B) = \frac{p(A \cap B)}{p(B)} \qquad (6\text{--}9)$$

Two sets, A and B, are independent if the conditional probability of A given B, $[\,p(A \mid B)\,]$ is equal to the probability associated with A, $[\,p(A)\,]$. In symbolic form, this definition says that A and B are independent if $p(A \mid B) = p(A \cap B)/p(B) = p(A)$. The reader can note that this definition of independence is equivalent to the multiplication rule. The equation $p(A \cap B)/p(B) = p(A)$ can be converted to the form $p(A) \times p(B) = p(A \cap B)$.

Just as a more general addition rule exists that does not assume mutual exclusiveness, there is also a more general multiplication rule that does not assume independence. This rule can be stated as follows:

$$p(A \cap B) = p(A)p(B \mid A) = p(B)p(A \mid B) \qquad (6\text{--}10)$$

The probability of the intersection of the outcomes A and B is equal to the product of the probability of A "times" the conditional probability of B, given A, and is also equal to the product of the probability of B "times" the conditional probability of A, given B.

Example Problems

We are now in a position to apply the background concepts and rules to the solution of problems.

EXAMPLE 1 (ONE PERMUTATION). The first example problem is a simple one involving only one permutation.

> Consider a four item true–false test. If a student does not know the answer to any of the items, what is the probability that he will get all four of them correct?

There are four considerations in this problem. First, since the student does not know the answer to any of the items, it will be assumed that the two alternatives "true" and "false" are equally likely. Thus, the classical model may be applied. According to the classical model, the probability of getting any one of the items correct is 1/2. It should be noted, however, that the equally likely assumption may, or may not, hold.

Second, since the problem is to calculate the probability of answering items 1 and 2 and 3 and 4 correctly, and not 1 or 2 or 3 or 4 correctly, the simple multiplication rule possibly is involved.

Third, since the probability of answering any item correctly is in no way affected by having answered any of the other items correctly, the probabilities are independent. Independence is, of course, required for application of the simple multiplication rule.

Fourth, since the problem is to determine the probability of getting all four items correct, only one permutation is involved. There is no need to be concerned with multiple permutations. Thus, the solution to the problem is to use the simple multiplication rule and raise 1/2 to the fourth power:

$$p(\text{all four items correct}) = \left(\frac{1}{2}\right)^4 = \frac{1}{16} = .0625$$

The probability of getting all four items correct is .0625.

EXAMPLE 2 (SEVERAL PERMUTATIONS). The second example problem involves several permutations.

> Consider the same four item true-false test. If a student does not know the answer to any of the items, what is the probability that he will get three of them correct?

Again, this problem involves four concerns. First, as before, on the basis of the classical model the probability of getting any item correct is 1/2. In addition, however, the classical model also tells us that the probability of getting any item incorrect is 1/2.

Second, as before, the simple multiplication rule is possibly involved since there is concern with the intersection of four outcomes.

Third, also as before, the probabilities are independent, thus giving further assurance that the simple multiplication rule is appropriate.

Fourth, unlike before, a number of permutations are involved. The problem does not concern the probability that some particular three items will be correct (e.g., items 2, 3, and 4), but the probability that any of the three items will be correct. Thus, it is necessary to determine the number of permutations of three correct items and one incorrect item. Applying formula 6–5:

$$_nP_{r, n-r} = \frac{n!}{r!(n-r)!}$$

$$= \frac{4!}{3!(4-3)!}$$

$$= \frac{(4)(3)(2)(1)}{(3)(2)(1)(1)}$$

$$= 4$$

TABLE 6.5. *Permutations and Probabilities of One Incorrect and Three Correct Responses to a Four Item Test.*

\multicolumn{4}{c}{ITEMS}	PROBABILITY
1	2	3	4	
Incorrect	Correct	Correct	Correct	.0625
Correct	Incorrect	Correct	Correct	.0625
Correct	Correct	Incorrect	Correct	.0625
Correct	Correct	Correct	Incorrect	.0625
				$\Sigma = .250$

Three correct items and one incorrect item will require four permutations. These permutations are shown in Table 6.5. The incorrect item may be the first item, the second item, the third item, or the fourth item.

Having specified the four permutations, we are now in a position to apply the simple multiplication rule to each permutation. What is the probability of the first permutation in Table 6.5—item 1 incorrect and items 2, 3, and 4 correct? Initially it might be thought that the answer can be obtained by multiplying the probability of getting item 2 correct by the probability of getting item 3 correct by the probability of getting item 4 correct; i.e., by raising 1/2 to the third power. Such a procedure, however, ignores the probability of getting item 1 incorrect. Notice that the intent is to obtain the probability of the entire permutation; i.e., the probability of getting item 1 incorrect *and* item 2 correct *and* item 3 correct *and* item 4 correct. Thus, the correct procedure involves raising 1/2 to the fourth power. One-half to the fourth power equals .0625, the probability of the permutation.

As Table 6.5 indicates, .0625 is the probability of each of the four permutations. Each of the permutations involves one incorrect and three correct items, and has a probability equal to 1/2 to the fourth power.

At this point it is instructive to consider exactly what is asked by the problem. What is asked is the probability of getting any three items correct (and thus, any one item incorrect). By way of restatement, what is asked is the probability of obtaining the first permutation *or* the second permutation *or* the third permutation *or* the fourth permutation. Applying the simple addition rule, the probabilities of each of the four permutations are added, giving .25. The probability of getting three of the four items correct is .25.

Binomial Formula

The procedures used in the above example problems can be symbolized by the following *binomial formula*:

$$\text{Prob.}\left(\frac{r}{n}\right) = \frac{n!}{r!\,(n-r)!}\,p^r q^{n-r} \qquad (6\text{–}11)$$

Computing Probabilities

Where: Prob. $(r \mid n)$ is the probability of obtaining r outcomes of one kind and $n - r$ of the other kind in n samplings or repetitions
p is the probability of obtaining one kind of outcome
q is the probability of obtaining the other kind of outcome.

In formula 6–11, the probability of $r \mid n$ is written Prob. $(r \mid n)$ rather than $p(r \mid n)$ in order to keep clear the distinction between the probability of $r \mid n$ and p in the right side of the equation—the probability of one of the two kinds of outcomes in the population set. The formula consists of two basic parts, $n! / r! (n - r)!$ and $p^r q^{n-r}$. First, consider $p^r q^{n-r}$. This expression gives the probability of each permutation. For the second of the above examples p is the probability of getting an item correct (1/2), q the probability of getting an item incorrect (1/2), n the number of items on the test (4), r the number of correct items (3), and $n - r$ the number of incorrect items (1). Thus:

$$p^r q^{n-r} = \left(\frac{1}{2}\right)^3 \left(\frac{1}{2}\right)^1 = \left(\frac{1}{2}\right)^4 = .0625$$

This is, of course, the probability that was calculated previously for each of the four permutations.

The other part of the binomial formula is formula 6–5, for the number of permutations of n elements when there are only two types of elements. The binomial formula requires the multiplication of the number of permutations, $n! / r! (n - r)!$, "times" the probability of any one permutation, $p^r q^{n-r}$. Since all of the permutations have the same probability, such multiplication is equivalent to the addition of the probabilities of each permutation. In terms of the example, $.0625 + .0625 + .0625 + .0625$ is equivalent to 4 "times" $.0625$. In both instances the result is $.25$, the probability of getting three out of four items correct. By way of summary:

$$\text{Prob.}\left(\frac{r}{n}\right) = \frac{n!}{r!(n-r)!} p^r q^{n-r}$$

$$= \frac{4!}{3!(4-3)!} \left(\frac{1}{2}\right)^3 \left(\frac{1}{2}\right)$$

$$= \frac{(4)(3)(2)(1)}{(3)(2)(1)(1)} \left(\frac{1}{8}\right)\left(\frac{1}{2}\right)$$

$$= .25$$

Assuming that p is 1/2, the probability of getting three out of four items correct is .25.

The binomial formula (formula 6–11) can also be used to solve the first example problem, which had only one permutation. Recall that the problem was to determine the probability that a student who did not know the answers to any of the items on a four-item, true-false test would get all four items correct. Using formula 6–11 (and recalling that 0! equals 1 and that any number raised to the zero power also equals 1):

$$\text{Prob.}\left(\frac{r}{n}\right) = \frac{n!}{r!(n-r)!} p^r q^{n-r}$$

$$= \frac{4!}{4!(4-4)!} \left(\frac{1}{2}\right)^4 \left(\frac{1}{2}\right)^{4-4}$$

$$= \frac{(4)(3)(2)(1)}{(4)(3)(2)(1)(1)} \left(\frac{1}{2}\right)^4 (1)$$

$$= \left(\frac{1}{2}\right)^4$$

$$= \frac{1}{16}$$

$$.0625$$

On the assumption that p equals 1/2, the probability of getting all four items correct is .0625. This is, of course, the same answer that was obtained above. The binomial formula provides a convenient summary of the procedures used in the calculation, regardless of the number of permutations involved.

PROBABILITY DISTRIBUTIONS

A probability distribution associates a probability number with each possible outcome in a set of exhaustive measurement classes. It is akin to a frequency distribution; however, a probability distribution differs from a frequency distribution in two respects. First, probabilities rather than frequencies are associated with each outcome in the measurement classes. Second, the measurement classes of a probability distribution are exhaustive (i.e., represent all of the possible outcomes to a procedure), while this may or may not be the case for a frequency distribution. Consider a 10 item true–false test. If a correct response is assigned a 1 and an incorrect response a 0, the range of possible scores is between 0 and 10. A frequency distribution of these scores need not consider all the possibilities. This would be the case, for example, if every individual got a score between 6 and 10. A probability distribution, however, would associate a probability with all 11 of the possible scores.

Table 6.6 illustrates a probability distribution. This distribution associates a probability with each of the 11 possible scores to a 10 item true–false test.

TABLE 6.6. *Probabilities Associated with All Possible Scores to a Ten Item, True–False Test.*

SCORES X	FREQUENCIES $f(X)$	PROBABILITIES $p(X)$
10	5	.05
9	8	.08
8	25	.25
7	30	.30
6	15	.15
5	12	.12
4	4	.04
3	1	.01
2	0	.00
1	0	.00
0	0	.00
	$\sum_{1}^{k} = 100$	$\sum_{1}^{k} = 1.00$

The probabilities in Table 6.6 have been empirically estimated by administering the test to 100 individuals. Thus, in accordance with the relative frequency model of probability, each probability is the ratio of the number of times a particular score was obtained to the total number of scores (100). Notice that the sum of the probabilities across all measurement classes is 1.

Discrete Probability Distributions

In Chapter 1, it was indicated that a discrete variable is a variable containing isolated numerical values—numerical values separated by gaps. The distribution in Table 6.6 is a discrete probability distribution. *In a discrete probability distribution, the probability numbers are associated with the isolated values of a discrete variable.*

Continuous Probability Distributions

By continuous is meant that between any two values, represented by the variable, there are no gaps, or values that cannot occur. Probabilities associated with numerical scores representing quantities of a property (e.g., amounts of anxiety, intelligence, conformity, and learning) are usually represented by continuous distributions rather than discrete distributions. *In a continuous probability distribution, probability numbers are associated with intervals of values of a con-*

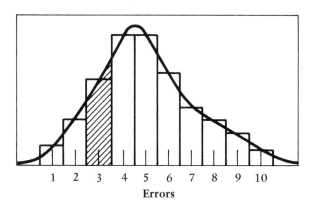

FIGURE 6-2. *Probability distribution of number of errors in a learning task represented by both a continuous curve and a histogram.*

tinuous variable. Figure 6-2 displays a continuous distribution superimposed on a histogram that describes a discrete distribution. Continuous distributions are represented graphically by a continuous line spanning, along the horizontal axis, the range of values of the continuous variable, and varying in height (roughly speaking) according to the probability of values on the horizontal axis.

The student may wonder why continuous distributions are used, since in Chapter 1 it was noted that measurement always results in discrete variables. The measurement process always consists of assigning observed outcomes to discrete scores. Furthermore, no real set of outcomes has members possessing every one of the infinite possible values of the particular property being measured.

In spite of these facts, several reasons dictate the use of continuous rather than discrete distributions:

1. Continuous probability distributions are easier to specify and manipulate mathematically than discrete probability distributions.
2. Particular continuous distributions, such as the normal distribution, are quite useful in inferential statistics.
3. It is possible to calculate, with minimal error, probabilities of discrete scores or intervals of scores from a continuous distribution that approximates the discrete distribution in shape.

Thus, while discrete distributions may represent reality more accurately, statistical inference typically makes use of continuous probability distributions.

A thorough understanding of continuous distributions demands knowledge of calculus. Since calculus is beyond the scope of this book, only those aspects of continuous distributions that can be understood without calculus and that are essential to comprehension of our subsequent uses of these distributions will be discussed.

> The student will note in Fig. 6-2 that the vertical axis of the graph is not labeled. Subsequent presentations of continuous probability distributions will continue this practice. For continuous distributions, the values represented along the vertical axis are not probabilities but are, instead, values called *probability densities*. Since an understanding of probability densities necessitates introducing calculus, these values will not be defined, and the vertical axis will not be labeled.
>
> Continuous distributions do not have probabilities represented along the vertical axis because each of the infinite number of values on a continuum has a probability so small as to be almost zero. On a continuum, any interval, however small, has an infinite number of values within its boundaries. Even if the probability associated with an interval is quite high, the probability of any one of the infinite number of values in the interval is close to zero. The outcome of dividing a probability number by infinity is a value very close to zero.

Probabilities as Proportions of Area

Probabilities in discrete distributions can be represented graphically by the height of a bar associated with each score. This height corresponds to a probability value represented on the vertical axis of the graph. Histograms so constructed also represent probabilities by the proportion of area in the histogram taken up by each bar. The probability associated with a given score equals the proportion of area in the histogram belonging to the bar associated with the score.

Continuous distributions represent probabilities in the latter manner. *The probability associated with any particular interval of values equals the proportion of area under the curve that lies within the interval.* This similarity between histograms and continuous distributions allows a visual demonstration of the fact that continuous distributions can accurately represent the probabilities of discrete scores.

In Fig. 6-2, where a continuous distribution is superimposed on a histogram, it can be seen that probabilities computed from the continuous distribution closely approximate probabilities represented by the histogram. For example, the area under the continuous curve between 2.5 and 3.5 (the lined area) closely approximates the area in the bar associated with an error score of 3. Since number of errors is a discrete variable, a score of 3 is the only score that falls in the interval 2.5-3.5. Thus, the probability of an error being in the interval 2.5-3.5 is equivalent to the probability of an error score being 3. For the discrete distribution, the probability of an error score of 3 is represented either by the height of the bar or by the ratio of the area in the associated bar to the total area. For the continuous distribution, the probability of a score falling in the interval 2.5-3.5 is the ratio of the lined area to the total area.

Mean, Variance, and Standard Deviation of a Probability Distribution

In Chapters 3 and 4, the arithmetic mean, variance, and standard deviation of a frequency distribution were defined and discussed. The mean, variance, and standard deviation of a probability distribution will now be defined.

MEAN. The mean of a discrete probability distribution is given by:

$$\mu = \sum_{1}^{k} Xp(X) \qquad (6\text{--}12)$$

Where: μ (the Greek letter mu) represents the mean
$p(X)$ represents the probability associated with a given X value
k represents the number of different values.

Calculation of the mean of the probability distribution in Table 6.6 is presented in the first three columns in Table 6.7. The mean of a probability distribution is called the *expected value* of the distribution. Computation of the mean of a

TABLE 6.7. *Calculation of the Mean of a Discrete Probability Distribution and of the Associated Frequency Distribution.*

X	$p(X)$	$Xp(X)$	$f(X)$	$Xf(X)$
10	.05	.50	5	50
9	.08	.72	8	72
8	.25	2.00	25	200
7	.30	2.10	30	210
6	.15	.90	15	90
5	.12	.60	12	60
4	.04	.16	4	16
3	.01	.03	1	3
2	.00	.00	0	0
1	.00	.00	0	0
0	.00	.00	0	0
	$\sum_{1}^{k} = 1.00$	$\sum_{1}^{k} = 7.01$	$\sum_{1}^{k} = 100$	$\sum_{1}^{k} = 701$

$$\mu = \sum_{1}^{k} Xp(X) = 7.01$$

$$\bar{X} = \sum_{1}^{k} Xf(X)/n = 701/100 = 7.01$$

continuous probability distribution requires calculus and is beyond the scope of this book.*

Due to the equivalence of the proportion of outcomes in a subset with the probability of the subset, the expected value of the discrete probability distribution in Tables 6.6 and 6.7 equals the arithmetic mean of the frequency distribution on which the probability distribution is based ($\mu = \bar{X}$). Table 6.7 contains the calculation of the arithmetic mean of the frequency distribution using formula 3–8 for the mean of an ungrouped frequency distribution. Notice that both μ and $\bar{X}$ equal 7.01.

VARIANCE AND STANDARD DEVIATION. The variance of a discrete probability distribution is given by:

$$\sigma^2 = \sum_{1}^{k} (X - \mu)^2 p(X) \qquad (6\text{--}13)$$

TABLE 6.8. *Calculation of the Variance and Standard Deviation of the Discrete Probability Distribution in Table 6.6.*

X	$p(X)$	$X - \mu$	$(X - \mu)^2$	$(X - \mu)^2 p(X)$
10	.05	2.99	8.94	0.4470
9	.08	1.99	3.96	0.3168
8	.25	0.99	0.98	0.2450
7	.30	0.01	0.00	0.0000
6	.15	−1.01	1.02	0.1530
5	.12	−2.01	4.04	0.4848
4	.04	−3.01	9.06	0.3624
3	.01	−4.01	16.08	0.1608
2	.00	−5.01	25.10	0.0000
1	.00	−6.01	36.12	0.0000
0	.00	−7.01	49.14	0.0000
	$\sum_{1}^{k} = 1.00$			$\sum_{1}^{k} = 2.1698$

$$\sigma^2 = \sum_{1}^{k} (X - \mu)^2 p(X) = 2.17$$

$$\sigma = \sqrt{\sum_{1}^{k} (X - \mu)^2 p(X)} = \sqrt{2.17} = 1.47$$

---

* For the reader who understands calculus, the mean of a continuous distribution is given by:

$$\mu = \int_{-\infty}^{+\infty} X p(X)\, dx$$

where $p(X)$ is a probability density rather than a probability.

Where: The symbol σ^2 (sigma squared) signifies the variance
$p(X)$ again represents the probability associated with a given X value
k represents the number of different values.

The standard deviation of a discrete probability distribution is simply the square root of the variance:

$$\sigma = \sqrt{\sum_{1}^{k} (X - \mu)^2 p(X)} \tag{6-14}$$

where σ (sigma) symbolizes the standard deviation.

Table 6.8 contains the calculation of the variance and standard deviation of the probability distribution for the ten item true–false test (Table 6.6). Calculation of the variance and standard deviation of the associated frequency distribution is contained in Table 6.9. Notice that the variance (s^2) and the standard deviation (s) of the frequency distribution have the values 2.17 and 1.47, respectively. These are the same values as the variance (σ^2) and standard deviation (σ) of the probability distribution.

TABLE 6.9. *Calculation of the Variance and Standard Deviation of the Frequency Distribution Associated with the Discrete Probability Distribution in Table 6.6.*

X	$f(X)$	$X - \bar{X}$	$(X - \bar{X})^2$	$(X - \bar{X})^2 f(X)$
10	5	2.99	8.94	44.70
9	8	1.99	3.96	31.68
8	25	0.99	0.98	24.50
7	30	0.01	0.00	0.00
6	15	−1.01	1.02	15.30
5	12	−2.01	4.04	48.48
4	4	−3.01	9.06	36.24
3	1	−4.01	16.08	16.08
2	0	−5.01	25.10	0.00
1	0	−6.01	36.12	0.00
0	0	−7.01	49.14	0.00
	$\sum_{1}^{k} = 100$			$\sum_{1}^{k} = 216.98$

$$s^2 = \sum_{1}^{k} (X - \bar{X})^2 f(X) / n = 216.98/100 = 2.17$$

$$s = \sqrt{\sum_{1}^{k} (X - \bar{X})^2 f(X) / n} = \sqrt{216.98/100}$$

$$= \sqrt{2.17} = 1.47$$

As was the case with the mean, computation of the variance and standard deviation of a continuous probability distribution requires calculus and is beyond the scope of this book.*

> The above illustrated equivalences of the mean, variance, and standard deviation of probability and associated frequency distributions can be shown by comparing formulas for the arithmetic mean, variance, and standard deviation of a frequency distribution (3–8, 4–10, and 4–12) with those for the mean, variance, and standard deviation of a probability distribution (6–12, 6–13, and 6–14). (As previously indicated, the arithmetic mean, variance, and standard deviation of an ungrouped frequency distribution are identical to the arithmetic mean, variance, and standard deviation of the set of scores represented by the frequency distribution.)
>
> The mean of a frequency distribution and the mean of a probability distribution are given respectively by:
>
> $$\bar{X} = \sum_{1}^{k} Xf(X) / n \qquad (3\text{–}8)$$
>
> and,
>
> $$\mu = \sum_{1}^{k} Xp(X) \qquad (6\text{–}12)$$
>
> Since n is a constant, it can be moved inside the summation sign, and formula 3–8 can be rewritten:
>
> $$\bar{X} = \sum_{1}^{k} X \frac{f(X)}{n}$$
>
> The arithmetic mean here is expressed as the summation of the product of each different value of X "times" the relative frequency or proportion of outcomes in the population set having that value.
>
> If the assumptions of the particular probability model are met, the probability of a subset equals the proportion of scores belonging to that subset. In this case the subsets are outcomes whose measurements have the same value of X. Thus, the relative frequency (proportion) of X, $f(X)/n$, equals the probability of X, $p(X)$; and formulas 3–8 and 6–12 are equivalent.
>
> In a like manner, it can be shown that if the assumptions of the probability model are met, the variance and standard deviation of a frequency distribution are equal to the variance and standard deviation of its probability distribution; that is:
>
> $$s^2 = \sum_{1}^{k} (X - \bar{X})^2 f(X) / n \qquad (4\text{–}10)$$
>
> is equivalent to:
>
> $$\sigma^2 = \sum_{1}^{k} (X - \mu)^2 p(X) \qquad (6\text{–}13)$$

---

* For those students who understand calculus, the variance of a continuous probability distribution is given by:

$$\sigma^2 = \int_{-\infty}^{+\infty} (X - \mu)^2 p(X)\, dx,$$

where $p(X)$ is a probability density rather than a probability. The standard deviation is the square root of the variance.

and,

$$s = \sqrt{\sum_{1}^{k}(X - \bar{X})^2 f(X) / n} \qquad (4\text{--}12)$$

is equivalent to:

$$\sigma = \sqrt{\sum_{1}^{k}(X - \mu)^2 p(X)} \qquad (6\text{--}14)$$

Population Probability Distributions

One of the more important types of probability distributions is a population probability distribution. A population probability distribution is a distribution that associates a probability with each of the numerical outcomes in the total set of the outcomes of concern. The term population was defined previously as the total set of elements of concern. Recall, however, that "elements" is a general term referring to anything in a set. Sometimes, the elements may be real objects—living or non-living. In psychology, of course, a population of objects is typically a population of living objects or organisms. For example, the population might consist of all residents of Berkeley, California. The elements, however, need not be real objects but may be the outcomes observed in connection with a procedure. Whenever outcomes are observed in connection with a procedure, these outcomes are assigned to measurement categories. Thus, it may be appropriate to refer to a population of numerical values. Following through with the above example, if we take height measurements of all the residents of Berkeley, the population would consist of the distribution of heights. Finally, as previously pointed out, all of the possible outcomes in a population need not have been observed; they may be only conceptual or theoretical possibilities. This would be the case if not all the residents of Berkeley were measured for height. One can conceptualize a distribution of numerical values for height, even though all of the residents of Berkeley have not been measured.

With a population probability distribution, the elements are numerical outcomes. They typically are not even the numerical values assigned through actual observation; rather, they are only theoretical possibilities. In such a situation, it becomes necessary for the researcher to make assumptions about the population probability distribution. Stated somewhat differently, it is necessary to assume that a certain model probability distribution accurately describes the population probability distribution. Fortunately, methods of statistical inference (to be discussed in subsequent chapters) do not require precise knowledge of the population probability distribution. It is usually enough to know or postulate the general shape of the probability distribution. The most important model probability distribution and one whose general shape is frequently assumed to accurately describe the population probability distribution is the continuous normal distribution (to be discussed in Chapter 8).

Finally, it is important for the student to understand how the term, "population distribution," will be used in the remainder of this book. "Population distribution" will refer to the population probability distribution, and not to the population frequency distribution. This is a simple matter of convenience in view of the fact that the researcher's concern is almost always with the population probability distribution rather than with the population frequency distribution.

HISTORICAL PERSPECTIVE ON PROBABILITY

The Astragalus and the Die

In discussing the history of probability, it is somewhat difficult to know where to begin. An intriguing possibility, however, is with the astragalus. The astragalus, variously referred to as the talus, hucklebone, or knuckle-bone, is the heel-bone of animals possessing developed feet. It is a hard bone containing no marrow and possessing four uneven sides. Archeologists frequently find sizeable quantities of astragali at various prehistoric sites. It seems quite likely that they were used for counting, as toys, or for games of some sort. To this day children in France and Italy play games with astragali. F. N. David (1962) in her book, *Games, Gods, and Gambling*, speculates that the term "knuckle-bone" may have arisen from a game in which four astragali were balanced on the knuckles of the hand, tossed, and then caught.

The Babylonians, Egyptians, Greeks, and Romans used the astragalus to play games. An Egyptian tomb painting has been found in which a nobleman is shown with an astragalus poised on a finger tip above a board containing "men" of some sort. Herodotus, the first Greek historian, writing around 450 B.C., credits the Greeks and allied peoples with the invention of various games of chance, some of which involved astragali and some of which involved dice. Thus, he writes that during the famine of 1500 B.C., the Lydians invented different games of chance to forget their hunger, and that other games of chance were invented by the Greeks during the seige of Troy to relieve the soldiers' boredom and bolster their sagging morale.

It is not known exactly when the die was first invented, but, as David (1962) points out, it seems quite likely that the die evolved from the astragalus. The first step in the evolution possibly involved a rubbing and flattening of the sides of the astragalus. Many such primitive dice have, in fact, been found. Nonetheless, in classical times the astragalus remained more popular than the die. Different numbers were assigned to the four different sides and four bones frequently were thrown at once. The best of all possible outcomes was the "venus-throw", when each of the four bones had a different side uppermost.

The Concept of Randomness

The concept of randomness is central to the theory of probability. This concept, however, had an extremely tardy intellectual development. The Greeks, who were innovative in so many respects, never achieved the concept. David (1962) generally discusses a number of possible contributing factors. Among these is the obvious fact that with the crude astragalus each of the four sides is not equally likely. A further possible factor is that the gods were believed to intervene and thus determine the results of die or astragalus casting. The priesthood, in fact, promoted the belief that the gods revealed their will by determining the outcome of lot casting. Thus, it might have been regarded as impious to dispute the priesthood, although it is difficult to believe that this would have deterred all of the bold and innovative Greek philosophers. A final factor contributing to the tardy development of the concept of randomness is the fact that the Greeks had less of an empirical orientation regarding natural phenomena of the earth than of the heavens. According to David, it was not part of the Greek cultural tradition to observe earthly phenomena, make generalizations about these phenomena, and check these generalizations through further observation.

Gambling was very popular among the Romans, and it was the Roman philosophers, most notably Cicero, who first demonstrated a grasp of the concept of randomness. In Book I of *De Devinatione*,

Cicero, advancing the argument that he later argues against, has Quintus say: "When the four dice produce the venus-throw, you may talk of accident; but, suppose you made a hundred casts and the venus-throw appeared a hundred times, could you call that accidental?" In Book II, Cicero argues against the notion that divine intervention determines the outcome of die casting:

> Nothing is so unpredictable as a throw of the dice; and yet, every man who plays often will at some time or other make a venus-cast. Now and then, indeed, he will make it twice and even thrice in succession. Are we going to be so feeble-minded, then, as to aver that such a thing happened by the personal intervention of Venus rather than by pure luck?

Cicero's ideas undoubtedly never were known to the uneducated and would have been unpopular with the priesthood. David does not speculate why the Romans succeeded in achieving the concept of randomness where the Greeks failed. Superficially, it would seem that the above mentioned factors should have deterred the Romans just as much as the Greeks. In any event, with the advent of Christianity, the idea of randomness was firmly rejected. St. Augustine, for example, argued that everything is controlled by the will of God and nothing occurs randomly. The appearance of randomness is due to man's limited understanding.

Combinations and Permutations

During the middle ages the Catholic Church opposed gambling and the casting of lots. David attributes this opposition to the prior association with pagan religion, and also notes that the continued preaching against dicing indicates that the Church was not successful in eliminating the behavior.

Around A.D. 960, Bishop Wibold of Cambrai, presumably against his superior's will, invented a moral dice game. He correctly enumerated the 56 possible combinations of three dice and assigned a virtue to each one. The order, or permutation, of the dice was ignored. A poem written in Latin, *De Vetula*, describes the number of ways in which three dice can fall, allowing for order. *De Vetula* was possibly written by Richard de Fournival (A.D. 1200–1250), Chancellor of Amiens Cathedral. This poem makes use of arabic numerals, "The wondrous system of notation having 9 digits and a cipher, with device of place." As David points out, the arabic numeral system, together with the growing interest in arithmetic and algebra, contributed to these developments. David also repeatedly emphasizes the importance of the developing empirical orientation—an epistemology that knowledge is acquired through observation and testable generalizations from such observations.

Girolamo Cardano

During the late middle ages an important figure in the history of probability was Girolamo Cardano (1501–1576). David characterizes Cardano as "one of the greatest eccentrics of all times." He was controversial in his time, and his place in history remains controversial to this day. Some of his biographers describe him as an unpleasant charlatan, and some, as a persecuted saint. David gives the following quote from Cardano's posthumously published autobiography:

> I was ever hot-tempered, single-minded, and given to women. From these cardinal tendencies there proceeded truculence of temper, wrangling, obstinacy, rudeness of carriage, anger, and an inordinate desire for revenge in respect of any wrong done to me. I am moreover truthful, mindful of benefits wrought to me, a lover of justice, a despiser of money, a worshipper of that fame that defies death, prone to thrust aside what is commonplace, and still more disposed to treat mere trifles in the same way.

The illegitimate son of Facio Cardano, a friend of Leonardo da Vinci, Cardano obtained his doctorate in medicine from the University of Padua in 1526. Thereafter, he practiced medicine and held successive positions at the Universities of Milan, Pavia, and Bologna. He was continually in one sort of trouble or another. These troubles resulted from his blatant attacks on the medical profession, his gambling losses, his feuds with contemporary mathematicians regarding the stealing of ideas, his grief at his favorite son's execution following the murdering of the son's wife—a former prostitute, and from rumors regarding the number and character of young men who visited his house, etc. In 1570, he was imprisoned by the Inquisition for, among other things, having cast the horoscope of Christ. After recanting and promising to abandon teaching, he went to Rome in 1571 where he obtained a lifetime annuity from Pope Pius V. He died in Rome September 20, 1576. According to

David, "The uncharitable gossip was that he had predicted he would die on that day and had starved himself for three weeks before to make it come true" (1962, p. 53).

Cardano possessed a universal mentality to which no area of learning was inaccessible. At the time of his death, he had 131 published works and 111 unpublished manuscripts dealing with such diverse topics as medicine, mathematics, physics, philosophy, religion, and music. He also claimed to have written and then burned an additional 170 manuscripts. How much of this prodigious output really represented Cardano's own ideas is not known. David points out that plagiarism was widely practiced and apparently had a kind of semi-legitimacy during this period.

Cardano's most famous work is *Artis Magnae Liber*, published in 1545. This book systematically presents many new and important ideas in algebra, such as the procedure for solving third-degree equations (equations involving a cubed unknown) and the fact that square roots may be either positive or negative. It is known that not all of the ideas in this book were original with Cardano, and it is possible that none were.

Cardano's place in the history of probability theory, however, is due to his book, *Liber de Ludo Aleae*, published in 1663, 87 years after his death. Historians speculate that the book was actually written in 1563 or 1564, while Cardano was at the University of Bologna. This book contains the first account of probability in terms of the ratio of specified outcomes to the total number of outcomes. The idea of combinations was introduced to enumerate the possible outcomes of die and astragalus casting. If we could be certain that these conceptions were, in fact, Cardano's, this contribution alone would be sufficient to establish his place in the history of ideas. Cardano's credibility, like the credibility of the "boy who cried wolf," is in doubt.

1550–1650

For the period of roughly 100 years after the writing of Cardano's book, *Liber de Ludo Aleae*, the historian is confronted by a curious lack of manuscripts treating probability. Except for a small fragment written by Galileo, no known records exist. As David comments, "Throughout these centuries, men continued to gamble with both cards and dice, and somehow among mathematicians the knowledge of how to calculate a probability percolated" (1962, p. 62). David believes that it is unlikely that the ideas spread from Cardano at the University of Bologna, since he was fairly isolated there and had no mathematical students. One likely possibility is that ideas were spread through the meetings of the more than seven hundred learned societies that existed in Italy during this period. It is hard to believe, however, that no mathematician wrote anything on probability in this era. The thought of some dusty, long forgotten manuscript waiting to be discovered is enough to make the historian's mouth water.

As indicated above, the only known record is a fragment written by Galileo. Galileo's main interest was, of course, astronomy, and it is unlikely that he would have had any concern with probability if his benefactor, Ferdinando dei Medici, the Grand Duke of Tuscany, had not asked him why it is that a cast of three dice is more likely to result in 10 rather than in 9. Galileo's answer, written sometime between 1613 and 1623, goes straight to the point. He notes that the total possible number of possibilities for three dice is 216. He gives a table for the possible throws yielding totals between 10 and 3 and notes that the numbers of possible throws for totals between 11 and 18 are symmetrical. Since this table shows that the number of possible throws yielding 9 is 25, and 10 is 27, the probability of obtaining 9 is 25/216 and the probability of obtaining 10 is 27/216. One remarkable thing about this is that "His Serenist Highness" had gambled enough to detect a difference in probability of only 2/216 or 1/108. David doubts that Galileo had developed anything original, since he wrote as if what he was saying was perfectly obvious. The procedure he used had, of course, already been described by Cardano.

The Fermat and Pascal Letters

The scene of the next noteworthy events in the history of probability shifted from Italy to France. As frequently seems to be the case, intellectual ascendancy is correlated with economic and military ascendancy. In this instance there may have been a direct causal relationship since successive French armies carried books and manuscripts, and, indeed, whole libraries, out of Italy.

The seventeenth century was a period of rapid development in mathematics. Reputedly, the greatest of the French mathematicians of this period was Pierre de Fermat (1601–1665). Fermat and Blaise Pascal (1623–1662) are frequently given credit for being the joint discoverers of the procedure for calculating probabilities. This credit is based on an exchange of letters between Fermat and Pascal in 1654. It is true that these letters contained an evolution toward and discovery of the calculus of probability. However, the basic procedure that they utilized had been previously set forth by Cardano. Cardano's book, *Liber de Ludo Aleae*, though, only existed in manuscript form at this time. It was not published until 1663—the year after Pascal's death.

In the summer of 1654, Pascal wrote to Fermat and posed a problem that previously had been proposed to Pascal by a gambler. The problem regarded how the stakes in a game of chance should be divided among the players if the game were prematurely ended. In discussing several such problems through a series of letters, Fermat and Pascal relied upon the principle that the expectation of each player should be based on the ratio of outcomes favorable to him to the total number of possible outcomes. Calculations based on the proportion principle were facilitated by the use of what has come to be called "Pascal's Triangle." Such a triangle is presented in Table 6.10. Each number "inside" the triangle is the sum of the two numbers immediately above it. Thus, the 2 in the third row is the sum of 1 and 1; 3 in the fourth row is the sum of 1 and 2, etc. The triangle simply represents the number of permutations of two kinds of outcomes (e.g., head and tail). Thus, the number of permutations of three coin tosses is represented in the third row, 1, 2, 1. There are 4 possibilities: HH, HT and TH, TT. If the probabilities of head and tail are equal, the probabilities of these outcomes are 1/4, 2/4 or 1/2, and 1/4.

TABLE 6.10. Pascal's Triangle.

```
                  1
               1     1
            1     2     1
         1     3     3     1
      1     4     6     4     1
   1     5    10    10     5     1
1     6    15    20    15     6     1
```

Subsequent Developments

Further developments in probability take us beyond the brief introduction to probability theory contained in this chapter, and thus, will be only briefly described.

After Fermat and Pascal the next important name in the history of probability is that of Christianus Huygens (1629–1695). Huygens was born to a family of wealth and position in Holland. In 1654, Huygens visited Paris and, although he did not meet Fermat or Pascal, heard through mutual acquaintances of the probability problems that had been the subject of the famous letters the year before. He, however, did not learn of the actual procedures that Fermat and Pascal had used in their calculations—possibly because Huygens's acquaintances were not competent to discuss them. On returning to Holland, Huygens set to work and quickly arrived at the same solutions that Fermat and Pascal had achieved. He did not leave the matter here, however, but went on to formalize a set of fourteen propositions regarding the theory of probability. Huygens's formalization of probability theory represents the first really new advance in the theory of probability since Cardano.

Huygens's treatise, *De Ratiociniis in Aleae Ludo*, was finally published in 1657. For almost half a century, it was the definitive work in the theory of probability. It was not superceded until the following century when further contributions were made by men like Montmort, Bernoulli, and de Moivre. Abraham de Moivre is remembered for his discovery of the normal distribution—the subject matter of Chapter 8.

For present purposes the most important of the later historical developments was the advancement of the long-run, relative frequency model of probability. Fermat and Pascal had approached probability in terms of the *proportion of specified outcomes to total number of outcomes*, i.e., they had relied upon the classical model of probability. Such a model, however, only applies in situations in which each of the outcomes are *equally likely*. Since any of the six sides of a fair die are equally likely, the probability of any one of them occurring is 1/6. What happens, however, when all of the possible outcomes are not equally likely? What, for example, is the probability that a person who is alive at age 20 will also be alive at age 40?

As previously indicated, the solution to the problem involves looking at probability in terms of *long-run, relative frequency.* Thus, if death and birth records indicate that of the 100,000 people alive at age 20, 80,000 are still alive at age 40, the ratio of these two numbers, 80,000/100,000 or .80, is taken as the probability that someone who is alive at age 20 will also be alive at age 40. Probability here is still a ratio or proportion, but not a proportion of specified to total number of outcomes. Probability is a relative frequency proportion.

EXERCISES

1. Define the following terms:
(a) inferential statistics
(b) procedure
(c) experiment
(d) long-run, relative frequency
(e) probability distribution
(f) population distribution
(g) population
(h) sample
(i) simple random sampling
(j) sampling with replacement
(k) sampling without replacement
(l) dependent sampling
(m) independent sampling
(n) independent outcomes
(o) dependent outcomes
(p) addition rule
(q) multiplication rule
(r) permutation

2. Assuming A and B are mutually exclusive outcomes, find the probability of $A \cup B$ for the following:

$p(A)$	$p(B)$
.5	.4
.2	.6
.3	.4

3. Assuming A and B are independent outcomes, find the probability of $A \cap B$ for the following:

$p(A)$	$p(B)$
.3	.8
.5	.4
.6	.9

4. Suppose a student takes a five item, multiple-choice test. If each item has four alternatives and the student performs at the chance level, what is the probability that he will get:

(a) all five items correct
(b) four items correct
(c) only the first item correct?

5. Compute μ and σ^2 for the following discrete probability distribution:

X	$p(X)$
2	.2
3	.5
5	.3

6. A population of 1,400 persons is composed of 150 Jews, 500 Catholics, 200 Protestants, and 550 persons not affiliated with any religious group. If one were to select randomly one person from the population, what is the probability that the person selected will not be affiliated with any religious group? What is the probability that after five selections, three of the five ($r/n = 3/5$) will not be affiliated with any religious group? Assume independent probabilities.

7. An elementary school class contains 30 students, 10 black and 20 white. The teacher selects 6 students with replacement. What is the probability that $r/n = 1/6$ will be black? What is the probability that 5/6 will be black? What is the probability that 4 or more will be black?

8. Consider a population of 100 individuals.
(a) If 60 of the individuals in the population are male, what is the probability of obtaining an r/n of 2/5, where r is the number of males in the sample?
(b) If 30 of the individuals are male, what is the probability of obtaining an r/n of 3/3, where r is the number of males in the sample?
(c) If 80 of the individuals are male, what is the probability of obtaining an r/n of 1/4, where r is the number of males in the sample?

9. Consider an urn containing five balls labeled with an "A", and five labeled with a "B."

(a) Sampling with replacement, what is the probability of obtaining one A and one B in two draws?

(b) Sampling without replacement, what is the probability of obtaining one A and one B on two draws?

10. At a large state university, the proportion of out-of-state freshmen is .3. If simple random sampling is used to assign students to a special section of a required freshman course, what is the probability that 9 of the 10 students assigned to fill the section will be from out of state? Assume independent probabilities.

11. Assuming the same conditions that exist in Exercise 10, what is the probability that 9 or more of the 10 students will be from out of state?

12. What is the rationale for the operations in the binomial formula (see formula 6–11)?

13. Use the binomial formula to compute the probability of obtaining 5 heads in 5 independent tosses of a coin.

14. Consider a four alternative, multiple-choice test with four items. What is the probability of getting:

(a) 0 items correct
(b) 1 item correct
(c) 2 items correct
(d) 3 items correct
(e) 4 items correct?

15. Explain the meaning of the term, "population distribution."

16. Explain why the mean of a discrete probability distribution equals the mean of the frequency distribution on which the probability distribution is based.

17. What is one disadvantage of the long-run, relative frequency model of probability?

18. What advantage does the long-run, relative frequency model of probability have over the classical model of probability?

19. What problem, if any, do you perceive with the classical model of probability as a possible definition of probability?

20. List the three axioms of probability and give the formal definition of probabilities or probability numbers.

Prologue to Chapter 7

Chapter 7 contains a discussion of sampling distributions—a matter of fundamental importance in statistical inferences from samples to population values. The bulk of the chapter is devoted to a methodical description of the sampling distributions of proportions. Since the sampling distribution of proportions is relatively simple, it provides a convenient means of illustrating sampling distributions without involving great amounts of higher mathematics. If the binomial formula, which was introduced in Chapter 6, is thoroughly understood, Chapter 7 should not pose any insurmountable problems. In fact, Chapter 7 contains less new information than Chapter 6.

7

Sampling Distributions

This chapter explores the characteristics of a particular type of probability distribution, a *sampling distribution*. As will subsequently become apparent, sampling distributions enable one to make inferences from samples to population values.

THE CONCEPT OF SAMPLING DISTRIBUTION

In Chapter 1, a statistic was defined as a summary measure of a set of scores. Means, variances, and standard deviations are statistics. *A sample statistic is a summary measure of a set of scores derived from a sample set of elements* (e.g., people). Sample means, sample variances, and sample standard deviations are sample statistics.

> *A sampling distribution is a function associating probabilities with all possible values of a sample statistic that could result from samples of a given size (n) drawn from a given population.*

To understand sampling distributions, the idea of repeatedly sampling and measuring n elements must be understood. Suppose that twenty students are sampled from a population, each student's IQ is measured, and the mean IQ for the sample of twenty is computed. Just as it is possible to repeatedly select and measure individual students, it is possible to repeatedly sample twenty students and compute the mean IQ for each sample. A sampling distribution associates probabilities with sample outcomes. For this example a sampling distribution of means will associate probabilities with the possible mean IQ scores that can result from computing mean IQ's for samples of twenty students.

In Chapter 6, certain assumptions and rules were used to calculate the probabilities of proportions; for example, the probability that a student would get three out of four true-false items correct. The next section of this chapter will extend the use of these assumptions and rules to describe the calculation of an entire sampling distribution of proportions. While the sampling distributions of other statistics are more important, they cannot be derived as simply. Thus, the more easily obtained sampling distribution of proportions will provide the initial illustration of the concept of sampling distributions.

SAMPLING DISTRIBUTIONS OF PROPORTIONS

Consider a population of first grade students divided at the median according to reading ability. Each student is placed into one of two mutually exclusive classes, above average reader or below average reader. Suppose that the general concern is with the probability of drawing a particular type of reader (above or below average) from this population. In order to apply the classical model, two assumptions must hold. First, the outcomes (or outcome sets) must be mutually exclusive, and second, all the particular instances of the outcomes must be equally likely. In this case the outcomes are above and below average readers, and, as already indicated, such classifications are mutually exclusive. Equally likely instances of the outcomes means that each reader in the population has an equal chance of being selected, and this is assured through simple random sampling. Given these two assumptions and the further knowledge that the population contains an equal number of above and below average readers (since the division is made at the median), the probability of either outcome is .5. Fig. 7-1 pictures the population distribution.

When the elements in a population are measured into only two classes, the probabilities associated with the two classes (in this case above average and below average readers) are referred to as the probability, p, *of a favorable outcome* and the probability, q, *of an unfavorable outcome*. As a matter of convention, p is used with reference to the favorable outcome (in this case an above average reader) while q is used with reference to the unfavorable outcome. The terms *favorable* and *unfavorable*, in this context, are not used necessarily to imply greater value to one outcome than the other. The favorable outcome is simply the outcome to which one is paying most attention.

It is, of course, possible to draw a sample of size n from a two-valued population. *The sample proportion is symbolized r/n, where n is the number of selections and r is the number of selections in which the favorable outcome occurs.* Care should be taken to distinguish between p of a favorable outcome and r/n. The p of a favorable outcome is the proportion of favorable outcomes in the population and thus, in fact, the probability of obtaining a favorable outcome.

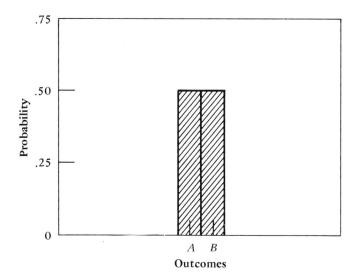

FIGURE 7-1. The population probability distribution for the population of above average readers (A) and below average readers (B).

For the above example, p of a favorable outcome is the proportion of above average readers in the population and thus the probability of obtaining an above average reader. The expression r/n defines a sample statistic—the proportion of favorable outcomes in a sample. For the above example, r/n is the proportion of above average readers in a sample of size n.

The concept of sampling distribution will be illustrated by developing a particular kind of sampling distribution, a sampling distribution of r/n. *The sampling distribution of r/n is a function associating probabilities with all possible values of r/n that could result from samples of a given size (n) drawn from a given two-valued population.* For present purposes, we will use the population of above average and below average readers whose probability distribution is pictured in Fig. 7-1. Sample sizes of 1, 2, and 5 will be used, and sampling with replacement will be assumed.

Sampling Distribution for $n = 1$

If n is 1, then r/n can be either 1 or 0. Selecting just once ($n = 1$), we can obtain either an above average or below average reader. Thus the proportion of above average readers in the sample will be either:

$$r/n = \frac{1}{1} = 1$$

Sampling Distributions of Proportions

TABLE 7.1. *Possible Outcomes and Their Probabilities for n = 1 Where p of Favored Outcome Equals .5.*

POSSIBLE OUTCOMES	r	r/n	$p(r/n)$
Above average reader (A)	1	1	0.5
Below average reader (B)	0	0	0.5

or:

$$r/n = \frac{0}{1} = 0$$

The probability of obtaining the proportion 1/1 is .5, [$p(1/1) = .5$], since the probability of obtaining an above average reader is .5, and the probability of obtaining the proportion 0/1 is .5, [$p(0/1) = .5$], since the probability of obtaining a below average reader is .5. Table 7.1 shows the possible outcomes to one selection, and the corresponding r, r/n, and $p(r/n)$. The sampling distribution of r/n for $n = 1$ is illustrated in Fig. 7–2. When $n = 1$, the sampling distribution of r/n is identical to the probability distribution for the population set illustrated in Fig. 7–1, since each sample consists of just one selection. There are just two possible outcomes, 1 and 0, corresponding to the two possible outcomes in the population, above average and below average readers.

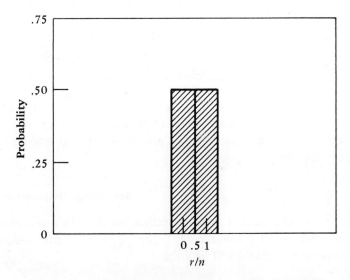

FIGURE 7–2. *Histogram description of sampling distribution of r/n for $n = 1$ where p of favored outcome $= .5$.*

TABLE 7.2. *Possible Outcomes and Their Probabilities for $n = 2$ Where p of Favored Outcome Equals .5.*

POSSIBLE OUTCOMES	r	r/n	PROBABILITY OF OUTCOMES
AA	2	1	0.25
AB	1	0.5	0.25
BA	1	0.5	0.25
BB	0	0	0.25

Sampling Distribution for $n = 2$

When n is two selections, we could obtain either an above average or below average reader on the first selection and either an above average or below average reader on the second selection. Using A and B to symbolize above and below average readers, respectively, there are four possible distinct samples (permutations) that could result from two selections, AA, BB, AB, and BA. Table 7.2 shows these possible outcomes to two selections, and the corresponding r, r/n, and probability. The probabilities were computed using the simple multiplication rule for independent selections discussed in Chapter 6. Assuming independent selections, the probability of each sample (AA, BA, AB, or BB) is the product of the probability of the outcome to the first selection times the probability of the outcome to the second selection. Since $p = q = .5$, the probability of each joint outcome is $(.5)(.5) = .25$.

Given this information, we can compute the sampling distribution of r/n for $n = 2$. From Table 7.2, it can be seen that the four possible samples yield three different proportions of above average readers, 1, .5, and 0. Table 7.3 describes the sampling distribution for r/n when $n = 2$, and Fig. 7–3 presents a histogram description of this distribution. In the case of the r/n of 0.5, the probability is calculated by summing $p(AB)$ and $p(BA)$ in accordance with the addition rule:

$$p(.5) = p(AB) + p(BA) = .50$$

TABLE 7.3. *Sampling Distribution of r/n for $n = 2$ Where p of Favored Outcome Equals .5.*

r/n	p(r/n)
1	0.25
0.5	0.50
0	0.25

Sampling Distributions of Proportions

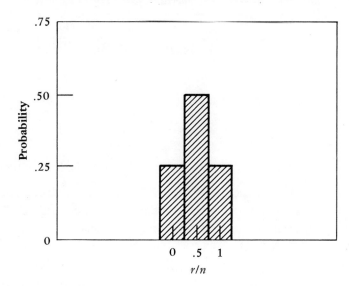

FIGURE 7-3. *Histogram description of sampling distribution of r/n for $n = 2$ where p of favored outcome = .5.*

USING THE BINOMIAL FORMULA. We could have determined the probability of each r/n by using the binomial formula (6-11) discussed in Chapter 6:

$$\text{Prob.}(r/n) = \frac{n!}{r!\,(n-r)!}\, p^r q^{n-r}$$

The probability of obtaining two above average readers ($r = 2$) on each of the two selections ($n = 2$) or an r/n of 1 can be easily calculated:

$$\text{Prob.}(r/n = 1) = \left(\frac{2!}{2!\,0!}\right)(.5)^2(.5)^0$$

$$= (1)(.25)(1)$$

$$= .25$$

When $r = 1$:

$$\text{Prob.}(r/n = .5) = \left(\frac{2!}{1!\,0!}\right)(.5)^1(.5)^1$$

$$= (2)(.5)(.5)$$

$$= .50$$

When $r = 0$:

$$\text{Prob.}(r \mid n = 0) = \left(\frac{2!}{0!\,2!}\right)(.5)^0(.5)^2$$
$$= (1)(1)(.25)$$
$$= .25$$

These are, of course, the same probabilities obtained above.

MEAN OF SAMPLING DISTRIBUTIONS OF PROPORTIONS. The sampling distributions for $n = 1$ and $n = 2$ represented in Figs. 7–2 and 7–3 are both symmetric and both centered around $p = .5$. In both cases, 50 percent of the distribution is above and 50 percent below .5; thus, the median for each distribution is .5. Further, since the median and mean of symmetric distributions are equivalent, it is apparent that the mean of each distribution is also .5. This fact can also be determined using formula 6–12:

$$\mu = \sum_1^k Xp(X)$$

for the mean of a probability distribution. In the present case X represents a sample proportion $(r \mid n)$, and $p(X)$ represents the probability of $r \mid n$, $[\,p(r \mid n)\,]$; thus, for $n = 1$:

$$\mu = \sum_1^k Xp(X)$$
$$= (1)(.5) + (0)(.5)$$
$$= .5$$

For $n = 2$:

$$\mu = \sum_1^k Xp(X)$$
$$= (1)(.25) + (.5)(.5) + (0)(.25)$$
$$= .5$$

These results indicate that the mean of both distributions equals p of an above average reader. In fact, the mean of any sampling distribution of $r \mid n$ equals p of the favored outcome.

Sampling Distribution for $n = 5$

Just as we can select once ($n = 1$) or twice ($n = 2$), computing the sample proportion ($r \mid n$) after each selection or each two selections, we can make five selections ($n = 5$) computing r/n for each sample of five. Table 7.4 shows the possible outcomes to five selections and the corresponding $r, r/n$, and probability. As above, the probabilities were computed using the simple multiplication rule for independent selections. Since $p = q = .5$, each of the outcomes has a probability of $(.5)(.5)(.5)(.5)(.5) = (.5)^5 = .03125$.

Table 7.5 presents the sampling distribution of r/n for $n = 5$, and Fig. 7–4 gives a histogram illustration of the distribution. The probabilities for each r/n value were obtained by applying the simple addition rule, summing the

TABLE 7.4. *Possible Outcomes and Their Probabilities for $n = 5$ Where p of Favored Outcomes Equals .5.*

POSSIBLE OUTCOMES	r	r/n	PROBABILITY OF OUTCOMES
AAAAA	5	1.0	.03125
AAAAB	4	0.8	.03125
AAABA	4	0.8	.03125
AABAA	4	0.8	.03125
ABAAA	4	0.8	.03125
BAAAA	4	0.8	.03125
AAABB	3	0.6	.03125
AABAB	3	0.6	.03125
AABBA	3	0.6	.03125
ABABA	3	0.6	.03125
ABBAA	3	0.6	.03125
ABAAB	3	0.6	.03125
BAAAB	3	0.6	.03125
BAABA	3	0.6	.03125
BABAA	3	0.6	.03125
BBAAA	3	0.6	.03125
AABBB	2	0.4	.03125
ABABB	2	0.4	.03125
ABBAB	2	0.4	.03125
ABBBA	2	0.4	.03125
BAABB	2	0.4	.03125
BABAB	2	0.4	.03125
BABBA	2	0.4	.03125
BBAAB	2	0.4	.03125
BBABA	2	0.4	.03125
BBBAA	2	0.4	.03125
ABBBB	1	0.2	.03125
BABBB	1	0.2	.03125
BBABB	1	0.2	.03125
BBBAB	1	0.2	.03125
BBBBA	1	0.2	.03125
BBBBB	0	0.0	.03125

TABLE 7.5. *Sampling Distribution of r / n for n = 5 Where p of Favored Outcome Equals .5.*

r/n	p(r/n)
1.0	.03125
0.8	.15625
0.6	.31250
0.4	.31250
0.2	.15625
0.0	.03125

probabilities of each of the outcomes having the same proportion (r / n) of above average readers.

USING THE BINOMIAL FORMULA. As above, rather than working out each distinct outcome and its probability and then summing the probabilities for outcomes having the same r / n, the probability of a particular r / n can be obtained by applying the binomial formula. For example, the probability of r / n = 2/5 is given by:

$$\frac{n!}{r!\,(n-r)!}\,p^r q^{n-r} = \left(\frac{5!}{2!\,3!}\right)(.5)^2(.5)^3$$

$$= (10)(.03125)$$

$$= .31250$$

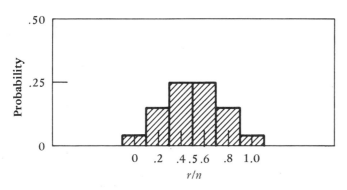

FIGURE 7-4. *Histogram description of sampling distribution of r / n for n = 5 and p of favored outcome = .5.*

Sampling Distributions of Proportions

MEAN OF SAMPLING DISTRIBUTION. The reader should note that this sampling distribution is also centered around .5; its mean is .5. Again this can be determined using formula 6–12 for calculating the mean of a probability distribution; in this case:

$$\mu = \sum_{1}^{k} Xp(X)$$
$$= (1.0)(.03125) + (.8)(.15625) + (.6)(.31250)$$
$$+ (.4)(.31250) + (.2)(.15625) + (.0)(.03125)$$
$$= .5$$

The proportions and corresponding probability values used in the above calculations are given in Table 7.5. The mean of the sampling distribution of r/n is always equal to p. In common sense terms this indicates that any particular (random) sample r/n is just as likely to be too large as too small, and on the average, will be correct.

Standard Error of r/n

The standard deviation of the sampling distribution of r/n is given by:

$$\sigma_{r/n} = \sqrt{\frac{pq}{n}} \qquad (7\text{–}1)$$

where p and q are the probabilities of the two outcomes and n is the number of selections or scores in each sample. *The standard deviation of a sample statistic is referred to as a standard error.* Thus, $\sigma_{r/n}$ is referred to as the *standard error of r/n*. The term *standard error* makes sense if the sample statistic, r/n, is considered an estimate of p. Since the mean of r/n equals p, the standard error is a measure of the degree to which the sample r/ns tend to deviate from p; that is, the degree to which r/ns tend to be in error as estimates of p.

EFFECT OF SAMPLE SIZE ON STANDARD ERROR. For the three binomial sampling distributions developed above, the standard errors of r/n are given respectively by:

$$\sqrt{\frac{pq}{n}} = \sqrt{\frac{.25}{1}}$$
$$= .5 \qquad \text{for } n = 1$$

$$\sqrt{\frac{pq}{n}} = \sqrt{\frac{.25}{2}}$$

$$= \frac{.5}{\sqrt{2}}$$

$$= .35 \quad \text{for } n = 2$$

$$\sqrt{\frac{pq}{n}} = \sqrt{\frac{.25}{5}}$$

$$= \frac{.5}{\sqrt{5}}$$

$$= .22 \quad \text{for } n = 5$$

Note that as n increases the standard error of r/n decreases. For $n = 1$, 2, and 5, the standard errors of r/n are respectively .50, .35, and .22. This observation is an illustration of the general fact that *when sampling from a population having only two subsets, the larger the sample size (n), the smaller the standard error of r/n.* Generally, the statistic r/n will be less in error as an estimate of p, the larger the number of selections (n).

Sample Size and Shape of Sampling Distributions of r/n

The reader should note also the change in shape of the sampling distributions of r/n as n increases from 1 to 5. The distribution of r/n approaches a bell shape. This observation is an illustration of the general fact that when sampling from a population that has only two subsets, the sampling distribution of r/n becomes more like a *normal distribution* as n, the sample size, increases. This fact will take on significance when we discuss the normal distribution in Chapter 8.

Computing Sampling Distributions of r/n

The binomial formula discussed in Chapter 6:

$$\text{Prob. } (r/n) = \frac{n!}{r!(n-r)!} p^r q^{n-r} \quad (6\text{--}11)$$

can be used to compute any particular sampling distribution of r/n, where samples are selected from a two-valued population. All we need to do is specify n and p and the probabilities for each possible r/n can be computed. Since the sum of the probabilities associated with the two values must equal 1, i.e., $p + q = 1$, the probability q must equal $1 - p$. Thus, by specifying p, q is also specified.

TABLE 7.6. *Computation of Sampling Distribution of r/n When $n = 4$ and p of Favored Outcome Equals .3.*

r/n	Prob. (r/n)	Computation of Prob. (r/n)
0.00	.2401	Prob. $(0/4) = \dfrac{4!}{0!\,4!}(.3)^0(.7)^4 = .2401$
0.25	.4116	Prob. $(1/4) = \dfrac{4!}{1!\,3!}(.3)^1(.7)^3 = .4116$
0.50	.2646	Prob. $(2/4) = \dfrac{4!}{2!\,2!}(.3)^2(.7)^2 = .2646$
0.75	.0756	Prob. $(3/4) = \dfrac{4!}{3!\,1!}(.3)^3(.7)^1 = .0756$
1.00	.0081	Prob. $(4/4) = \dfrac{4!}{4!\,0!}(.3)^4(.7)^0 = .0081$
	$\Sigma = 1.0000$	

Table 7.6 illustrates the computation of a particular binomial sampling distribution using formula 6–11. Figure 7–5 displays a histogram representation of this distribution. The p of .3 and the n of 4 provide the information needed to compute this sampling distribution. This example could represent the results of sampling at random from a population of 100 persons of whom 30 are females. Because sampling is random, the probability of selecting a female is .3. The resulting distribution represents probabilities associated with the various possible

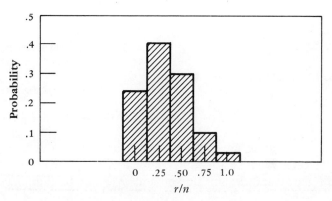

FIGURE 7–5. *Histogram description of sampling distribution of r/n for $n = 4$ and p of favored outcome $= .3$.*

proportions of females that could result from sampling four times from the population.

Note that this sampling distribution is not symmetric. The sampling distribution of r/n is symmetric only when p is .5. For ps less than .5, the sampling distribution of r/n is skewed to the right, and for ps greater than .5, the sampling distribution of r/n is skewed to the left, the degree of skew being less the closer p is to .5. Also, for any particular value of $p \neq .5$, the larger the n comprising a sample, the less skewed the distribution of r/n.

The degree of skewness of the distribution of r/n has no bearing on the fundamental relationship:

$$\mu_{r/n} = p \qquad (7\text{--}2)$$

The mean of a sampling distribution of r/n always equals p.

SAMPLING DISTRIBUTIONS OF MEANS

Sampling distributions of r/n serve as a good focus for the introduction of sampling distributions. This is because the sampling distributions of r/n can be easily calculated (using algebraic rules) once we know or postulate the population probability distribution. The psychologist most often refers, however, to sampling distributions of means, differences between means and sampling distributions of special statistics called *test statistics*. The remainder of this chapter will be devoted to a discussion of sampling distributions of means. The other distributions mentioned will be treated in subsequent chapters.

Just as it is possible to make n selections from a population set having just two values and compute a proportion, it is possible to make n selections from a population set having two or more numerical values and compute a mean. *A sampling distribution of means is a function associating probabilities with all possible means that could result from samples of a given size (n) drawn from a given population.*

An Illustrative Population

In order to illustrate a sampling distribution of the mean, it is necessary to have a population distribution. An illustrative population is contained in Table 7.7. The population consists of the score values 1, 2, 3, 4, and 5, each of which has an associated probability of 0.20. Such a population could result from a frequency distribution of 20 1s, 20 2s, 20 3s, 20 4s, and 20 5s—if the probabilities were computed on the assumption of random sampling with replacement. Random sampling guarantees that each of the five possible score values is equally likely,

TABLE 7.7. *A Population (Probability) Distribution and Its Mean.*

X	$p(X)$	$Xp(X)$
5	0.20	1.00
4	0.20	0.80
3	0.20	0.60
2	0.20	0.40
1	0.20	0.20
	$\sum_{1}^{k} = 1.00$	$\sum_{1}^{k} = 3.00$

$$\mu = \sum_{1}^{k} Xp(X) = 3.00$$

and thus, enables use of the classical model of probability to determine the probability of each score value. Sampling with replacement guarantees that the probability of any score value does not change with repeated sampling.

Table 7.7 also contains the calculation of the mean, μ, by the use of formula 6–12. In this case the mean of the population, μ, is equal to 3.00.

Previously, μ was used as the symbol for the mean of any probability distribution. In the present context, the probability distribution of concern is the population distribution. Thus, μ is the mean of the population distribution. Subsequently, μ without a subscript will be used mainly as the symbol for the mean of a population distribution.

Table 7.8 contains the calculation of the variance of the population distribution, σ^2, by the use of formula 6–13. In this case the variance of the population,

TABLE 7.8. *Calculation of the Variance and Standard Deviation of the Population in Table 7.7.*

X	$p(X)$	$X - \mu$	$(X - \mu)^2$	$(X - \mu)^2 p(X)$
5	0.20	2	4	0.80
4	0.20	1	1	0.20
3	0.20	0	0	0.00
2	0.20	−1	1	0.20
1	0.20	−2	4	0.80
	$\sum_{1}^{k} = 1.00$			$\sum_{1}^{k} = 2.00$

$$\sigma^2 = \sum_{1}^{k} (X - \mu)^2 p(X) = 2.00$$

$$\sigma = \sqrt{\sigma^2} = \sqrt{2.00} = 1.414$$

σ^2, is equal to 2.00. Also, as Table 7.8 indicates, the standard deviation, σ, is the square root of the variance, or 1.414.

Previously, σ^2 was used as the symbol for the variance of any probability distribution; and σ, as the symbol for the standard deviation of any probability distribution. In the present context, the probability distribution of concern is the population distribution. Thus, σ^2 is the variance of the population distribution, and σ is the standard deviation of the population distribution. Subsequently, σ^2 without a subscript will be used mainly as the symbol for the variance of the population distribution; and σ without a subscript, as the symbol for the standard deviation of the population distribution.

One final matter relating to the population distribution in Table 7.7 is that it does not have a normal, bell shape. Since all the score values have an equal probability of occurrence, the "top" of the distribution is flat. Such a distribution is said to be rectangular.

A Sampling Distribution of the Mean

The next step in illustrating a sampling distribution of the mean is to draw all possible samples of a given size from the population distribution in Table 7.7. In order to keep the calculations as simple as possible, a sample size of 2 ($n = 2$) will be used. Furthermore, sampling with replacement will be assumed. The first column of Table 7.9 provides the possible outcomes of two samplings from the population in Table 7.7. Thus, a 5 and a 5, a 5 and a 4, a 5 and a 3, etc., may be obtained. The second column gives the means for the possible outcomes. The mean of 5 and 5 is 5, the mean of 5 and 4 is 4.5, the mean of 5 and 3 is 4, etc.

Finally, the third column gives the probabilities of all the outcomes. According to Table 7.7, the probability of any one score value is 0.20. In the present case, the concern is with the probability of two score values; for example, 5 and 4. Using the multiplication rule, the probability of the outcome 5,4 (5 and 4) is 0.20 "times" 0.20 or 0.04. As the third column of Table 7.9 indicates, the probability of each of the possible outcomes is 0.04.

In Table 7.10, each of the possible sample means is associated with its probability. The mean of 5.0 occurs only once in Table 7.9. Hence, its probability is indicated as 0.04 in Table 7.10. The mean of 4.5, however, occurs twice in Table 7.9. A mean of 4.5 results from outcomes 5,4 and 4,5. In both cases the sum is 9, and the mean is 4.5. As Table 7.9 indicates, each of these outcomes has an associated probability of 0.04. Using the addition rule, the probability of 5,4 or 4,5 (each with a mean of 4.5) is 0.04 "plus" 0.04, or 0.08. The second column in Table 7.10, therefore, associates a probability of 0.08 with the mean of 4.5.

TABLE 7.9. *Possible Outcomes and Their Probabilities for Samples of n = 2 Drawn from the Population in Table 7.7.*

POSSIBLE OUTCOMES	$\bar{X}$	PROBABILITY OF OUTCOMES
5,5	5.0	0.04
5,4	4.5	0.04
5,3	4.0	0.04
5,2	3.5	0.04
5,1	3.0	0.04
4,5	4.5	0.04
4,4	4.0	0.04
4,3	3.5	0.04
4,2	3.0	0.04
4,1	2.5	0.04
3,5	4.0	0.04
3,4	3.5	0.04
3,3	3.0	0.04
3,2	2.5	0.04
3,1	2.0	0.04
2,5	3.5	0.04
2,4	3.0	0.04
2,3	2.5	0.04
2,2	2.0	0.04
2,1	1.5	0.04
1,5	3.0	0.04
1,4	2.5	0.04
1,3	2.0	0.04
1,2	1.5	0.04
1,1	1.0	0.04

TABLE 7.10. *Sampling Distribution of $\bar{X}$ for Population in Table 7.7 and n = 2.*

$\bar{X}$	$p(\bar{X})$
5.0	0.04
4.5	0.08
4.0	0.12
3.5	0.16
3.0	0.20
2.5	0.16
2.0	0.12
1.5	0.08
1.0	0.04

All the probabilities of means, $p(\bar{X})$, in Table 7.10 were calculated in an analogous way. For example, since a mean of 3.0 occurs 5 times in Table 7.9:

$$p(3.0) = 0.04 + 0.04 + 0.04 + 0.04 + 0.04$$
$$= 0.20$$

Table 7.10 contains a sampling distribution of the mean. As previously indicated, a sampling distribution of the mean is a function associating probabilities with all possible means that could result from samples of a given size (n) drawn from a given population. In the present case the sampling distribution was constructed by drawing all possible samples of size 2 from the population in Table 7.7.

Mean of Sampling Distribution of Mean

The first three columns in Table 7.11 contain the calculation of the mean of the sampling distribution of the mean, $\mu_{\bar{X}}$. Here, μ is written with an $\bar{X}$ subscript to distinguish the mean of a sampling distribution of the mean, $\mu_{\bar{X}}$, from the mean of a population, μ. The calculations in Table 7.11 were accomplished with a variation of formula 6–12, the formula for the mean of any probability

TABLE 7.11. *Calculation of the Mean, Variance, and Standard Deviation for the Sampling Distribution of the Mean in Table 7.10.*

$\bar{X}$	$p(\bar{X})$	$\bar{X}p(\bar{X})$	$\bar{X} - \mu$	$(\bar{X} - \mu)^2$	$(\bar{X} - \mu)^2 p(\bar{X})$
5.0	0.04	0.20	2.0	4.00	0.16
4.5	0.08	0.36	1.5	2.25	0.18
4.0	0.12	0.48	1.0	1.00	0.12
3.5	0.16	0.56	0.5	0.25	0.04
3.0	0.20	0.60	0.0	0.00	0.00
2.5	0.16	0.40	−0.5	0.25	0.04
2.0	0.12	0.24	−1.0	1.00	0.12
1.5	0.08	0.12	−1.5	2.25	0.18
1.0	0.04	0.04	−2.0	4.00	0.16
$\sum_{1}^{k} = 1.00$		$\sum_{1}^{k} = 3.00$			$\sum_{1}^{k} = 1.00$

$$\mu_{\bar{X}} = \sum_{1}^{k} \bar{X}p(\bar{X}) = 3.00$$

$$\sigma_{\bar{X}}^2 = \sum_{1}^{k} (\bar{X} - \mu_{\bar{X}})^2 p(\bar{X}) = 1.00$$

$$\sigma_{\bar{X}} = \sqrt{\sigma_{\bar{X}}^2} = \sqrt{1.00} = 1.00$$

distribution. Since the concern in the present context is with a probability distribution of means, μ is written with an $\bar{X}$ subscript, and $\bar{X}$ is substituted for X; i.e., formula 6–12:

$$\mu = \sum_{1}^{k} Xp(X)$$

is written,

$$\mu_{\bar{x}} = \sum_{1}^{k} \bar{X}p(\bar{X})$$

As Table 7.11 indicates, $\mu_{\bar{x}}$ is 3.00.

The mean of the sampling distribution in Table 7.11 is 3.00. It is, furthermore, the case that the mean of the population distribution in Table 7.7 is also 3.00. This equality of the two means illustrates an important fact. *The mean of a sampling distribution of means equals the mean of the probability distribution for the population from which the samples were drawn*:

$$\mu_{\bar{x}} = \mu \qquad (7\text{–}3)$$

Thus, just as the mean of r/n equals p of the favorable outcome, the mean of sample means equals the mean of the population distribution.

The fact that $\mu_{\bar{x}}$ equals μ provides a partial basis for drawing inferences about a population mean from a particular sample mean. Suppose somone wishes to know the probability of obtaining a sample mean within ± 1 of the population mean. Since the population mean equals the mean of the sampling distribution, what is being asked is the probability of obtaining a sample mean of 2.0 or 2.5 or 3.0 or 3.5 or 4.0. Summing the probabilities for these means (Table 7.10) gives:

$$0.12 + 0.16 + 0.20 + 0.16 + 0.12 = 0.76$$

The probability that a sample mean based on any two random scores from the population in Table 7.7 will be within ± 1 of the population mean is 0.76.

The sampling distribution in Table 7.7 is discrete. (There are gaps between the values for adjacent sample means; for example, 5.0 and 4.5.) More typically, however, the concern is with sampling distributions of the mean that are theoretically continuous. Such sampling distributions are worked out with the aid of calculus. Figure 7–6 displays a continuous sampling distribution of means of need-aggression scores for samples of 20 persons. Suppose someone wishes to known how probable it is that an obtained sample mean will be within ± 1 of the population mean. This can be determined by calculating the proportion of area within the limits $\mu_{\bar{x}} \pm 1$ in the sampling distribution. In Fig. 7–6 the sampling distribution mean is 3. While exact calculation of the area one unit on either side of 3 requires calculus, it is apparent that about 2/3 of the

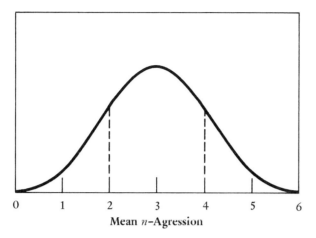

FIGURE 7-6. *Sampling distribution of means for n-aggression scores.*

distribution in Fig. 7-6 is within the limits ±1 of the sampling distribution mean of 3. The probability is about .67 that the mean of a sample of 20 scores will be in the range 2–4. Further, the fact that $\mu_{\bar{x}}$ equals μ allows us to form expectations about the proximity of a sample mean to the population mean. In this particular case, if someone repeatedly sampled 20 scores, each time computing the sample mean, roughly 67/100 sample means would be expected to be within ±1 of the population mean.

> Calculus is necessary for an exact computation of the probability associated with an interval of values along the domain of a continuous distribution. It is not possible to compute the probability of each value within an interval and then sum these probabilities as one can for intervals of a discrete variable. It is not possible to enumerate all of the values in an interval on a continuous dimension, because there are an infinite number of values in any interval. To facilitate computation of probabilities for intervals of a continuous variable, tables of probability values for certain distributions, such as the normal distribution, are presented in Appendix C. We will discuss the use of these tables in subsequent chapters.

Standard Error of the Mean

The last three columns in Table 7.11 contain the calculation of the variance and standard deviation of a sampling distribution of the mean, $\sigma_{\bar{x}}^2$ and $\sigma_{\bar{x}}$. Here σ^2 and σ are written with $\bar{X}$ subscripts to distinguish the variance and standard deviation of a sampling distribution of the mean from the variance and standard deviation of a population (σ^2 and σ). The calculations in Table 7.11 were accomplished with a variation of formula 6–13, the formula for the variance of any probability distribution. Since the present concern is with a probability distribution of means, σ^2 and μ are written with $\bar{X}$ subscripts, and $\bar{X}$ is substituted for X; i.e., formula 6–13:

$$\sigma^2 = \sum_{1}^{k} (X - \mu)^2 p(X)$$

is written,

$$\sigma_{\bar{X}}^2 = \sum_{1}^{k} (\bar{X} - \mu_{\bar{X}})^2 p(\bar{X})$$

As Table 7.11 indicates, $\sigma_{\bar{X}}^2$ is 1.00. Further, since the square root of 1.00 is 1.00, the standard deviation of this sampling distribution of the mean, $\sigma_{\bar{X}}$, is also 1.00.

The standard deviation of a sampling distribution of means, $\sigma_{\bar{X}}$, is called the standard error of the mean. A standard deviation is a measure of the degree of deviation of scores from the mean of their distribution. Because $\mu_{\bar{X}}$ equals μ, deviation of sample means from the sampling distribution mean is equivalent to deviation from the population mean. Thus, the larger the standard deviation of a sampling distribution of means, the more the sample means tend to deviate from population mean, i.e., the more in error they tend to be as estimates of the population mean. This accounts for the term *standard error of the mean* as a name for the standard deviation of a sampling distribution of means.

In Table 7.11 the standard error of the mean was directly calculated from the sampling distribution, and a value of 1.00 was obtained. The standard error of the mean can also be obtained by formula 7–4:

$$\sigma_{\bar{X}} = \frac{\sigma}{\sqrt{n}} \tag{7–4}$$

the standard deviation of the population divided by the square root of the sample size. As is indicated in Table 7.8, the standard deviation of the population in Table 7.7 is 1.414. The sampling distribution in Table 7.10 is for $n = 2$; thus:

$$\sigma_{\bar{X}} = \frac{\sigma}{\sqrt{n}}$$
$$= \frac{1.414}{\sqrt{2}}$$
$$= \frac{1.414}{1.414}$$
$$= 1.00$$

This is, of course, the same answer that was previously obtained in Table 7.11.

Formula 7–4 makes clear the effect of sample size, n, on the variability of the sampling distribution of the mean. For example, if someone samples from a population whose standard deviation is 8, the standard error of the sampling distribution of means for samples of 16 scores is 2:

$$\sigma_{\bar{x}} = \frac{8}{\sqrt{16}} = \frac{8}{4} = 2$$

If the sample size is 9:

$$\sigma_{\bar{x}} = \frac{8}{\sqrt{9}} = 2.67$$

and if the sample size is 4, $\sigma_{\bar{x}}$ equals 4. Figure 7–7 illustrates the effect of sample size on the variability of sample means. The larger the n, the narrower the distribution; that is, the less the variability of sample means.

The effect of sample size on the standard error has the further implication that the larger the size of the sample, the greater the probability that an obtained

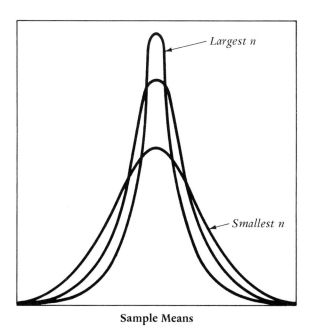

FIGURE 7–7. *Decreasing variability of sampling distributions as n increases.*

sample mean will be close to the population mean. In this sense, sample means become better and better estimates of their respective population means as the number of scores comprising the sample increases.

Sample Size and Shape of Sampling Distribution

It is of interest to examine the shape of the sampling distribution of the mean in Table 7.10. Notice that the distribution has a rough bell shape. The most probable mean, 3.0, is in the center; and the other means on either side of 3.0 progressively decline in probability. This bell shape is intriguing because the population from which the sample means were obtained does not have a bell shape. As previously indicated, the population in Table 7.7 has a so-called rectangular shape; i.e., all the population values have an equal probability of occurrence. The reason why the samples drawn from the rectangular population in Table 7.7 produce the bell shaped distribution of means in Table 7.10 can be understood by examining the possible sample outcomes in Table 7.9. Notice that only one possible outcome can produce a mean of 5.0: 5 and 5. On the other hand, five possible outcomes can produce a mean of 3.0: 5 and 1, 4 and 2, 2 and 4, 3 and 3, and 1 and 5. In view of this fact, 3.0, the mean at the center of the distribution, has a greater probability of occurrence than 5.0, the mean at one tail of the distribution.

While we have not as yet precisely defined a normal distribution, it is important to mention at this juncture that, *no matter what the shape of the population*

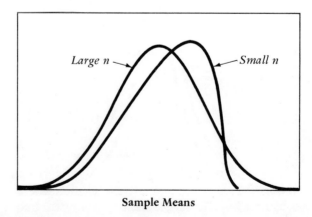

FIGURE 7–8. *Change in shape of sampling distribution of means as n increases.*

distribution, as sample size increases the sampling distribution of means approaches (becomes more like) a normal (bell shaped) distribution. This relationship between the size of *n* and the shape of the sampling distribution of means is called the *central limit theorem*. Figure 7–8 pictures the difference in shape of sampling distributions of means based on differing size samples from the same negatively skewed population distribution. The distribution based on the largest sample size is closest to a bell shaped curve.

SAMPLING DISTRIBUTIONS AND STATISTICAL INFERENCE

As noted at the beginning of the chapter, sampling distributions enable one to make inferences from samples to population characteristics. A brief example at this point will introduce an inference procedure called *hypothesis testing*. Chapter 8 contains a more detailed discussion of hypothesis testing.

Suppose someone questions the fairness of a particular coin; i.e., he questions whether the probability of obtaining a head to a toss of the coin is .5. If the coin is fair and the coin is tossed five times, the sampling distribution in Table 7.5 describes the probabilities of the various proportions of heads (favorable outcomes) that could be obtained. This sampling distribution is based upon samples from a population containing just two outcomes where the favored outcome has a probability (p) equal to .5. In this case the population is theoretical, consisting of outcomes to single tosses of the coin.

If the coin is biased ($p \neq .5$) then a sample proportion (r / n) in the direction of the bias will be more probable than indicated in Table 7.5. For example, if the true probability of obtaining a head is .7, then r / ns of .8 and 1.0 are more probable than indicated.

Suppose the coin is tossed five times and five heads are obtained. If the coin is fair, an outcome as extreme as five heads in five tosses ($r / n = 5/5 = 1$) has a very low probability of occurrence. According to Table 7.5 this probability is only .03125.

Hypothesis testing proceeds by assuming that a characteristic of a population distribution has a certain value (in this case that $p = .5$), and then noting whether the obtained sample statistic falls into an improbable, extreme interval of the sampling distribution for that statistic. If so, the original assumption regarding the value of the characteristic of the population distribution is rejected. In this case, if p is assumed to be .5, the obtained sample outcome (five heads in five tosses) falls at the extreme of the sampling distribution of proportions and has a probability of only .03125. Thus we may reject the assumption that the coin is fair ($p = .5$) and conclude that the coin is probably biased.

EXERCISES

1. Define the following terms:
   (a) statistic
   (b) sample statistic
   (c) sampling distribution
   (d) p of a favorable outcome
   (e) standard error
   (f) sampling distribution of means
   (g) standard error of the mean
   (h) central limit theorem
   (i) sampling distribution of variance

2. Discuss the relationship among the following concepts: sample, population distribution, and sampling distribution.

3. Compute the sampling distribution of r/n where $p = .8$ and $n = 3$.

4. In Exercise 3, what is the probability that:
   (a) $r/n \leq 1/3$
   (b) $r/n \leq 2/3$

5. Compute the sampling distribution of r/n where:
   (a) $p = .6, n = 4$
   (b) $p = .6, n = 6$

6. Compare the sampling distributions in Exercise 5.
   (a) How are they similar?
   (b) How are they different?
   (c) What general principle does the difference illustrate?
   (d) Compute the standard errors of these two distributions.

7. Solve for:
   (a) $\sigma_{\bar{x}}$, when $\sigma = 4, n = 5$
   (b) $\sigma_{\bar{x}}$, when $\sigma^2 = 10, n = 2$
   (c) $\sigma_{\bar{x}}^2$, when $\sigma^2 = 6, n = 7$
   (d) $\sigma_{\bar{x}}^2$, when $\sigma = 4, n = 15$
   (e) σ, when $\sigma^2 = 4, n = 10$
   (f) $\sigma_{\bar{x}}$, when $\sigma = 6, n = 5$
   (g) σ, when $\sigma_{\bar{x}}^2 = 3, n = 7$
   (h) $\sigma_{\bar{x}}^2$, when $\sigma^2 = 12, n = 15$
   (i) n, when $\sigma_{\bar{x}} = 2, \sigma^2 = 16$

8. In a sampling distribution of the proportion r/n, when $p = .3$ and $n = 5$, what is the probability that after n selections r/n will be:
   (a) greater than 3/5
   (b) less than or equal to 3/5

9. What is the relationship between $\sigma_{\bar{x}}$ and σ?

10. Why does it make sense to use the term "standard error" for the standard deviation of a sampling distribution? Use the example of the sampling distribution of r/n.

11. Suppose that you were trying to explain the fact that $\mu_{\bar{x}} = \mu$ to someone who had not taken a course in statistics. How would you do it?

12. Explain the central limit theorem as if you were discussing it with someone who had never taken a course in statistics.

13. Under what circumstances is the median of the sampling distribution of r/n not equal to the mean of the sampling distribution of r/n?

14. What happens to the sampling distribution of the mean as n increases? Relate your answer to formula 7–4 for the standard error of the mean.

15. Explain how sampling distributions are an aid to statistical inference.

Prologue to Chapter 8

Chapter 8 begins by describing the normal distribution and then illustrates how the normal distribution may be used to test hypotheses concerning population means. Along the way the various technical terms and considerations that relate to hypothesis testing are introduced. These include "critical intervals," "significance level," "two-tailed test," and "null hypothesis." Students typically find such matters difficult to grasp. The same is true of the material on error and power, which is discussed in the final section of the chapter. The material is difficult, not because of complex mathematics, but because of the unfortunately high level of abstraction involved. It may be small reassurance, but you should recognize that psychologists of the previous generation also had to struggle with this material.

8

Normal Distribution and Hypothesis Testing

The normal distribution was referred to in prior chapters simply as a symmetrical bell shaped distribution. It was not stated whether it was a frequency or a probability distribution, or whether the involved variable was discrete or continuous. In fact, the normal distribution is a *probability distribution* of a continuous variable, or more simply, a continuous probability distribution. The importance of the normal distribution derives from its use as a descriptive model of population probability distributions and sampling distributions.

In this chapter we will initially define and discuss the normal distribution and then use the normal distribution for testing hypotheses concerning population means.

NORMAL DISTRIBUTION

Relation of the Normal Distribution to the Sampling Distribution of Proportions

A convenient way to approach the normal distribution is from the perspective of the sampling distribution of proportions—frequently referred to as the binomial (i.e., two-valued) distribution. The sampling distribution of proportions was discussed in some detail in Chapter 7. Recall that a population of above and below average readers was used as an illustrative example. From this population the sampling distributions of r/n for varying sample sizes was constructed. Figure 7–2 pictures the sampling distribution for $n = 1$, Fig. 7–3 for $n = 2$, and Fig. 7–4 for $n = 5$. For present purposes the important thing is to notice that visual comparison of the figures reveals an interesting tendency for the graphs

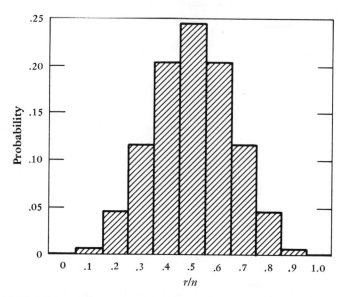

FIGURE 8–1. *Histogram description of sampling distribution of r / n for n = 10, where p of favored outcome = .5.*

to take on more of a bell shape as the sample size increases from 1 to 5. This tendency continues if the sample size is increased from 5 to 10. Figure 8–1 pictures the sampling distribution of proportions for $n = 10$. This distribution is most conveniently calculated through use of the binomial formula—as was illustrated in Chapter 7 for smaller ns. As indicated above, such sampling distributions are, in fact, frequently referred to as binomial distributions.

As sample size increases beyond 10, the tendency to approximate a smooth bell shape continues. If the total amount of space on the horizontal axis between proportions of 0 and 1 is held constant, the "stair steps" on the histogram become smaller and smaller until, with a sufficiently large n, these discontinuities become imperceptible. It can be, in fact, mathematically proven that *the normal distribution is the limit of the sampling distribution of proportions (or binomial distribution) where p equals .5 and n becomes indefinitely large.*

Equation for Probabilities in a Normal Distribution

The following formula is used to compute the probability associated with any interval of values along the domain of a normal distribution. Use of this formula requires calculus, and thus, is beyond the scope of this text. Presentation of the formula, however, will be useful for describing certain characteristics of a normal distribution. Probabilities for intervals of values along the domain of a normal distribution are given by formula 8–1:

$$\text{Prob.} (a < X < b) = \int_a^b \frac{1}{\sqrt{2\pi\sigma^2}} e^{-(X-\mu)^2/2\sigma^2} \, dX \qquad (8\text{--}1)$$

where Prob. $(a < X < b)$ is read "probability that an X score is between a and b."

Recognizing that the right side of this equation looks imposing, we will approach the equation by labeling and describing some of its elements. The character $\int$ is an integral sign symbolizing the process of *integration,* an operation in integral calculus. The reader might think of the *integration* operation as a summation operation and thus think of the integral sign, $\int$, as a summation sign, Σ. The operation $\int_a^b$ can be thought of as Σ_a^b, the summation across all values of a variable between a and b. Since the normal distribution represents a continuous variable, the process of integration can be thought of as summation across an infinite number of values. It will be recalled from prior mathematics courses that the symbol π stands for the constant, 3.1416, and the symbol e stands for the constant, 2.718. The other symbols, σ^2 and μ, represent respectively the variance and mean of a normal distribution. Finally, X represents any particular value of the variable described by the distribution, and dX an infinitesimal interval of X values.

The product to the right of the integral sign—that is,

$$\frac{1}{\sqrt{2\pi\sigma^2}} e^{-(X-\mu)^2/2\sigma^2} \, dX$$

can be thought of as the probability associated with an infinitesimal interval of X around a particular X value. Therefore,

$$\int_a^b \frac{1}{\sqrt{2\pi\sigma^2}} e^{-(X-\mu)^2/2\sigma^2} \, dX$$

can be thought of as the summation of probabilities across all of the infinitesimal intervals between values a and b.

Parameters of a Normal Distribution

Before we can integrate formula 8–1 to obtain the probability associated with a given interval of X, two values need to be known or postulated—the *variance* and *mean* of the particular normal distribution. If these two values are known, the values of the X variable that bound the interval of interest can be specified; and integral calculus can be used to compute the probability. As stated in Chapter

7, we will use special tables to ascertain probabilities on a normal distribution. The probabilities in these tables were computed using integral calculus.

The mean and variance of a normal distribution are called *parameters* of the distribution. Technically speaking, *a parameter is an arbitrary constant specified in the equation for a probability distribution.* The mean and variance of a normal distribution fit this definition. *The term parameter also is used more generally, however, to refer to any summary characteristic of a probability distribution* (whether a population distribution or a sampling distribution).

Family of Normal Distributions

As may be apparent by now, there is no one, "single," normal distribution. An infinite number of different normal distributions are specified by formula 8–1, each distribution resulting from a particular combination of the infinite variety of means and variances that can enter into formula 8–1. Thus, formula 8–1 is said to represent a *family* of normal distributions. The value for the mean will determine the exact location of the high point of the curve along the X, or horizontal, axis. The larger the mean is, the further the central tendency of the curve is moved to the right along the X axis. Likewise, the value of the variance will determine the extent to which a large proportion of the total area is either close to the mean or more distant from the mean. Normal distributions are always symmetrical and unimodal. However, the exact nature of any particular normal distribution is determined by the values for the mean and variance in formula 8–1. As long as the probability for any interval of values can be computed with formula 8–1, a given probability distribution is a normal distribution.

Standard Normal Distribution

One member of the family of normal curves is referred to either as the *standard normal distribution* or the *unit normal distribution*. *This distribution has a mean of zero and a variance of one as its two parameters.* The standard normal distribution is of particular importance because *any* normal probability distribution of X scores becomes a standard normal distribution when the X scores are transformed to z scores by formula 8–2:

$$z = \frac{X - \mu}{\sigma} \qquad (8\text{–}2)$$

Recall that z scores have a mean of zero, and a variance and standard deviation of one.

In addition to having a mean of zero and a standard deviation of one, the standard normal distribution always has a bell shape. Figure 8–2 portrays a

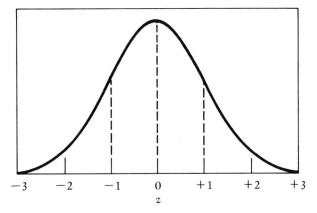

FIGURE 8-2. *Standard normal distribution.*

standard normal distribution. In Fig. 8-2, the interval +1 to −1 marks off 1 standard deviation to either side of the distribution mean. About 68 percent of the area under the curve lies within this interval. Likewise, the interval −2 to +2 marks off 2 standard deviations to either side of the mean. About 96 percent of the area under the curve lies within this interval.

Regardless of the values of the mean and variance of a normal probability distribution of X scores, transformation of the X scores to z scores will result in the probability distribution of z pictured in Fig. 8-2.

Up to this point we have defined the standard normal distribution as if the z scores were always transformed X scores. However, just as a normal probability distribution of X scores may be transformed to a standard normal distribution, a normally distributed sampling distribution of means can be transformed to a standard normal distribution. The latter transformation is given by formula 8-3:

$$z = \frac{\bar{X} - \mu_{\bar{x}}}{\sigma_{\bar{x}}} \qquad (8\text{-}3)$$

Note that this transformation follows directly from the definition of z in formula 8-2. We begin with a normally distributed score, in this case the sample mean ($\bar{X}$), subtract the mean of its distribution ($\mu_{\bar{x}}$), and divide by the standard deviation of its distribution ($\sigma_{\bar{x}}$). The same standard normal distribution which describes probabilities associated with intervals of z, where the z scores are transformed X scores from a population set, will describe probabilities associated with intervals of z where the z scores are transformed sample means from a sampling distribution of means.

Since the mean of a sampling distribution of means ($\mu_{\bar{x}}$) equals the mean of the population distribution (μ) (formula 7–3), and the standard deviation of the same sampling distribution equals the population standard deviation divided by $\sqrt{n}$ (formula 7–4):

$$\sigma_{\bar{x}} = \frac{\sigma}{\sqrt{n}}$$

Formula 8–3 can be restated as:

$$z = \frac{\bar{X} - \mu}{\sigma / \sqrt{n}} \tag{8-4}$$

For example, a sample mean of 6 ($\bar{X} = 6$) can be converted to a z score if we know the population mean (μ), standard deviation (σ), and the sample size (n). Given a μ of 4, a σ of 5, and an n of 16:

$$z = \frac{\bar{X} - \mu}{\sigma / \sqrt{n}}$$

$$= \frac{6 - 4}{5 / \sqrt{16}}$$

$$= 1.6$$

This z score is normally distributed if $\bar{X}$ is normally distributed.

ASCERTAINING PROBABILITIES FOR INTERVALS OF A NORMAL DISTRIBUTION

As previously noted, we will refer to tables in order to ascertain probabilities for intervals on a continuous distribution. Because every normal distribution assumes the same shape, mean, and standard deviation when it is transformed to a z distribution, we need only one table, a table of probability values for the standard normal distribution, in order to ascertain the probability associated with an interval of values from a particular normal distribution. Such a table is located in Appendix C, Table I.

Table I has three columns. Column A presents the various positive z scores (up to $z = 4$), col. B the area under the curve that is between the mean of the standard normal distribution ($\mu_z = 0$) and the adjacent z score in col. A, and col. C the area under the curve that is beyond the particular z score. For example, the area between the mean (a z score of 0) and a z score of 1 (col.A) is .3413 (col. B), and the area to the right of a z score of 1 is .1587 (col. C). Figure 8–3 displays the areas under the curve that are given in cols. B and C.

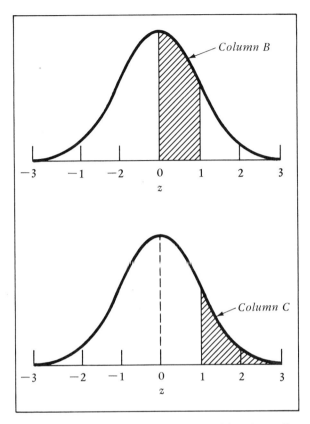

FIGURE 8–3. Illustration of areas represented in columns B and C of Table I, Appendix C.

Table I can be used to ascertain the probability associated with any interval of z scores between $+\infty$ (plus infinity) and $-\infty$ (minus infinity). To illustrate such usage, let us compute the area within the interval -1 to $+1$, an interval we have previously stated to include about 68 percent of the total area. This interval can be divided into two smaller intervals, -1 to 0 and 0 to $+1$. Since the standard normal distribution is a symmetric distribution, the interval -1 to 0 intersects the same amount of area as the interval 0 to $+1$. This is illustrated in Fig. 8–4. Referring to Table I, we see that the interval 0 to $+1$ intersects .3413 of the area. Thus, the interval -1 to 0 also intersects .3413 of the area. The interval -1 to $+1$, being the union of these two mutually exclusive intervals, intersects the area .3413 + .3413 = .6826, or about 68 percent of the total area.

Because the standard normal distribution is symmetric, only areas for

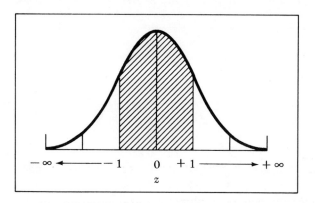

FIGURE 8-4. *Standard normal distribution showing the equivalence of areas cut off by -1 to 0 and 0 to $+1$; $-\infty$ to -1 and $+1$ to $+\infty$.*

positive z scores need be presented in Table I. In order to obtain the areas between negative z scores, simply find the areas for the positive z scores having the same absolute values. For example, the area between 0 and $-.85$ equals the area between 0 and $+.85$. Both equal .3023. And the area beyond -2.18 (i.e., between -2.18 and $-\infty$) equals the area beyond $+2.18$ (i.e., between $+2.18$ and $+\infty$). Both equal .0146.

As long as the mean and variance of a particular normal distribution are known, Table I can be used to compute probabilities for intervals of the variable. The probability associated with any interval of X scores is equivalent to the probability associated with the interval of z scores that results when the X scores are transformed to z scores.

As an example, suppose we desire to compute the probability that a student's IQ will be greater than 120. Assume that a population distribution of student IQ scores has a mean of 100, a standard deviation of 20, and that the probabilities for intervals of IQ values are represented quite accurately by a normal distribution. The IQ value of 120 can be easily converted to a z score using formula 8-2:

$$z = \frac{X - \mu}{\sigma}$$

which gives:

$$z = \frac{120 - 100}{20} = 1.0$$

Referring to Table I, we find that the probability associated with the z score interval of 1.0 to ∞ is .1587. Thus, the probability of a student having an IQ greater than 120 is .1587.

USING THE STANDARD NORMAL DISTRIBUTION TO TEST HYPOTHESES CONCERNING POPULATION MEANS

Hypothesis testing is a procedure for drawing inferences about a population parameter(s) from a sample set(s) of data. The procedure will be illustrated in the context of inferring the mean of a population distribution from the mean of a sample. Discussion will be limited to those situations in which the normal distribution can be used in the inference process.

An Example Problem

Consider the following problem. The dean of students at a given university suspects that the average IQ of entering freshman students has changed markedly during the past ten years and wants to check his hunch. When the university was smaller, it used to be routine practice to administer an IQ measure to all entering freshmen and tabulate the mean and standard deviation of these IQ measures. Rapid increase in the number of freshman admissions has subsequently rendered the practice too cumbersome to continue. The mean IQ and standard deviation of IQ scores for the total freshman class is, however, available for the entire freshman class of ten years ago.

Examination of data from ten years ago reveals a mean IQ of 110 with a standard deviation of 10 for incoming freshmen. There were 500 students in that class, and the relative frequency histogram (estimated probability distribution) of IQ scores is approximated quite closely by a continuous normal distribution.

Given this data, the following question is posed for research: Is the mean IQ of incoming freshman students different than it was ten years ago? The dean suspects that the mean IQ may be higher than ten years previous, but he wants his research to allow for the possibility that his judgment is incorrect, that the mean IQ has actually decreased. A method is needed that allows a conclusion to be reached but does not require testing each freshman student. The hypothesis testing procedures described below provide such a method.

The actual calculations used to provide information relevant to the dean's concern are, in fact, quite simple. The difficulty lies in understanding the rationale for such calculations. This rationale is described in the following subsections beginning with the alternative and null hypotheses and ending with completion of the test through transformation to the standard normal distribution.

A final subsection recapitulates the bare outline of the entire hypothesis testing procedure.

Stating the Alternative and Null Hypotheses

The initial step in hypothesis testing is stating two hypotheses—the *alternative* hypothesis and the *null* hypothesis. For the present example, these hypotheses are stated as follows:

1. *Alternative hypothesis*: the mean IQ is not equal to 110: *i.e., mean IQ has changed*
   $H_1: \mu \neq 110$
2. *Null hypothesis*: the mean IQ is 110; i.e., the mean IQ has not changed from 110:
   $H_0: \mu = 110$

The alternative hypothesis specifies the situation that the investigator believes to be true. In this case he believes that the mean IQ is presently different from 110. *The null hypothesis is the logical contradiction of the alternative hypothesis.* It is assumed to describe the true situation until evidence to the contrary is encountered. In this case it is the hypothesis of no change, i.e., the hypothesis that the mean IQ is 110 just as it was ten years before. The alternative hypothesis is regarded as descriptive of the true state of affairs if the null hypothesis is rejected.

Specifying a Sampling Distribution

After stating the null and alternative hypotheses, the next step in hypothesis testing is to specify a sampling distribution. Since the dean's concern is with mean IQ, the relevant sampling distribution is the sampling distribution of the mean. Actually, "the" sampling distribution of the mean is a family of sampling distributions. Before one particular member of this family can be specified, it is necessary to know or assume its mean, $\mu_{\bar{x}}$.

How can the mean of the sampling distribution of the mean be specified or assumed? In Chapter 7, it was pointed out that the mean of a sampling distribution of the mean equals the mean of the population from which the samples are drawn. This equality was stated in formula 7–3:

$$\mu_{\bar{x}} = \mu$$

Assumption of the population mean allows for specification of the mean of the sampling distribution. The population mean was, of course, assumed above by the null hypothesis, H_0, to be 110. It is, thus, the null hypothesis regarding the population mean that allows for specification of the mean of the sampling distribution of the mean.

In the present example the mean of the specified sampling distribution of the mean is 110. It is, furthermore, possible to determine the variance of the sampling distribution of the mean. The square root of a variance is a standard deviation; and the standard deviation of the sampling distribution of the mean—the standard error of the mean—can be determined from formula 7-4:

$$\sigma_{\bar{x}} = \frac{\sigma}{\sqrt{n}}$$

The use of formula 7-4 requires knowledge of σ, the standard deviation of the population distribution. Assuming σ to be the same as ten years prior and setting the sample size (n) at 50 gives $\sigma_{\bar{x}}$ a value of 1.414:

$$\sigma_{\bar{x}} = \frac{\sigma}{\sqrt{n}} = \frac{10}{\sqrt{50}} = 1.414$$

The square of the standard error of the mean is the variance of the sampling distribution of the mean:

$$\sigma_{\bar{x}}^2 = (1.414)^2 = 2.00$$

Use of the Normal Distribution as a Theoretical Model

At this point it might be felt that a reasonable next step would involve the calculation of the sampling distribution of the mean. In fact, however, this is not done. It is not done because the population values upon which the sampling distribution would be based are only theoretical possibilities. Not all of the IQ scores for the freshmen in the current class have been measured. Since these values do not exist, we obviously cannot take all possible samples of a given size and construct the sampling distribution. What is done, therefore, is to use the normal distribution as a theoretical model of the sampling distribution of the mean. Provided sample size is large (e.g., 30 or more) it can be assumed that the sampling distribution of means is well described by a normal distribution. As indicated in Chapter 7, this assumption follows from the *central limit theorem*, which states that, regardless of the form or shape of the population distribution, the sampling distribution of means approaches a normal distribution as sample size (n) increases.

The central limit theorem thus provides the rationale for using the normal distribution as a model for the sampling distribution of the mean. As stated above, however, the normal distribution is actually a family of distributions. Before any particular member of the family can be selected, it is necessary to know or assume values for the two parameters, the mean of the distribution and the variance of the distribution. Which member of the family of normal distributions should be selected? Stated differently, what values should be assigned to the mean and the variance of the distribution? Since the selected normal distribution is to be used as a model for the particular sampling distribution of the mean specified above, it is necessary and reasonable that the values for the mean and variance of the selected normal distribution be the same as those in the specified sampling distribution of the mean. For the above example these are 110 and 2.00. *By specifying the mean as 110 and the variance as 2.00, a particular normal distribution is selected as a model for the sampling distribution.*

Specifying Critical Interval(s) and Significance Level

In the example problem, the more deviant the obtained sample mean is from the population mean assumed in H_0 ($\mu = 110$), the more suspect the truth of H_0. But just how deviant must the sample mean be from the mean assumed in H_0 to justify the decision that the mean assumed in H_0 is not true, and the alternative hypothesis is true?

This question is answered by specifying a probability level called *significance level*. *The significance level is a probability specifying just how unlikely an extreme interval(s) of sample values must be before an obtained sample statistic falling in the extreme interval(s) will lead to rejection of the null hypothesis.* The particular significance level chosen is symbolized as α. Significance levels frequently used in psychology are .05 and .01, with .05 being the most frequent. Some of the factors affecting choice of significance level are discussed below.

In Fig. 8–5 for the example problem, α equals .05, the sum of .025 + .025, the lined areas. Two *critical intervals* are also labeled in Fig. 8–5. *Critical intervals are those intervals of sample mean values at the extremes of the sampling distribution for H_0 that have a combined probability of α. They define the set of sample values which, if obtained, will serve to reject the null hypothesis.* Each critical interval extends from a *critical value*, the value at the limit of the interval toward μ_{H_0}, to infinity.

The logic behind the choice of critical intervals becomes clear if we consider sampling distributions that might result if the alternative hypothesis H_1: $\mu \neq 110$ is true. The alternative hypothesis is true if the true population mean is 111, 113, 96, 109, etc. For purposes of discussion we will choose two mean values that might represent the true state of affairs if the alternative hypothesis is true, 114 and 106. Figure 8–6 displays three different sampling distributions, each assumed to describe the true state of affairs if its mean is the true population mean. Note that in constructing the sampling distributions for μ equals 114 and μ equals 106 it has been assumed that each distribution has exactly the same shape. *When testing a hypothesis concerning a mean, it is assumed that the mean is the only aspect of the population in question. The population's shape and the standard error of the mean are assumed constant no matter which of the hypothesized population*

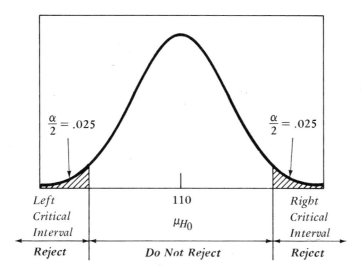

FIGURE 8-5. Illustration of α and critical intervals.

means is in fact true. Examination of Fig. 8-6 makes it apparent just why intervals at the extremes of the sampling distribution for H_0 are chosen as intervals for rejection of the null hypothesis. If the null hypothesis is true, the probability is very low (.05) that an obtained sample mean will fall within these critical intervals. However, if the true population mean is either greater than 110 (e.g.,

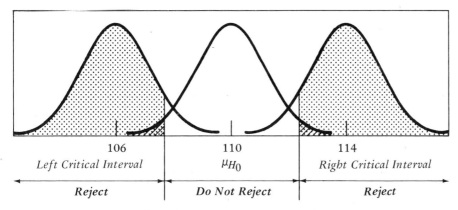

FIGURE 8-6. Sampling distributions for three different population means.

Using the Standard Normal Distribution to Test Hypotheses

114) or less than 110 (e.g., 106), then it is more probable that the obtained sample mean will fall within one of these critical intervals. If the true population mean is 114, then the probability of a sample mean falling within a critical interval is about .80. It can be seen in Fig. 8–6 that about 80 percent of the area of the sampling distribution for μ equals 114 covers a critical interval. Likewise, about 80 percent of the area of the sampling distribution for μ equals 106 covers a critical interval. If either of these alternative hypotheses is true, the probability is quite high that an obtained sample mean will fall within a critical interval and the null hypothesis will be rejected.

Figure 8–6 also helps clarify the reason for splitting the critical interval between the two tails of the sampling distribution. Stating the null and alternative hypotheses respectively as $H_0: \mu = 110$ and $H_1: \mu \neq 110$ means that the researcher wants to reject the null hypothesis if the true mean is either greater than or less than 110.

ONE- AND TWO-TAILED TESTS. When the null and alternative hypotheses are stated as they are in the present example and thus the critical interval is divided, the statistical test is called, reasonably enough, a *two-tailed test*. A two-tailed test is used whenever the experimenter thinks that the true mean may be either greater or less than the mean assumed in the null hypothesis and thus wants the statistical test to be sensitive to deviations in either direction. Two circumstances may lead to use of a one-tailed test: (1) reasonable confidence that the true mean differs from the null hypothesis in only one direction; (2) concern about results only if the true mean differs in a particular direction from the null hypothesis. In a one-tailed test the critical interval is in just one tail of the sampling distribution.

Figure 8–7 displays once again the sampling distribution for μ equal to 110 in the example experiment. In this case the critical interval is in just the right tail of the distribution. Such a one-tailed critical interval would follow if the researcher had designated the null hypothesis as $H_0: \mu \leq 110$ and the alternative hypothesis as $H_1: \mu > 110$. Here the researcher is interested only in the possibility that the true mean is greater than 110. The researcher believes that the mean IQ has increased in the past decade and wants to check out this hunch, not being interested in detecting change in the negative direction. Notice that the null hypothesis for a one-tailed test is specified as an interval of values, μ less than or equal to 110, rather than a specific value. In performing a one-tailed test, the specific value μ equals 110 is assumed and tested (just as in a two-tailed test). However, the null hypothesis is stated as an interval because rejection of $H_0: \mu = 110$ automatically implies rejection of all values of μ less than 110. Suppose that a sample mean large enough to fall within the critical interval in Fig. 8–7 were obtained. This same sample mean also would fall within the critical interval if μ were less than 110, i.e., 105. Any obtained sample mean falling within the critical region of the sampling distribution for $\mu = 110$ at a

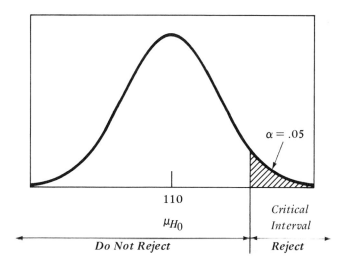

FIGURE 8–7. *Critical interval for one-tailed test.*

given α level would also fall within the critical regions of sampling distributions for values of μ less than 110.

In summary, when the significance level is set (for example, at .05), a critical interval then is specified at the tail(s) of the distribution for H_0 such that the set of values covered by the critical interval has a probability equal to α. If the test is a two-tailed test, the critical interval is typically split evenly between the two tails of the distribution. Thus, in Fig. 8–5, where two critical intervals are specified and $\alpha = .05$, each interval has a probability of .025 if H_0 is true. The null hypothesis may be rejected by the sample mean falling within either critical interval. The probability of rejecting the null hypothesis, if, in fact, it is true, is .05, the sum of the probabilities associated with the two critical intervals (.025 + .025 = .05).

If a one-tailed test is specified and the significance level is set at .05, as illustrated in Fig. 8–7, one critical interval is located at the tail corresponding to the direction of the alternative hypothesis. The range of values covered by the single critical interval has a probability of .05 if H_0 is true.

Completing the Test

TRANSFORMATION TO THE STANDARD NORMAL DISTRIBUTION. Our hypothetical dean previously decided to use a normal distribution with a mean of 110 and a standard deviation of 1.414 as a model for his sampling distribution. With such a distribution it is possible to compute the probabilities for various intervals of possible mean values. Such calculation, however, involves calculus (formula 8-1). The easiest procedure is to use the standard normal distribution for which the probabilities for various intervals of z scores have already been computed (see Appendix C, Table I). As was previously indicated, formulas 8-3 or 8-4 can be used to transform a sample mean from a normal distribution into a z score from the standard normal distribution.

COMPLETING A TWO-TAILED TEST. With the above terminology as background we can now follow through with the example problem and test $H_0: \mu = 110$ against $H_1: \mu \neq 110$. The hypotheses are so stated as to imply a two-tailed test and the researcher has previously selected $\alpha = .05$ as his significance level. Now he must specify the critical intervals. When $\alpha = .05$ and the test is two-tailed, critical intervals are typically placed at each tail of the sampling distribution for H_0 such that the two intervals each have a probability of .025. Examination of Table I reveals that the z score interval $+1.96$ to ∞ intersects .025 of the standard normal distribution in the right tail; the interval $-\infty$ to -1.96 intersects .025 of the standard normal distribution in the left tail. Thus the two critical intervals of z scores are $+1.96$ to ∞ and -1.96 to $-\infty$.

The dean selects a *random sample* of 50 students from the population of 5000. He tests each of the 50 students and finds the mean IQ for the sample to be 113.25. Transformed to a z score (formula 8-4) this mean becomes 2.3, as illustrated in Table 8.1. This value is within the right critical interval. Thus H_0 is rejected in favor of H_1. The dean concludes that the mean IQ for incoming freshmen is no longer 110.

COMPLETING A ONE-TAILED TEST. Consider now the same example problem with H_0 and H_1 restated so as to imply a one-tailed test. The null hypothesis is stated $H_0: \mu \leq 110$ and the alternative hypothesis $H_1: \mu > 110$. Specifying α to be .05, the critical interval of z scores is 1.65 to ∞ according to Table I. This interval intersects 5 percent of the area of the standard normal distribution in the right

TABLE 8.1. *Transformation of Obtained Sample Mean to a z Score According to Formula 8-4.*

$$z = \frac{\bar{X} - \mu}{\sigma / \sqrt{n}}$$

$$= \frac{113.25 - 110}{10 / \sqrt{50}}$$

$$= 2.3$$

tail. The obtained sample mean IQ of 113.25 transforms to a z score of 2.3, which falls within the critical interval. Thus $H_0: \mu \leq 110$ is rejected in favor of $H_1: \mu > 110$, at the .05 level of significance.

Recapitulation: Why Have a Null Hypothesis?

At this point it may be helpful to state succinctly why it is necessary to have a null hypothesis. Some students may feel that to insist on the use of the null hypothesis is outright perverseness. Since the main interest is typically the alternative hypothesis, "why state things backwards?"

Null hypotheses are formulated in order to develop or specify probability distributions that associate probability with particular sample outcomes or intervals of sample outcomes. If the probabilities are sufficiently small, the null hypothesis may be rejected, and the alternative hypothesis accepted.

In Chapter 7, the sampling distribution of proportions was developed and illustrated for a binomial population of above and below average readers. Development of this sampling distribution was dependent upon the knowledge or assumption of the proportion of above and below average readers in the population. This proportion was set at .5. In the context of hypothesis testing, we could say that a null hypothesis was formulated. This hypothesis could be stated simply as $H_0: p = .5$. The knowledge or the assumption of this population parameter was essential to the construction of the sampling distribution of proportions, or of the r/n statistic.

Once the sampling distribution of r/n for a given sample size was constructed, it indicated the probability of obtaining a particular sample proportion—on the H_0 assumption that the population contained an equal number of above and below average readers. Thus, as Table 7.5 and Fig. 7-4 indicate, the probability that the random sampling of 5 outcomes from such a population would result in 5 above average readers (or that $r/n = 1.0$) is .03125. Since this probability is fairly small, a sample r/n of 1.0 provides some evidence that the population does not contain an equal number of above and below average readers. Under some circumstances this null hypothesis might be rejected.

In the present chapter the concern is not with proportions but with means. The probability distribution of concern is, thus, the sampling distribution of the mean. The sampling distribution, however, usually cannot be developed because the population distribution is typically unknown. In Chapter 7, a particular sampling distribution of the mean for a particular population was developed. If the population is unknown, though, it is not possible to develop an appropriate sampling distribution of the mean. What is done, therefore, is to use a normal distribution as a theoretical model for the sampling distribution of the mean. As previously indicated, this use of the normal distribution as a theoretical model is justified by the central limit theorem. Also, as previously indicated, there is not one normal distribution but a family of normal distributions. At this point the importance of the null hypothesis becomes apparent. It is the null hypothesis that allows for specification of the mean of the sampling distribution and, thus, the appropriate mean for the theoretical normal distribution. In order to select one member of the family of normal distributions, it is necessary to specify values for the two parameters—the mean and the variance. *It is the null hypothesis that allows for specification of one of the two parameters of the theoretical normal distribution.*

If the population standard deviation is known, it is also possible, with the aid of formula 7–4, to specify a value for the standard deviation and, thus, for the variance of the theoretical normal distribution. In this way values for both parameters are specified, and one particular member of the family of normal distributions is selected. With such a distribution it is possible, with the aid of calculus, to calculate the probabilities of certain intervals of sample outcomes. More typically, however, such probabilities are ascertained by using formulas 8–3 or 8–4 to transform a sample mean from a normal distribution into a z score from the standard normal distribution. The probabilities of various intervals of z then are looked up in Table I of Appendix C.

ERROR AND POWER

Types of Error

When testing a null hypothesis, one of two decisions is made: either (1) reject the null hypothesis and accept the alternative hypothesis, or (2) do not reject the null hypothesis. Which conclusion is reached depends on the sample mean outcome. Because a judgment is being made from a sample mean rather than by determining the actual population mean, the judgment may be in error or it may be true.

Table 8.2 illustrates the possible outcomes to the decision process for the example problem. In two of the four possibilities, a correct decision is made; and in the other two, an incorrect decision or an error is made. Consider the first column in which the population mean, μ, equals 110. Recall that the dean's null hypothesis was that μ equals 110. Thus, the first column represents a situation in which the null hypothesis is correct. Therefore, rejection of this true null hypothesis is an error, and a decision not to reject is correct. The error in this case is called a *Type I error*. *A Type I error is the error of rejecting a true null hypothesis.* The probability associated with making such an error is equal to the significance level, α (alpha). If the significance level is .05 and the null hypothesis

TABLE 8.2. *Decisions and Possible Outcomes Given the Null and Alternative Hypotheses.*

	True Mean	
	$\mu = 110$	$\mu \neq 110$
Reject Null Hypothesis	Type I Error	Correct
Do Not Reject Null Hypothesis	Correct	Type II Error

is true, the probability is .05 that an obtained sample mean will fall within a critical interval. On the other hand, the probability of correctly not rejecting the null hypothesis is $1 - \alpha$. Thus, if the significance level, or α, is .05, the probability of correctly not rejecting a true null hypothesis is .95.

The situation existing for the first column of Table 8.2 is graphically represented in Fig. 8–8. Figure 8–8 pictures two identical distributions superimposed on each other. Since the distributions are identical, a total degree of overlap exists. The first of these distributions is the hypothesized distribution of

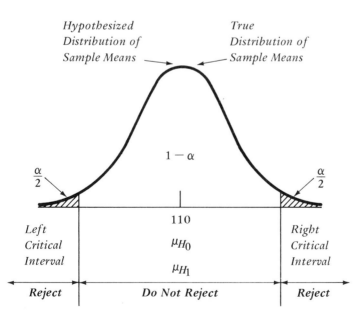

FIGURE 8–8. *Illustration of Type I error, α, and correct acceptance of the null hypothesis, $1 - \alpha$—complete overlap of hypothesized and true sampling distributions of the mean.*

sample means (the hypothesized sampling distribution of the mean). Its mean equals 110. The second, or other, distribution is the true distribution of sample means (the true sampling distribution of the mean). Its mean equals 110, and 110 is also the true population mean. Thus, the mean of the hypothesized sampling distribution of the mean, the mean of the true sampling distribution of the mean, and the population mean all equal 110. The lined areas in the tails of the distribution represent α, the probability of a Type 1 error. Notice that α is split between the two tails, as is appropriate for a two-tailed test. For a one-tailed test, the total area, of course, would all be in one tail. The remaining, or unlined, area represents $1 - \alpha$, the probability of not rejecting a true null hypothesis.

The second column of Table 8.2 represents a situation in which the population mean, μ, does not equal 110. Since the null hypothesis is that μ does equal 110, the second column describes the situation in which the null hypothesis is not true. Thus, rejecting this false null hypothesis is a correct decision, and deciding not to reject is an error. The error in this case is called a *Type II error*. *A Type II error is the error of failing to reject a false null hypothesis.* The probability associated with a Type II error is symbolized β (beta).

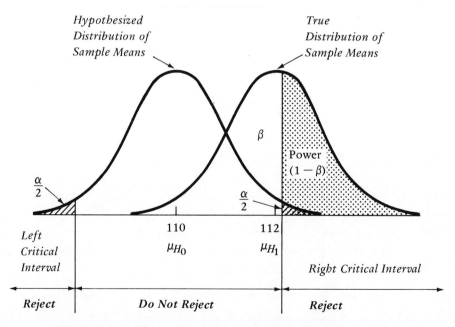

FIGURE 8–9. *Illustration of Type II error, β, and power, $1 - \beta$—non-overlapping, hypothesized, and true sampling distributions of the mean.*

With an inexact alternative hypothesis such as $H_1: \mu \neq 110$, β cannot be specified. It is possible, however, to determine β for specific values of μ within the range of values delimited by H_1. Suppose that the true sampling distribution of the means has a mean equal to 112, as is indicated in Fig. 8-9. In Fig. 8-9, the distribution on the left is the hypothesized sampling distribution of the means, and the distribution on the right is the true sampling distribution of the means. As determined by the null hypothesis, the distribution on the left has a mean of 110. The distribution on the right has a mean of 112, and the true mean of the population is 112. Thus, Fig. 8-9 illustrates one specific situation in which $\mu \neq 110$, a situation in which $\mu = 112$. In Fig. 8-8, the population mean did equal 110, and the two distributions completely overlapped.

In Fig. 8-9, the undotted area in the true distribution represents β, the probability of a Type II error. Recall that a Type II error is the error of not rejecting a false null hypothesis. Notice that the undotted area in the true distribution is the area that does not fall within the critical interval for the hypothesized distribution. Even if the true population mean is 112, it is still possible to select a sample whose mean is considerably less than 112. If this mean is sufficiently small so as not to fall within the critical interval, a Type II error will be committed. The undotted area in the true distribution represents the probability of this occurrence.

Power

In Fig. 8-9, the dotted area represents power. Power equals $1 - \beta$, and is *the probability of correctly rejecting a false null hypothesis*. Since investigators wish to reject false null hypotheses, power is desirable.

Factors Affecting Type II Error and Power

Since the probability of a Type II error is β and power is $1 - \beta$, any factor that affects power will also affect the probability of committing a Type II error. Factors that increase power will decrease the probability of committing a Type II error. These factors, which are discussed below, include: the discrepancy between the true population mean and the hypothesized population mean, the level of significance, the population standard deviation, and the sample size.

DISCREPANCY BETWEEN THE TRUE POPULATION MEAN AND THE HYPOTHESIZED POPULATION MEAN. The upper graph in Fig. 8-10 pictures the distribution of sample means specified by the dean's null hypothesis and a possible true distribution of sample means. The hypothesized distribution has a mean of 110, and the true distribution has a mean of 112. Thus, the hypothesized population mean

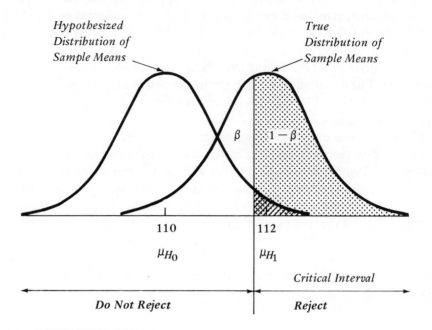

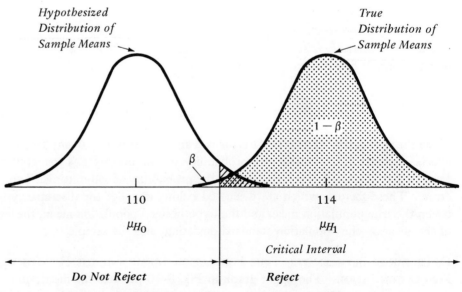

FIGURE 8–10. Illustration of an increase in power, $1 - \beta$, as the discrepancy between the hypothesized population mean and the true population mean increases.

is 110, and the true population mean is 112. The lower graph in Fig. 8-10 pictures the same situation except for the fact that the mean of the true distribution of sample means (and the true population mean) is now represented as 114. The discrepancy between the hypothesized and the true mean has increased from 2 to 4. Comparison of the upper and lower graphs indicates that power, or $1 - \beta$, has increased; and the probability of a type II error, or β, has decreased. *In general, the greater the discrepancy between the true population mean and the hypothesized population mean, the greater the power and the less the probability of a Type II error.*

LEVEL OF SIGNIFICANCE (PROBABILITY OF A TYPE I ERROR, α). The upper and lower graphs in Fig. 8-11 picture the hypothesized and the true distribution of sample means for $\mu_{H_0} = 110$ and $\mu_{H_1} = 112$. The upper and lower graphs differ only by virtue of the fact that $\alpha = .01$ in the upper graph, and .05 in the lower. Notice that this change produces an increase in power, or $1 - \beta$; and a decrease in the probability of a Type II error, or β, for the lower graph. In general, *the larger the level of significance, or α, the greater the power, and the less the probability of a Type II error.*

As previously stated, power is desirable. Investigators wish to be able to reject false null hypotheses. Thus, why should an investigator not increase power by selecting a large level of significance, α? The problem here is that increasing α increases the probability of a Type I error, the error of rejecting a true null hypothesis.

To some extent an investigator can determine α on the basis of situational requirements. If the study is one in which rejection of the null hypothesis will lead to large expenditures of money and/or investment of time, a Type I error is to be guarded against. In this situation the investigator may be willing to sacrifice power just to lower the probability of a Type I error. Thus, a small level of significance, .01 or even .001, may be desirable. On the other hand, if the study is only a pilot investigation to ascertain whether or not further research is justified, power may be much more important, and a Type I error is less important. In this situation an investigator may wish to increase power even at the increased risk of a Type I error. Thus, a large level of significance, .10 or even .20, may be desirable. As previously indicated, for most situations psychologists use the .05 level of significance. While this particular value for α is to some extent arbitrary, it does fall within a range of values that seems reasonable.

POPULATION STANDARD DEVIATION AND SAMPLE SIZE. Power and the probability of a Type II error are also affected by the population standard deviation and the sample size. *The smaller the population standard deviation the greater the power and the less the probability of a Type II error.* Also, *the greater the sample size, the greater the power and the less the probability of a Type II error.*

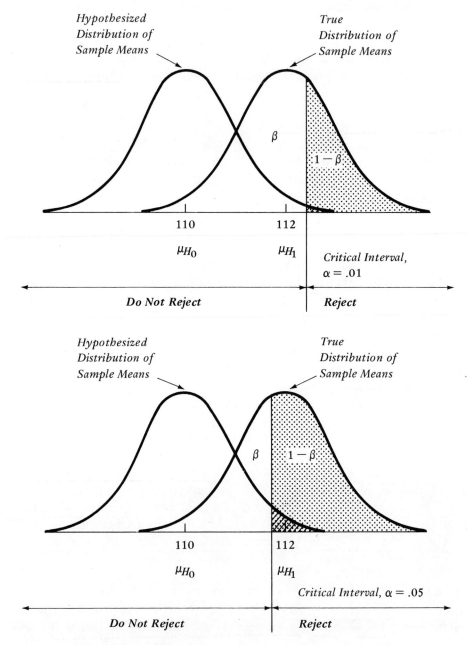

FIGURE 8–11. Illustration of an increase in power, $1 - \beta$, with an increase in α from .01 to .05.

Normal Distribution and Hypothesis Testing

The population standard deviation, σ, and the sample size, n, have an effect on power because of their effect on the standard error of the mean. The formula for the standard error of the mean was previously given (formula 7–4) as:

$$\sigma_{\bar{x}} = \frac{\sigma}{\sqrt{n}}$$

From formula 7–4, it can be seen that as σ decreases, the standard error decreases. It can also be seen that as n increases, the standard error decreases. Thus, the standard error decreases with either a decrease in the population standard deviation or an increase in the sample size.

The effect of a decrease in the standard error of the mean is illustrated in Fig. 8–12. The standard error decreases from the upper to the lower graph. This decrease is associated with an increase in power, $1 - \beta$, and a decrease in the probability of a Type II error, β.

How does knowledge of the effect of the population standard deviation and the sample size help the investigator increase power? Superficially, it would seem that nothing can be done about the population standard deviation. To some extent this may be true. However, it is possible that some of the variability in the population is due to unreliable or inconsistent measurement. For example, to the extent that raters interpret a rating scale differently on different occasions the ratings are unreliable. Thus, one way to increase power may be to improve the reliability of measurement. This may be accomplished, for example, by giving the raters more extensive training, or by giving the raters more explicit instructions.

Increasing the sample size is in some respects one of the simplest procedures for increasing power. Of course, the cost in time and effort has to be balanced against the value of an increased probability of rejecting a false null hypothesis.

ONE- AND TWO-TAILED TESTS. Finally, power and the probability of a Type II error are affected by the use of a one-tailed or a two-tailed test. *Assuming that the direction of the true mean from the mean specified under the null hypothesis has been correctly anticipated and that other things are equal, there is an increase in power and a decrease in the probability of a Type II error with the use of a one-tailed, as opposed to a two-tailed, test.*

Figure 8–13 provides an illustration of the increase in power with the use of a one-tailed test. In the upper graph is a two-tailed test, and in the lower graph, a one-tailed test. Notice that the dotted, or $1 - \beta$, area is larger in the lower graph. Thus, an investigator may increase power by the use of a one-tailed test.

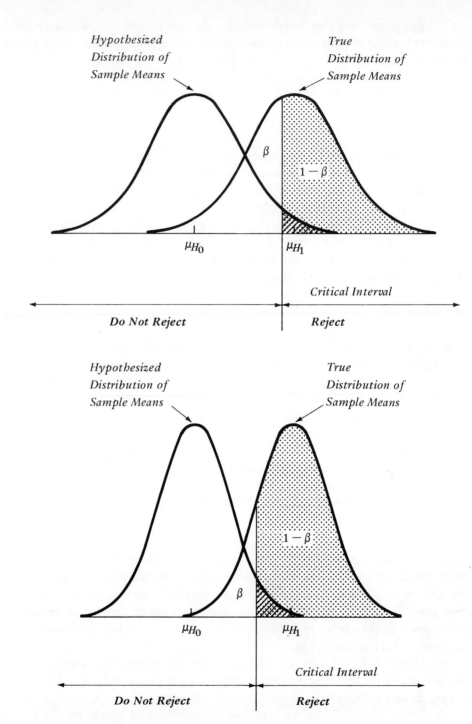

FIGURE 8–12. Illustration of an increase in power, $1 - \beta$, with a decrease in standard error.

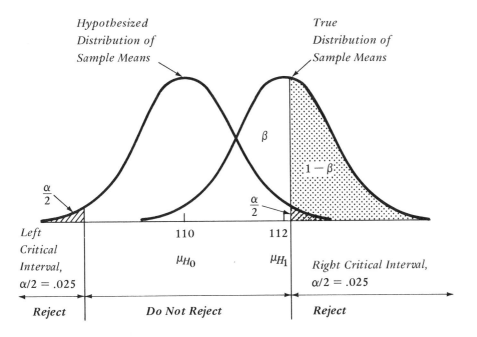

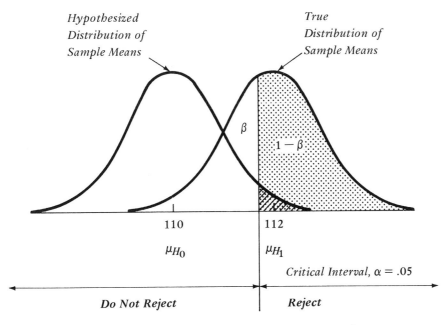

FIGURE 8–13. Illustration of an increase in power, $1 - \beta$, with a change from a two-tailed to a one-tailed test.

TWO CONTROVERSIAL ISSUES

Finally, we will end the consideration of hypothesis testing with a brief discussion of two issues that have been the source of some controversy among psychologists, and are frequently misunderstood by introductory statistics students. These issues are the arbitrariness of the significance level and one- versus two-tailed tests.

The Arbitrariness of the Significance Level

As indicated above, there are some general considerations bearing upon the use of a given significance level. These relate to the relative importance of Type I and Type II errors in the specific context of the research problem. The smaller the significance level, the less the probability of a Type I error (rejecting a true null hypothesis) and the greater the probability of a Type II error (not rejecting a false null hypothesis). Type I errors are most important in a context in which rejecting a true null hypothesis would lead to a considerable waste of time, effort, and/or money. A good example of this type of situation is in consumer preference studies, which are carried out to determine whether or not sufficient demand exists to justify the manufacturing and marketing of a new product—like an Edsel. Thus, a .01 or even a .001 level of significance may be selected. Type II errors, on the other hand, are most important in a context in which not rejecting a false null hypothesis would prevent some important discovery. A good example of this type of situation is in the pilot-testing of an experiment in order to determine whether or not the study of a larger sample of subjects is justified. In this situation a .10 or even a .20 level of significance may be selected.

Thus, many considerations can have a bearing upon the relative importance of Type I and Type II errors. In most psychological research, however, both types of error are considered important. Partly for this reason, a tradition of using the .05 level of significance has developed. But why specifically .05, it may be asked? Are not .03, .04, .06, or .07 just as reasonable? Do not these values also represent reasonable middle values? Yes, most certainly. Tradition has selected .05 possibly for no better reason than that we possess five fingers on each hand. Thus, there is a sense in which the specific .05 level of significance is arbitrary.

To this point we have been discussing determination of the level of significance *before* completion of the test. (The level of significance should be chosen before completion of the test.) Now, however, consider the situation *after* completion of the test. Suppose that a researcher had initially decided to use the .05 level of significance, and that after completion of the test he or she is unable to reject the null hypothesis. Suppose, also, that the z is very close to the critical value of 1.96 required for a two-tailed test at the .05 level, e.g., 1.89. A

z of 1.89 would enable rejection of the null hypothesis at the .06 level. Thus, since the .05 is "arbitrary," why not use the .06 level and reject the null hypothesis? The reason why such a procedure is unacceptable is that it makes it impossible to determine exactly the probability of a Type I error. *If an investigator is always willing to alter the significance level after completion of the test, the stated level of significance does not accurately reflect the probability of committing a Type I error.* The probability of committing a Type I error is partially determined by the experimenter's willingness to alter the level of significance; and the extent of such willingness is unknown. Such willingness may vary from investigator to investigator, or from time to time for the same investigator. Maybe the true probability is .06, or .07, or .08, or .09, or .10—there is no way of knowing for sure. Thus, after completion of the test there is a very important sense in which the level of significance is *not arbitrary*.

The student should appreciate the pressures upon some researchers to reject null hypotheses. Rejected null hypotheses mean accepted alternative or research hypotheses, and thus, hopefully, the advancement of knowledge and understanding. Rejected null hypotheses, however, may also mean accepted publications, tenure, increased salaries, and job offers from prestigious universities.

In view of all this pressure to reject null hypotheses, there is a definite advantage in having a publicly sanctioned level of significance; i.e., a level of significance used by journal editors to help determine the acceptability of research manuscripts for publication. It is certainly true that an investigator may legitimately use the .06 level of significance if this level is specified before completion of the test. However, the use of the .05, instead, would mitigate any doubt regarding the investigator's having altered the stated significance level after completion of the test. There is thus a real advantage in having some publicly sanctioned level of significance. This advantage does not relate as directly to the logic of hypothesis testing as it does to human trust. Thus, given that the level of significance has been publicly set at .05 (rather than at some other level), there is an additional sense in which the .05 level of significance is not arbitrary. This is even true *before* completion of the test.

From the standpoint of publicly establishing the probability of a Type I error, the .05 level of significance is not arbitrary. There is, however, another important probability—a probability that does not bear as directly upon Type I errors and the acceptability of manuscripts for publication as it does upon an experimenter's future research activity. This is the probability of a z equal to or beyond the obtained one. In the above example the probability of $z \geq +1.89$ or ≤ -1.89 is .06. What should an investigator do if he or she obtains a z that intersects an interval having a probability somewhat larger than .05, say .06 or even .10? The hypothesis testing procedure, as described above, indicates that the null hypothesis cannot be rejected. There are, however, additional considerations that this procedure does not include. Two of these considerations are the probability associated with the interval intersected by the obtained z, and a judgment as to the importance of the alternative or research hypothesis. If the probability associated with the obtained z is reasonably small and if the alternative hypothesis is judged important, the experiment should be repeated with modifications designed to increase the power of the test. These modifications might, for example, include an improvement in the reliability of measurement or an increase in the sample size. However, if an investigator chooses not to repeat the experiment, he or she is not justified in complaining about the arbitrariness of the significance level.

One- Versus Two-Tailed Tests

It was previously indicated that two considerations allow for the use of a one-tailed test: (1) reasonable confidence that the true mean differs from the null hypothesis in only one direction and (2) concern about results only if the true mean differs in a particular direction from the null hypothesis. Each of these considerations will be discussed in turn.

The argument for the first consideration can be stated as follows: If an investigator is confident of the direction of an experimental result, the use of a two-tailed test will reduce his or her stated significance level by one-half (e.g., from .05 to .025). Thus, the probability of a Type I error will not be the stated significance level (.05), but half of that value (.025). In order to make the probability of a Type I error agree with the significance level, it is necessary to place the entire critical interval within one tail; i.e., to do a one-tailed test.

A rejoinder to this argument is that we are never very confident or certain about the direction of the true mean. Of course, confidence is highly variable. And in psychological research, unfortunately, there may not be many circumstances in which we are reasonably confident of the direction of the true mean. This is partially because little research effort is devoted to replicating previous findings, and partially because of the lack of well-established theories.

What happens if results are obtained in the opposite direction from a one-tailed test? No matter how extreme the results are, the null hypothesis cannot be rejected. This quite simply results from the fact that the entire critical interval was placed in the "wrong" tail. It is the fear of this occurrence that has led some investigators to avoid the use of one-tailed tests, no matter how confident they are regarding the direction of an anticipated result. If results opposite in direction to a one-tailed test are obtained, the investigator's only recourse is to replicate the experiment.

What would be the situation, however, if results opposite to expectation were obtained—even if a one-tailed test had not been formulated? Given the reasons for expecting the opposite, such results should be viewed with considerable suspicion. Therefore, a replication may be advisable whenever highly unexpected results are obtained. Thus, it is not really the use of a one-tailed test that introduces the possibility of replication; rather, it is the obtaining of highly unexpected results.

Whether or not an investigator replicates an experiment may, of course, depend on his or her judgment as to the importance of the "opposite" results. This takes us to the second consideration that dictates the use of a one-tailed test. If an investigator is only concerned about results in a given direction, it seems only sensible to maximize power by placing the entire critical interval in one tail. It is sometimes felt, however, that investigators should always be interested in results opposite to prediction. From our perspective, though, this argument is fallacious. While there are many circumstances in which results in either direction are important, there are many in which results in only one direction are important. As an example of the former, consider research on the efficacy of some teaching method. Results indicating either that the method increased learning performance or decreased learning performance would certainly be considered important. On the other hand, suppose that an investigator was interested in testing a new theory of social change—a theory that among other things implies that the male undergraduates at the University of North Carolina had shorter hair than the male undergraduates at the University of South Carolina during a certain time period. The investigator really was not interested in hair length, per se, particularly the hair length of UNC and USC male graduates, but was interested in the theory of social change. Thus, the only issue was whether the prediction was or was not confirmed. Under this circumstance it would not matter to the investigator whether hair length was longer at UNC than at USC, or whether there was no difference in hair length at UNC and USC. The prediction would not be confirmed in either case.

A possible reply is that it it always possible that some future investigator may be interested in results opposite to prediction. It is possible, for example, that in the future someone may be interested in the fact that during a certain academic year hair length was longer at UNC than at USC. Yes, this is certainly possible; how probable, though, is another matter. It is, of course, also possible that in the future someone may question the reliance on the .05 level of significance. All an investigator should be required to do is to proceed according to his or her best considered judgment at the time of the investigation.

Finally, we turn again to the matter of human trust. Suppose that an investigator obtains results that enable rejection of the null hypothesis at the one-tailed .05, but not at the two-tailed .05. This would be the case, for example, with a z of 1.89. Should a journal editor accept an investigator's assertion that he or she initially formulated a one-tailed test and did not formulate the test after having seen the results? In our opinion the answer to this question is yes—if the investigator can provide a compelling justification for the use of a one-tailed test. This justification should include reference to one or both of the above considerations relating to confidence of direction of results, or concern with only one direction of results.

HISTORICAL PERSPECTIVE ON THE NORMAL DISTRIBUTION

Abraham De Moivre

The normal distribution, or what was originally referred to as the "curve of error," was independently discovered by several different men. Writing in 1924, Karl Pearson points out that at that time it was fashionable to credit Carl Fredrich Gauss (1777–1855) with the discovery of the normal distribution. To this day the normal distribution is sometimes referred to as the "Gaussian distribution." Pearson points out, however, that Gauss was actually preceded by another mathematician-astronomer, Pierre-Simon, Marquis De Laplace (1749–1827). Then, Pearson further notes that "...in studying De Moivre, I have come across a work that long antedates both Laplace and Gauss" (p. 402).

Abraham De Moivre (1667–1754) published in 1718 *The Doctrines of Chances; or, A Method of Calculating the Probabilities of Events in Play*. This work is a gambler's manual that lays out the arithmetic principles for certain "practical" problems relating to size of wager, and so forth. Subsequent editions shifted to a consideration of life insurance problems and mortality tables. In 1730, De Moivre published *Miscellanea Analytica*, and then three years later, *Approximatio ad Summam Terminorum Binomii a + b^n in Seriem Expansi*. This latter paper of only seven pages is the one discovered by Pearson. Pearson notes that only a few copies of the *Miscellanea Analytica* have the *Approximatio* supplement attached to them. He speculates that it was appended only to those copies of the *Miscellanea* sold three years after the original issue. The supplement was dated November 12, 1733.

November 12, 1733, then, is the birthday of the normal distribution. As the title of the seven page paper suggests, the normal distribution was derived from an expansion of the binomial; i.e., by computing the sampling distribution of r/n when $p = .5$ and n is indefinitely large. An appreciation of the significance of the discovery of the normal distribution can be obtained from the following comment by Helen Walker.

> In this obscure treatise on abstract mathematics, written in Latin nearly two centuries ago, and supposed by its author to have no practical implications outside the realm of games of chance; in this brief supplement now so rare that only two copies have been reported extant; we have the first formulation of the momentous concept of a law of errors. We like to imagine De Moivre sitting at a table with a small group of cronies in some London coffee house, imparting to them the discovery that he had made through his knowledge of the properties of infinite series and his interest in the fortunes of the gambler. And we like to imagine the astonished incredulity with which they would have greeted a prophecy that there would come a time when that theorem would powerfully affect the thinking of the world on all its social problems; when it would enter the schools and shape the policies of educators, and when its aid would be invoked in thousands of investigations in sciences whose very names were then unknown (1929, p. 14).

As Pearson (1924) points out, De Moivre's basic concern was not really to understand games of chance but with a theological problem. De Moivre wanted to show that the apparent disorder in the universe was consistent with an "Original Design." Walker notes that the subsequent editions of De Moivre's *Doctrine of Chances* strove "...to free the subject of probability from the opprobrium attaching to it because of its intimate connection with the art of gambling, and also to establish the theological doctrine of a divine order working through human affairs and exhibiting itself in the regularity of statistical ratios" (1929, p. 17).

Laplace and Gauss

Pierre-Simon, Marquis De Laplace and Carl Frederich Gauss are two of the most illustrious names in the history of mathematics and astronomy. Laplace made important contributions to calculus, differential equations, geodesy, celestial mechanics, and probability. Gauss made contributions to the theory of numbers, algebra, geometry, probability, celestial mechanics, surveying, geodesy, geomagnetism, electromagnetism, mechanics, optics, and the design of scientific equipment. Gauss is considered to have been one of the greatest scientific geniuses ever to have lived. He reputedly learned to calculate before he could talk, and according to one well authenticated story, at the age of 3 corrected an error in his father's wage calculations. One frequently quoted story relates to his formulation, at the age of 8, of the equation for obtaining the sum of the first 100 integers. His teacher had given the class the problem of summing the first 100 integers just to keep them busy. Gauss quickly calculated the answer and laid it on his teacher's desk—much to the teacher's astonishment.

Gauss made numerous important contributions to probability and statistics. He formulated the concept of the standard error of the mean (formula 7-4), developed the theory of least squares, and, of course, the curve of error. His development of least squares had been partially anticipated by Laplace, as had the curve of error. It was Gauss, however, who was responsible for drawing major attention to the curve of error. Gauss reached the curve of error through a consideration of errors in astronomical observations. He assumed that each observation was contributed to by a large number of elementary errors—different physiological conditions of the observer, vibrations of the telescope, variations in the atmosphere, and so forth. He further assumed that each error was independent of all others, and thus regarded each observational error as the algebraic sum of the contributing errors. From this beginning point he was able to demonstrate mathematically that as the number of elementary errors approaches infinity the distribution of observations approaches a normal distribution. It is also historically important that he regarded the mean of the distribution as the value with the greatest probability of being correct.

Quetelet and Galton

Adolphe Quetelet (1796–1874), like Laplace and Gauss, was a mathematician-astronomer. He was the director of the royal observatory at Brussels. Quetelet was influenced by both Laplace and Gauss. After studying probability under Laplace, his attention was increasingly directed toward statistics. In 1829, he visited Gauss at the latter's observatory in Göttingen, Germany.

Quetelet's importance in the history of probability theory flows from his discovery that the "Law of Frequency of Error" worked well for "anthropological" measures such as the weight, height, chest diameter, etc. of soldiers in the Belgian army. Quetelet also had an important influence on the advancement of the discipline of statistics through his correspondence, travel, personal contacts, influence on the collection of census data, and role in the creation of various statistical societies. It was, for example, at his instigation that the first International Statistical Congress met in Brussels in 1853.

Quetelet had a profound influence on Francis Galton—an influence that Galton frequently acknowledged, as illustrated by the following quote, "I need hardly remind the reader that the Law of Error upon which these Normal Values are based, was excogitated for the use of astronomers and others who are concerned with extreme accuracy of measurement, and without the slightest idea until the time of Quetelet that they might be applicable to human measures" (1889, pp. 54–55). Galton, like Quetelet, was fascinated by the possibility of applying the Law of Error to human and social phenomena, and thus, helped to further the development of what has been called the "Normal Mystique." Galton's excitement regarding the normal distribution was, as usual, expressed with a brilliant literary style.

> I know of scarcely anything so apt to impress the imagination as the wonderful form of cosmic order expressed by the "Law of Frequency of Error." The law would have been personified by the Greeks and deified, if they had known of it. It reigns with serenity and in complete self-effacement amidst the wildest confusion. The more huge the mob and the greater the apparent anarchy, the more perfect is its sway. It is the supreme law of Unreason. Whenever a large sample of chaotic elements are taken in hand and marshalled in the order of their magnitude, an unsuspected and most beautiful form of regularity proves to have been latent all along (1889, p. 66).

In his book, *Natural Inheritance*, from which we have been quoting, Galton presents a beautifully simple illustration of how random errors can generate a normal appearing distribution. Figure 8–14 is a reproduction of Galton's Figure 7, which pictures the apparatus. Galton describes it as follows:

> It is a frame glazed in front, leaving a depth of about a quarter of an inch behind the glass. Strips are placed in the upper part to act as a funnel. Below the outlet of the funnel stand a succession of rows of pins stuck squarely into the backboard, and below these again, are a series of vertical compartments. A charge of small shot is inclosed. When the frame is held topsy-turvy, all the shot runs to the upper end; then, when it is turned back into its working position, the desired action commences. Lateral strips, shown in the diagram, have the effect of directing all the shot that had collected at the upper end of the frame to run into the wide mouth of the funnel. The shot passes through the funnel, and issuing from its narrow end, scampers deviously down through the pins in a curious and interesting way; each of them darting a step to the right or left, as the case may be, every time it strikes a pin. The pins are disposed in a quincunx fashion, so that every descending shot strikes against a pin in each successive row. The cascade issuing from the funnel broadens as it descends, and, at length, every shot finds itself caught in a compartment immediately after freeing itself from the last row of pins. The outline of the column of shot that accumulate in the successive compartments approximates to the Curve of Frequency..., and is closely of the same shape, however often

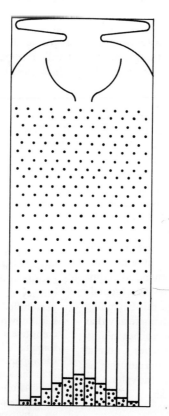

FIGURE 8–14. Galton's mechanical illustration of the "Curve of Error."

the experiment is repeated. The outline of the columns would become more nearly identical with the Normal Curve of Frequency, if the rows of pins were much more numerous, the shot smaller, and the compartments narrower; also, if a larger quantity of shot were used.

The principle on which the action of the apparatus depends is that a number of small and independent accidents befall shot in its career. In rare case, a long run of luck continues to favor the course of a particular shot towards either outside place, but in the large majority of instances the number of accidents that cause Deviation to the right, balance in a greater or less degree those that cause Deviation to the left. Therefore, most of the shot finds its way into the compartments that are situated near to a perpendicular line drawn from the outlet of the funnel, and the Frequency with which shots stray to different distances to the right or left of that line diminishes in a much faster ratio than those distances increase. This illustrates and explains the reason why mediocrity is so common (pp. 63–65).

It is easy to see why subsequent generations, from Pearson on, fell under Galton's spell. As indicated earlier, it was Karl Pearson who introduced the term "normal curve" for the previous term, "curve of error."

EXERCISES

1. Define the following terms:

(a) integration
(b) parameter
(c) standard normal distribution
(d) z score
(e) alternative hypothesis
(f) null hypothesis
(g) significance level
(h) critical interval
(i) two-tailed test
(j) one-tailed test
(k) Type I error
(l) Type II error
(m) power

2. Assuming that z is normally distributed, with $\mu = 0$ and $\sigma = 1$, find the following probabilities:

(a) $z \geq .8$
(b) $z \geq 1.8$
(c) $z \geq .6$ or $\leq .4$
(d) $z \leq .2$
(e) $.8 \leq z \leq 1.2$
(f) $-2.4 \leq z \leq -.2$
(g) $z \leq .4$ or ≥ 2.2

3. Assuming that X and $\bar{X}$ are normally distributed, solve for:

(a) z, when $X = 130$, $\sigma = 30$, $\mu = 90$.
(b) $z_{\bar{x}}$, when $\bar{X} = 20$, $\sigma = 40$, $n = 50$, $\mu = 45$.
(c) $z_{\bar{x}}$, when $\bar{X} = 3$, $\sigma_{\bar{x}} = 1.27$, $\mu_{\bar{x}} = 2.22$.

4. If $\bar{X}$ is normally distributed, solve for z:

(a) when $\sigma = 4$, $\mu = 8$, and $X = 9$. What is the probability that $X \geq 9$?
(b) when $\mu = 4$, $X = 2$, and $\sigma = 3$. What is the probability that $X \geq 2$?
(c) when $\sigma^2 = 16$, $\mu = 5$, and $X = 8$. What is the probability that $X \geq 8$?

5. Assuming the standard normal distribution, test the following hypotheses:

(a) $H_0: \mu \leq 4$ against $H_1: \mu > 4$, given $\bar{X} = 6$, $\sigma = 4$, $n = 9$, and setting α at .01.
(b) $H_0: \mu = 8$ against $H_1: \mu \neq 8$, given $\bar{X} = 12$, $\sigma^2 = 25$, $n = 10$, and setting α at .05.
(c) $H_0: \mu \geq 20$ against $H_1: \mu < 20$, given $\bar{X} = 21$, $\sigma = 13$, $n = 25$, and setting α at .01.
(d) $H_0: \mu \leq 45$ against $H_1: \mu > 45$, given $\bar{X} = 30$, $\sigma = 20$, $n = 35$, and setting α at .01.
(e) $H_0: \mu = 6$ against $H_1: \mu \neq 6$, given $\bar{X} = 7.75$, $\sigma = 10$, $n = 100$, and setting α at .05.

6. A researcher wishes to compare two tests of visual-motor coordination with regard to mean score in a population of first grade school children. Test 1 has been given to the entire population; the mean score is 12 with a standard deviation of 4. Because of limitations of time and money, the researcher decides not to give Test 2 to the entire population, but to a random sample of 100 children. He assumes that both tests yield normally distributed scores, and the standard deviation for Test 2 is the same as the standard deviation for Test 1. The sample mean for Test 2 was 12.96. Can the null hypothesis that the population means for Test 2 equals 12 be rejected? Let $\alpha = .05$.

7. Relate the sampling distribution of proportions to the normal distribution.

8. What does it mean to refer to a "family of normal distributions"?

9. Consider the following formulas:

$$z = \frac{X - \bar{X}}{s}, \quad z = \frac{X - \mu}{\sigma}$$

$$z = \frac{\bar{X} - \mu_{\bar{x}}}{\sigma_{\bar{x}}}, \quad z = \frac{\bar{X} - \mu_{\bar{x}}}{\sigma / \sqrt{n}}$$

How can all four expressions be equated to z?

10. How does the central limit theorem provide a rationale for use of a normal distribution as a theoretical model for the sampling distribution of the mean?

11. The table of proportions of area under the standard normal curve is used as an aid in solving problems regarding the mean. Why is such an aid convenient?

12. List the factors affecting Type II error and power.

13. Why is it important to specify a significance level before, rather than after, completion of a statistical test?

14. Explain the role of the null hypothesis in hypothesis testing.

15. Some individuals have argued against the use of one-tailed tests—preferring to always use two-tailed tests. What basis can you perceive for such a position?

Prologue to Chapter 9

Chapter 9 presents one of the better known statistical tests, the t test. The t test can be used to test for the significance of means resulting from the classical experimental design of one experimental group and one control group. Psychology journals contain numerous reports of t test results. The chapter also introduces the F test for variance. Since variances are seemingly less important than means, there may be a tendency to underestimate the importance of the F test. The F test, when used in conjunction with a technique known as analysis of variance, provides a procedure for obtaining information about means in complex experimental designs that involve more than two groups or conditions. Analysis of variance is introduced in Chapter 10.

9

Inferences Concerning Means and Variances

The normal distribution provides an excellent context within which to introduce the essential aspects of hypothesis testing. This chapter will continue discussion of procedures designed to test hypotheses concerning population means. The discussion of testing hypotheses about a single population mean will be extended, and considerations concerning hypotheses about differences between two population means will be added. Further, there will be a brief discussion of procedures for testing hypotheses concerning the equality of two population variances.

HYPOTHESES CONCERNING A SINGLE POPULATION MEAN

Since Chapter 8 dealt at length with hypotheses about a population mean, why should additional space be devoted to this matter? The reason centers around the difficulty in specifying the standard error of the sampling distribution of means. Knowledge of the standard error is required both to specify one of the parameters of the normal distribution and to transform the sample mean, $\bar{X}$, to a standard normal z score. Formula 8–4 was used in Chapter 8 to transform the obtained value of $\bar{X}$ to a standard normal z score:

$$z = \frac{\bar{X} - \mu}{\sigma / \sqrt{n}}$$

To use this z transformation, it is necessary to know the value of σ, the population standard deviation. Recall that the denominator of formula 8–4 is the standard error of the mean from formula 7–4:

$$\sigma_{\bar{x}} = \frac{\sigma}{\sqrt{n}}$$

Unless the population standard deviation is known or accurately estimated, the standard error of the mean cannot be calculated; and the z formula and standard normal distribution (Appendix C, Table I) cannot be used to test the null hypothesis. Fortunately, it is possible to use another statistic, t, when confronted with this problem.

The t Statistic

The t statistic is defined by formula 9–1:

$$t = \frac{\bar{X} - \mu}{\text{est. } \sigma_{\bar{x}}} \tag{9-1}$$

The formula for t is identical in the numerator to that of z (formula 8–4), but differs in the denominator in that it uses an *estimate* of the standard error of the mean rather than the actual standard error.

It was pointed out in Chapter 4 that the standard deviation of a population distribution is estimated by computing the standard deviation of a sample of scores with $n - 1$ in the denominator:

$$\text{est. } \sigma = \hat{s} = \sqrt{\frac{\sum (X - \bar{X})^2}{n - 1}} \tag{4-5}$$

An estimate of the standard error of the mean is obtained by dividing $\hat{s}$ by the square root of the sample size n:

$$\text{est. } \sigma_{\bar{x}} = \frac{\hat{s}}{\sqrt{n}} \tag{9-2}$$

Since the t statistic is not designed to be descriptive of a sample, as is the sample mean, and its primary value rests with its use in testing hypotheses, it is referred to as a *test statistic*. The majority of statistics with which we will be concerned in the remainder of the book are test statistics.

t Distribution

In Chapter 7, a sampling distribution was defined as a function associating probabilities with all possible values of a sample statistic that could result from samples of a given size (n) drawn from a given population. This concept was illustrated by discussing in detail the sampling distribution of proportions, or of the statistic r/n. In constructing the sampling distribution of r/n, it was first necessary to specify both a population and a sample size, and then to work out all of the possible sample values of r/n. The task was made fairly easy, particularly for small sample sizes, by virtue of the fact that a binomial population only contains two values, 0 and 1.

Just as it is possible to construct a sampling distribution of the r/n statistic, it might also be possible to construct a sampling distribution of the t statistic. The procedure would involve first specifying the population (with its mean, μ, and distribution of scores) and the sample size (or more exactly, the degrees of freedom), and then drawing all possible samples of the specified size from the specified population. In actuality, however, this is not done. Why not? The reason, quite simply, is that in the typical case the distribution of scores for the entire population is not known. Thus, statisticians employ a theoretical model, known as Student's t distribution, which has been worked out with the aid of integral calculus. When certain assumptions are made (assumptions discussed below), the sampling distribution of t follows Student's t distribution, the values and probabilities of which are given in Table II, Appendix C. In subsequent discussion the term t distribution is used to mean sampling distribution of t, and the sampling distribution of t is assumed to follow Student's t distribution.

A t distribution can be compared to the standard normal distribution. When sample size is small, the t distribution is narrower in the center and broader in the tails than the standard normal distribution. Figure 9-1 shows this relationship. As with z, the proportion of area under the t distribution between any two

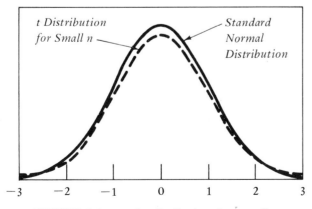

FIGURE 9-1. t and z distributions for a small n.

points on the horizontal axis equals the probability of a t falling in that interval. Thus, one can observe in Fig. 9–1 that the probability of an extreme t is somewhat greater than the probability of an extreme z. That is, for small ns the t distribution has a larger variance than the standard normal distribution. As sample size increases the variance of t decreases and the t distribution changes shape, becoming more closely approximated by a normal distribution. When sample size is infinity (an impossibility, of course), the t distribution is identical to the standard normal distribution.

The decrease in variance and approach to normality make intuitive sense if t and z are examined side by side:

$$z = \frac{\bar{X} - \mu}{\sigma_{\bar{X}}}, \quad t = \frac{\bar{X} - \mu}{\text{est. } \sigma_{\bar{X}}}$$

There is only one variable in the z formula, $\bar{X}$; μ and $\sigma_{\bar{X}}$ are constants. The t formula, on the other hand, depends on two variables, $\bar{X}$ and est. $\sigma_{\bar{X}}$. For any given $\bar{X}$ there is only one corresponding z. However, for a given $\bar{X}$ there are a variety of possible ts depending on the particular sample standard deviation obtained to estimate $\sigma_{\bar{X}}$. In Chapter 7 it was noted that the variability of estimates of the (population) variance and standard deviation decreases as the number of observations on which the estimates are based increases. The estimate of the standard error of the mean (est. $\sigma_{\bar{X}}$) is partially based on the estimate of the (population) standard deviation ($\hat{s}$):

$$\text{est. } \sigma_{\bar{X}} = \frac{\hat{s}}{\sqrt{n}}$$

Thus, when n is small the estimate of the standard error of the mean varies considerably from sample to sample. This means that with a small n the t distribution can also be expected to have a large variance because the variability of t is directly affected by the variability of its denominator, est. $\sigma_{\bar{X}}$. As n increases, however, the variability of the estimate of the standard deviation ($\hat{s}$) decreases and the likelihood increases that $\hat{s}$ will closely approximate the true population standard deviation (σ). Since formula 9–2 for the estimate of the standard error of the mean (est. $\sigma_{\bar{X}}$) and formula 7–4 for the standard error of the mean ($\sigma_{\bar{X}}$) differ only by virtue of $\hat{s}$ and σ, an increasing n results in the increasing likelihood that est. $\sigma_{\bar{X}}$ will closely approximate $\sigma_{\bar{X}}$. Hence the t distribution approaches the normal z distribution as $n \to \infty$.

Table II in Appendix C gives ts that cut off various proportions of a t distribution at the right or upper tail. The specific proportions of area for which ts are given are .10, .05, .025, .01, and .005. The table gives only ts as the upper tail of the distributions because t distributions are symmetric and thus negative ts, corresponding in absolute value to the positive ts in the table, cut off the same proportions of area in the lower tail of the distribution.

As noted above, the precise shape of a t distribution is directly related to the size of the sample from which the estimates of mean and standard error are derived. More specifically, the t distribution depends on the degrees of freedom, *df*, associated with the estimated standard error of the mean. For the estimated standard error of the mean:

$$df = n - 1$$

that is, degrees of freedom equal the sample size minus 1. There is a different *t* distribution for each value of *df*. Each row in Table II, corresponding to a different value of *df*, gives *t*s that cut off certain portions of area of a particular *t* distribution.

The fact that the variance of *t* decreases as sample size increases, and also that as $n \to \infty$, *t* approaches normality, is clearly illustrated in Table II. In any column corresponding to a given proportion of area, the *t*s decrease as *df* increases. For example, when the proportion is .05, the critical value of *t* is 1.812 for *df* equals 10, but 1.671 for *df* equals 60. When *df* equals ∞, the *t*s are identical to *z*s from the standard normal distribution. When *df* equals ∞, the *t* distribution is identical to the standard normal distribution.

Assumptions Underlying Use of *t*

The theoretical *t* distribution is a continuous probability distribution just as is the normal distribution. We use the theoretical *t* distribution as a model to describe the sampling distribution of *t*s that result from actual samples of scores. The theoretical *t* distribution is an accurate model as long as certain conditions are met:

1. The sample set (from which $\bar{X}$ and est. $\sigma_{\bar{x}}$ are computed) is selected by independent random sampling.
2. A normal distribution accurately describes the probability distribution for the population from which the sample set is selected.
3. The population mean is known or can be assumed (e.g., the population mean specified in the null hypothesis).

> A primary reason for the second assumption lies in the requirement of the *t* distribution that the numerator and denominator of *t* be independent. The *t* distribution will provide an accurate model as long as such independence exists. *The mean and variance of a sample are independent if the population (distribution) from which the sample is obtained is normally distributed.* Thus, the numerator and denominator of *t* will be independent if the population distribution is accurately described by a normal probability distribution.

The assumption of normality can actually be violated without severely affecting accuracy of the *t* distribution as a model. As long as sample size is moderate (say 20), slight deviations from normality in the population (e.g., skewness) will not markedly change the accuracy of probability estimates obtained from the theoretical *t* distribution. The more marked the deviations from normality, the larger the sample size necessary to guarantee the accuracy of the theoretical t distribution.

Example Problem: Two-Tailed Test

In the last chapter, the process of testing the hypothesis that the mean IQ for incoming freshmen at a given university differed from 110 was described. The null hypothesis $H_0: \mu = 110$ was tested by sampling 50 incoming freshmen and measuring their IQs. In this example, our hypothetical dean assumed the population distribution standard deviation to be 10. It could be argued that since the school enrollment has increased tenfold and a wider variety of persons now attend the university, there is no good reason to assume that the standard deviation of 10, obtained ten years previously, still holds. Thus, the dean might decide to use a statistic that does not require specification of the population standard deviation. Such a statistic is t.

When the dean sampled 50 incoming freshmen and measured their IQs, the estimate of the standard deviation ($\hat{s}$) was found to be 15, and the sample mean 113.25. The null hypothesis dictates a two-tailed test, and, as before, a significance level (α) of .05 is selected, thus splitting the critical region equally between the two tails of the t distribution.

Figure 9–2 displays the critical regions for this statistical test. Since α is

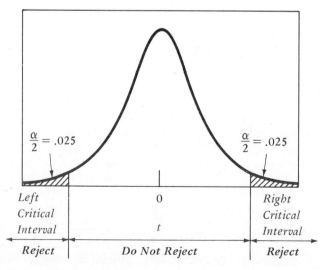

FIGURE 9–2. t distribution with critical regions designated for a two-tailed test where $\alpha = .05$.

TABLE 9.1. Computation of t Value for Sample Data in the Two-Tailed Example Problem.

$$t = \frac{\bar{X} - \mu}{\text{est. } \sigma_{\bar{x}}}$$

$$= \frac{\bar{X} - \mu}{\hat{s} / \sqrt{n}}$$

$$= \frac{113.25 - 110}{15 / \sqrt{50}}$$

$$= \frac{3.25}{2.12}$$

$$= 1.533$$

.05, critical values of t are chosen that cut off .025 of the t distribution in both tails. Referring to Table II, Appendix C, we find that for:

$$df = n - 1$$
$$= 50 - 1$$
$$= 49$$

the critical t at the upper tail of the distribution is between 2.021 (the t for 40 df) and 2.000 (the t for 60 df). Since 49 is between 40 and 60, the desired critical t is between 2.021 and 2.000, or about 2.011. In view of the fact that a t distribution is symmetric, the critical t that cuts off .025 of the lower end of the distribution will be about -2.011. In other words, if the absolute value of t obtained from the sample of 50 exceeds 2.011, it will be concluded that H_0 is not true.

As in Chapter 8, the null hypothesis is tested by assuming that it is true. Computation of t for the example experiment is illustrated in Table 9.1. The obtained t of 1.533 does not exceed the critical value of 2.011 and therefore the null hypothesis is not rejected.

Example Problem: One-Tailed Test

A psychology professor administered a questionnaire measuring degree of self-disclosure to many different college students over a long period of time and found the mean self-disclosure score to be 30. He then had the opportunity to administer the questionnaire to a randomly selected sample of 25 semi-skilled employees from a local factory. The professor's expectation was that the mean self-disclosure score for college students is higher than the mean for semi-skilled workers. In order to check out his expectation, he decided to test the null hypothesis:

$$H_0: \mu \geq 30$$

against the alternative:

$$H_1: \mu < 30$$

that the population mean self-disclosure for the semi-skilled workers is less than 30.

Answers to the questionnaire revealed a sample mean self-disclosure score of 23 ($\bar{X} = 23$) and an estimated population standard deviation of 6.36 ($\hat{s} = 6.36$) for the semi-skilled workers. The null hypothesis dictated a one-tailed test with the critical region in the left, or lower, tail. The left tail was chosen because it was expected that $\bar{X}$ would be less than μ. If $\bar{X}$ is less than μ, the numerator of the t formula (see Table 9.2) would be negative, and the resulting t also negative. As previously indicated, negative ts are in the left tail. The df associated with the test was:

$$df = n - 1$$
$$= 25 - 1$$
$$= 24$$

Since the professor had decided to use a significance level of .01, the critical region of t scores (see Appendix C, Table II; $df = 24$, $\alpha = .01$) was -2.492 to $-\infty$.

Computation of t for this problem is given in Table 9.2. The resultant t, -5.51, falls in the critical region; thus the null hypothesis is rejected in favor of the alternative, that semi-skilled workers have a lower mean self-disclosure score than college students.

TABLE 9.2. Computation of t Value for Sample Data in the One-Tailed Example Problem.

$$t = \frac{\bar{X} - \mu}{\text{est. } \sigma_{\bar{x}}}$$

$$= \frac{\bar{X} - \mu}{\hat{s}/\sqrt{n}}$$

$$= \frac{23 - 30}{6.36/\sqrt{25}}$$

$$= \frac{-7}{1.27}$$

$$= -5.51$$

A reasonable question concerning the calculations in Table 9.2 asks why μ was set equal to 30 when the null hypothesis is that μ is equal to or greater than 30. The null hypothesis does not specify a particular value for μ, but a range of values. Examination of the calculations in Table 9.2, however, should make it apparent that for all values of μ greater than 30 the calculated t will be even more negative than the obtained t of -5.51. Thus, with a μ of 31:

$$t = \frac{-8}{1.27} = -6.30$$

If specification of μ as 30 results in a significant t, the same result will occur for all values of μ that are greater than 30.

Error and Power

Type I errors, Type II errors, and power are relevant concepts for a t test just as they are for a z test. The probability of making a Type I error (the error of rejecting a true null hypothesis) is determined by α, the level of significance. The smaller the α is, the less the probability of a Type I error. The probability of making a Type II error (the error of failing to reject a false null hypothesis or β) is affected by a number of factors: (1) discrepancy between true and hypothesized population means, (2) level of significance (α), (3) population standard deviation (σ), (4) degrees of freedom (df), and (5) one- and two-tailed tests. This was exactly the situation with z except that β was affected by n instead of df. For hypotheses relating to a single mean, df is, of course, $n - 1$.

Power (the probability of rejecting a false null hypothesis) is $1 - \beta$ for t, as for z. The less the probability of making a Type II error is, the greater the power.

HYPOTHESES CONCERNING DIFFERENCES BETWEEN POPULATION MEANS

So far our discussion of hypothesis testing has been restricted to the so-called one-sample case. In the one-sample case a statistic computed from a single sample is compared with a hypothesized population characteristic. For example, a sample proportion (r/n) is compared with a hypothesized population proportion (p); or a sample mean ($\bar{X}$) is compared with a hypothesized population mean (μ).

We will now deal with the two-sample case, or, more specifically, the two-sample case for means. The two-sample case includes the classic experimental design of an experimental condition and a control condition. Subjects sampled are assigned to one or both of the two conditions. In the two-sample case for means, the investigator is interested in determining the difference, if any, between the mean of the population under the experimental condition (μ_1) and the mean of the population under the control condition (μ_2). An inference regarding the difference between μ_1 and μ_2 is made on the basis of the difference between the mean of a sample in the experimental condition ($\bar{X}_1$) and the mean of a sample in the control condition ($\bar{X}_2$).

The two-sample case has two subcases, depending upon the type of sampling employed. Hypotheses concerning the difference between μ_1 and μ_2 may be tested with either independent samples or dependent samples. Independent samples are achieved by randomly assigning each subject to one of the two conditions. Dependent samples are achieved either by assigning every subject to both conditions or by matching the subjects in one condition with subjects in the other condition. Matching may be accomplished by using pairs of subjects with similar IQs, sex, etc. Actually, assigning the same subjects to both conditions can be regarded as a type of matching. Matching will tend to produce a correlation between the scores in the two samples. For that reason dependent samples are sometimes referred to as correlated samples. We will first consider the t test for independent samples and then the t test for dependent samples.

Sampling Distribution of Differences Between Independent Sample Means

Suppose an experimental treatment is applied to one sample of subjects (experimental condition) and not to another sample of subjects (control condition). As indicated above, this is the classic experimental design. Thus, for example, an experimental sample of subjects could be injected with a drug, and a control sample injected with a placebo. The experimenter is interested in comparing the mean of the experimental sample with the mean of the control sample. Although it is typically not done, the experiment could be repeated many times. Theoretically, an experimenter could repeatedly select n subjects and apply the experimental treatment, on a random basis, to one-half of the subjects and the control treatment to the other half. In other words, the experimenter could repeat the experiment over and over again, each time generating a sample mean for the experimental condition and a sample mean for the control condition.

It is possible to derive a sampling distribution that directly refers to the probabilities of various possible differences between sample means. Rather than plotting separate sampling distributions for the experimental and control conditions, a sampling distribution for differences between sample means can be plotted. This is a plot of probabilities of the various possible differences between experimental and control sample means resulting from an infinite number of repetitions of the experiment. Each repetition of the experiment results in a difference score—the mean of the sample in the experimental condition minus the mean of the sample in the control condition.

Mean of Sampling Distribution of Differences Between Independent Sample Means

The mean of the sampling distribution of differences between sample means equals the difference between the population means under the two conditions. If the two conditions are labeled 1 and 2:

$$\mu_1 - \mu_2$$

is the mean of the sampling distribution of differences.

This situation is analogous to the above described one-sample case where the concern was not the sampling distribution of differences but the sampling distribution of the mean. Recall that the mean of the sampling distribution of the mean, $\mu_{\bar{x}}$, is equal to the population mean, μ (formula 7–3).

Normality of Sampling Distribution of Differences Between Independent Sample Means

It turns out that *if two scores comprising a difference are each randomly sampled from a normal distribution, the difference score is also normally distributed.* Thus the difference between two sample means is a normally distributed score if the sample means are normally distributed.

Sampling from normally distributed populations assures that the distributions of sample means will be normal, and thus that the sampling distribution of differences between means will be normal. In addition, we known from the central limit theorem that regardless of the shape of the population distribution, a sampling distribution of means approaches a normal distribution as the sample size increases. Thus, a sampling distribution of differences between sample means also approaches a normal distribution as the sample size for the two conditions increases.

Standard Error of Sampling Distribution of Differences Between Independent Sample Means

The standard error of a sampling distribution of differences between independent sample means is given by formula 9–3:

$$\sigma_{\bar{x}_1 - \bar{x}_2} = \sqrt{\sigma_{\bar{x}_1}^2 + \sigma_{\bar{x}_2}^2} \tag{9-3}$$

If subjects are randomly assigned to two conditions, the standard error of the sampling distribution of differences between pairs of sample means equals the square root of the sum of variances of the sampling distributions for the two conditions.

Squaring both sides of formula 9–3 gives:

$$\sigma^2_{\bar{x}_1 - \bar{x}_2} = \sigma^2_{\bar{x}_1} + \sigma^2_{\bar{x}_2} \tag{9-4}$$

The variance of differences between means equals the sum of the variances of the sampling distributions for the two conditions. It makes sense that the variance of a difference score should be larger than the variance of either of the component scores. If sampling from two distributions is completely independent and random, the two component scores might come from anywhere in their respective distributions. Thus, the difference $\bar{X}_1$ minus $\bar{X}_2$ will be a large positive number if $\bar{X}_1$ is large and $\bar{X}_2$ is small, or a large negative number if $\bar{X}_1$ is small and $\bar{X}_2$ is large.

Estimating the Standard Error of the Sampling Distribution of Differences Between Independent Sample Means

Formula 9–3 for the standard error of the difference between means requires knowledge of the population standard deviation in both of the two conditions. This can be readily seen by making some simple substitutions in formula 9–3. According to formula 7–4:

$$\sigma_{\bar{x}} = \frac{\sigma}{\sqrt{n}}$$

Squaring both sides:

$$\sigma^2_{\bar{x}} = \frac{\sigma^2}{n}$$

Assigning appropriate subscripts and substituting σ^2 / n for $\sigma^2_{\bar{x}}$ in formula 9–3:

$$\sigma_{\bar{x}_1 - \bar{x}_2} = \sqrt{\frac{\sigma^2_1}{n_1} + \frac{\sigma^2_2}{n_2}} \tag{9-5}$$

Since the standard deviations of the populations in either the experimental or control conditions are rarely known, the typical procedure is to estimate the standard error of the difference between sample means. In order to estimate the standard error of the difference, formula 9–6, or its algebraic equivalent, formula 9–7, is used:

$$\text{est. } \sigma_{\bar{x}_1 - \bar{x}_2} = \sqrt{\text{est. } \sigma^2_{\bar{x}_1} + \text{est. } \sigma^2_{\bar{x}_2}} \tag{9-6}$$

$$= \sqrt{\frac{\hat{s}^2_1}{n_1} + \frac{\hat{s}^2_2}{n_2}} \tag{9-7}$$

Formula 9-7 for the estimate of the standard error of the difference differs from formula 9-5 solely in the presence of $\hat{s}_1^2$ and $\hat{s}_2^2$ rather than σ_1^2 and σ_2^2.

Homogeneity of Variance Assumption

Typically, the estimate of standard error of the difference is not calculated by formula 9-7. Usually experimenters assume that the variances of the populations under the two conditions do not differ. This is referred to as the *homogeneity of variance assumption*.

Assuming the same population variance under experimental and control conditions (i.e., σ_1^2 equals σ_2^2), the formula for the standard error of the difference between means can be modified as follows:

$$\sigma_{\bar{x}_1 - \bar{x}_2} = \sqrt{\sigma_{\bar{x}_1}^2 + \sigma_{\bar{x}_2}^2} \tag{9-3}$$

$$= \sqrt{\frac{\sigma_1^2}{n_1} + \frac{\sigma_2^2}{n_2}}$$

$$= \sqrt{\sigma^2 \left(\frac{1}{n_1} + \frac{1}{n_2}\right)} \tag{9-8}$$

In formula 9-8, σ^2 is the common population variance equal to σ_1^2 and to σ_2^2.

By virtue of the homogeneity of variance assumption, formula 9-7 for the estimate of the standard error of the difference can also be modified:

$$\text{est. } \sigma_{\bar{x}_1 - \bar{x}_2} = \sqrt{\frac{\hat{s}_1^2}{n_1} + \frac{\hat{s}_2^2}{n_2}}$$

$$= \sqrt{\hat{s}_p^2 \left(\frac{1}{n_1} + \frac{1}{n_2}\right)} \tag{9-9}$$

In formula 9-9, $\hat{s}_p^2$ is the *pooled estimate of the common population variance*. Assuming homogeneity of variance, $\hat{s}_1^2$ and $\hat{s}_2^2$ are estimates of the same population variance. In such a situation the two estimates can be pooled (or averaged) to obtain a more accurate estimate of the common population variance, $\hat{s}_p^2$.

The following formula can be used to combine the two sample estimates into a single pooled estimate:

$$\hat{s}_p^2 = \frac{(n_1 - 1)\hat{s}_1^2 + (n_2 - 1)\hat{s}_2^2}{n_1 + n_2 - 2} \tag{9-10}$$

Formula 9-10 can then be substituted into formula 9-9 for estimating the standard error of the difference between sample means:

$$\text{est. } \sigma_{\bar{x}_1 - \bar{x}_2} = \sqrt{\frac{(n_1 - 1)\hat{s}_1^2 + (n_2 - 1)\hat{s}_2^2}{n_1 + n_2 - 2} \left(\frac{1}{n_1} + \frac{1}{n_2}\right)} \tag{9-11}$$

The pooled variance estimate has $n_1 + n_2 - 2$ degrees of freedom (df) associated with it.

Formula 9–10 will, perhaps, be more comprehensible if we take a close look at its components. The forms $(n_1 - 1)s_1^2$ and $(n_2 - 1)s_2^2$ are actually the sums of squares for the two samples:

$$(n_1 - 1)s_1^2 = (n_1 - 1)\frac{\sum(X_1 - \bar{X}_1)^2}{n_1 - 1}$$
$$= \sum(X_1 - \bar{X}_1)^2$$
$$= \sum x_1^2$$

Similarly:

$$(n_2 - 1)s_2^2 = \sum x_2^2.$$

Thus, the numerator of formula 9–10 is the sum of two sums of squares, each representing the variability of scores in one of the two conditions.

As was indicated in Chapter 4, the degrees of freedom associated with a sum of squared deviations from the mean equals $n - 1$. (Recall that $n - 1$ of the deviations are free because the sum of deviations from the mean must equal zero.) The degrees of freedom associated with a pooled sum of squares equals the simple sum of degrees of freedom associated with each of the sums of squares in the pool:

$$df_p = df_1 + df_2$$
$$= (n_1 - 1) + (n_2 - 1)$$
$$= n_1 - 1 + n_2 - 1$$
$$= n_1 + n_2 - 2$$

The pooled sum of squares has associated degrees of freedom equal to the total number of scores in the two sample sets minus one degree of freedom for each sample mean. Thus, the denominator of formula 9–10 is the sum of degrees of freedom associated with each of the sums of the squares.

Since the division of sums of squares by appropriate degrees of freedom results in a variance estimate, the division of the sum of the two sums of squares by the sum of the degrees of freedom associated with each of the sums of squares gives a pooled variance estimate. This pooled variance is mathematically equivalent to the weighted mean of the variance estimates for the two samples. When the two variance estimates have the same degrees of freedom they may be simply added together and divided by two. When the two variance estimates have different degrees of freedom they must be differently weighted in a manner analogous to that specified in formula 3–7 for the mean of a set of scores from the mean of its subsets.

t for Differences Between Independent Sample Means

We now have the elements necessary to construct a t for differences between independent sample means. The relevant t is given in formula 9–12:

$$t = \frac{(\bar{X}_1 - \bar{X}_2) - (\mu_1 - \mu_2)}{\text{est. } \sigma_{\bar{x}_1 - \bar{x}_2}} \qquad (9\text{--}12)$$

Substituting formula 9–11:

$$t = \frac{(\bar{X}_1 - \bar{X}_2) - (\mu_1 - \mu_2)}{\sqrt{\frac{(n_1 - 1)\hat{s}_1^2 + (n_2 - 1)\hat{s}_2^2}{n_1 + n_2 - 2}\left(\frac{1}{n_1} + \frac{1}{n_2}\right)}} \qquad (9\text{–}13)$$

The t defined in formula 9–12 fits the definition of t given in formula 9–1:

$$t = \frac{\bar{X} - \mu}{\text{est. } \sigma_{\bar{x}}} \qquad (9\text{–}1)$$

A t is composed of a normally distributed score minus the mean of the distribution from which it is drawn divided by an estimate of the standard deviation of the distribution. This is the case for both formula 9–1 and formula 9–12 (and 9–13). The difference between formulas 9–1 and 9–12 is due to the fact that they were designed for two different sampling distributions. In the case of formula 9–12, the relevant sampling distribution is the sampling distribution of differences between independent sample means. In the case of formula 9–1, the relevant sampling distribution is the sampling distribution of the mean. Consider first the numerators of the two formulas. In formula 9–12, we have a normally distributed score from the sampling distribution of differences, $\bar{X}_1 - \bar{X}_2$, minus the mean of the sampling distribution of differences, $\mu_1 - \mu_2$. In formula 9–1, we have a normally distributed score from the sampling distribution of the mean, $\bar{X}$, minus the mean of the sampling distribution of the mean, μ. Now, examine the denominators of the two formulas. In both cases the denominators are estimates of the standard error of a sampling distribution. For formula 9–12, the distribution is the sampling distribution of the difference; for formula 9–1, the distribution is the sampling distribution of the mean.

Testing Hypotheses Concerning Differences Between Independent Sample Means

Suppose a researcher applies an experimental treatment to one sample of subjects and not to another. With the obtained scores on the dependent variable he computes a mean for each sample and takes the difference between sample means. He then asks the question, "Can it be concluded that the population mean under the experimental condition is different from the population mean under the control condition?" To answer this question he uses the hypothesis testing procedures discussed in Chapter 8 and (since the standard error must be estimated) the t distribution discussed in the present chapter.

ASSUMPTIONS. As noted above, certain assumptions are necessary before it can be assumed that the theoretical t distribution provides an adequate model for the sampling distribution of the t statistic. These assumptions are:

1. The sample sets are selected by independent random sampling.
2. A normal distribution accurately describes the population distributions.
3. The difference between population means, μ_1 minus μ_2, is known or can be assumed (e.g., assumed as specified in the null hypothesis).
4. The population distributions have equal variances, σ_1^2 equals σ_2^2 (homogeneity of variance assumption).

> The major reason for the second assumption lies in the fact that the theoretical t distribution is accurate only if the numerator and denominator of the t are independent. As discussed earlier, only if a population is normally distributed are the sample mean and variance independent. Thus, only if the populations are normally distributed are the numerator and denominator of t for differences between sample means independent.

HYPOTHESIS TESTING. Having made the necessary assumptions for the use of t, hypothesis testing procedures require that a number of steps be taken before any data are collected.

1. The experimenter first decides upon an alternative hypothesis. The alternative hypothesis will be either two-tailed ($\mu_1 \neq \mu_2$) or one-tailed ($\mu_1 > \mu_2$ or $\mu_2 > \mu_1$).
2. The experimenter states the null hypothesis that is to be tested. (The null hypothesis is the logical contradictory of the alternative hypothesis.)
3. An α level is chosen.
4. The critical region(s) is determined. In the case of a one-tailed test, the critical region is placed in one tail, and in the case of a two-tailed test, the critical region is divided (evenly in the typical case) between the two tails.

Example Problem: One-Tailed Test

Suppose a psychologist is interested in the effects of food deprivation on the amount of food intake during a subsequent time period. He wants to examine, first of all, whether a very short period of food deprivation increases the mean food intake by rats during a subsequent but equivalent test period. Thus, he randomly selects two sample sets of ten rats each from his rat colony and randomly assigns the experimental treatment of two hours food deprivation to one sample set and the control treatment of zero hours food deprivation to the other sample set. The rats in the two hours food deprivation condition go two hours without food, and the rats in the zero hours food deprivation condition go zero hours without food.

Prior to performing the experimental manipulations, he decides upon an alternative hypothesis stating the expectation that two hours food deprivation will increase subsequent food intake:

$$H_1: \mu_1 > \mu_2$$

that is, the population mean food intake under the condition of two hours deprivation, μ_1, is greater than the population mean food intake under the condition of zero hours food deprivation, μ_2. The null hypothesis, the logical contradictory of H_1, is stated:

$$H_0: \mu_1 \leq \mu_2$$

that is, the population mean food intake under the condition of two hours deprivation, μ_1, is equal to or less than the population mean food intake under the condition of zero hours food deprivation, μ_2. These hypotheses dictate a one-tailed test.

Finally the experimenter chooses an α level of .05, and specifies the critical region in the upper tail of the t distribution. This means that the critical value of t cuts off .05 of the t distribution in the upper tail. The upper, or right, tail is chosen since an $\bar{X}_1$ larger than an $\bar{X}_2$ would result in a positive value for $\bar{X}_1 - \bar{X}_2$ in the numerator of t (formula 9-13); and hence, a positive rather than a negative t. Since the degrees of freedom associated with a t for differences between sample means equals $n_1 + n_2 - 2$, the df for this experiment is $10 + 10 - 2 = 18$. Referring to Table II, the critical t value for a one-tailed test at $\alpha = .05$ and 18 df is 1.734.

After completing the experimental and control procedures, the experimenter measures the grams of food intake for each rat during an immediately subsequent two-hour period. Table 9.3 displays the scores obtained for each rat

TABLE 9.3. Grams of Food Intake for Experimental and Control Rats.

TWO HOURS DEPRIVATION		ZERO HOURS DEPRIVATION	
X_1	X_1^2	X_2	X_2^2
2.0	4.00	1.3	1.69
3.1	9.61	0.9	0.81
2.6	6.76	2.3	5.29
2.8	7.84	1.5	2.25
4.0	16.00	1.7	2.89
3.3	10.89	0.7	0.49
2.3	5.29	2.5	6.25
1.9	3.61	1.6	2.56
3.5	12.25	1.4	1.96
3.2	10.24	2.1	4.41
$\Sigma = 28.7$	$\Sigma = 86.49$	$\Sigma = 16.0$	$\Sigma = 28.60$
$n_1 = 10$		$n_2 = 10$	
$\bar{X}_1 = 28.7/10$		$\bar{X}_2 = 16.0/10$	
$= 2.87$		$= 1.60$	

and the calculations of the mean for each condition. Table 9.4 contains the calculations leading to an estimate of the standard error of the difference between sample means according to formula 9–11. The resultant estimate serves as the denominator of the t (formula 9–12) used to test the null hypothesis.

Assuming the null hypothesis to be true (i.e., $\mu_1 \leq \mu_2$), the t for differences between sample means can be calculated. The null hypothesis is tested by assuming that the difference between population means and thus the mean of the sampling distribution of differences between means equals zero, i.e., $\mu_1 = \mu_2$ or $\mu_1 - \mu_2 = 0$. In a one-tailed test, if the resultant t leads to rejection of the hypothesis of no difference, $\mu_1 - \mu_2 = 0$, then all other values for the difference between population means included in the null hypothesis, i.e., $\mu_1 - \mu_2 < 0$, would also be rejected at the same α level. The values for $\bar{X}_1$ and $\bar{X}_2$ (2.87 and 1.60) were calculated in Table 9.3, and the value for est. $\sigma_{\bar{x}_1 - \bar{x}_2}$ (.2813) was calculated in Table 9.4. Table 9.5 illustrates the calculation of t for the above experiment. Since the resultant t, 4.5147, exceeds the critical t value of 1.734, the

TABLE 9.4. *Calculation of Estimate of Standard Error of the Difference Between Means.*

$$\hat{s}_1^2 = \frac{\sum (X_1 - \bar{X}_1)^2}{n_1 - 1} \qquad \hat{s}_2^2 = \frac{\sum (X_2 - \bar{X}_2)^2}{n_2 - 1}$$

$$= \frac{n_1 \sum X_1^2 - (\sum X_1)^2}{n_1(n_1 - 1)} \qquad = \frac{n_2 \sum X_2^2 - (\sum X_2)^2}{n_2(n_2 - 1)}$$

$$= \frac{10(86.49) - (28.7)^2}{10(10 - 1)} \qquad = \frac{10(28.60) - (16.00)^2}{10(10 - 1)}$$

$$= \frac{864.9 - 823.7}{90} \qquad = \frac{286.00 - 256.00}{90}$$

$$= .458 \qquad = .333$$

$$\text{est. } \sigma_{\bar{x}_1 - \bar{x}_2} = \sqrt{\frac{(n_1 - 1)\hat{s}_1^2 + (n_2 - 1)\hat{s}_2^2}{n_1 + n_2 - 2} \left(\frac{1}{n_1} + \frac{1}{n_2}\right)}$$

$$= \sqrt{\frac{9(.458) + 9(.333)}{18} \left(\frac{2}{10}\right)}$$

$$= \sqrt{\frac{(4.12 + 3.00)(2)}{180}}$$

$$= \sqrt{\frac{14.24}{180}}$$

$$= \sqrt{.07911}$$

$$= .2813$$

TABLE 9.5. *Calculation of t for Food Deprivation Problem.*

$$t = \frac{(\bar{X}_1 - \bar{X}_2) - (\mu_1 - \mu_2)}{\text{est. } \sigma_{\bar{x}_1 - \bar{x}_2}}$$

$$= \frac{(\bar{X}_1 - \bar{X}_2) - 0}{\text{est. } \sigma_{\bar{x}_1 - \bar{x}_2}}$$

$$= \frac{(2.87 - 1.60) - 0}{.2813}$$

$$= \frac{1.27}{.2813}$$

$$= 4.5147$$

experimenter concludes that the null hypothesis is not true and that two hours food deprivation does increase mean food intake in the population distribution. The null hypothesis is rejected because, if it were to be true, the obtained outcome would fall into a very unlikely interval of scores.

Example Problem: Two-Tailed Test

Some of the data from the Stroebe *et al.* experiment can be used to illustrate a two-tailed test. Recall that in this experiment subjects from the University of North Carolina initially spent some time examining materials regarding another University of North Carolina student. These materials included a picture of the other person and an attitude questionnaire supposedly filled out by this person. For female subjects the other person was male; and for male subjects, the other person was female. In the high-similarity condition, the questionnaire responses indicated that the other person had attitudes that were highly similar to those of the subjects toward such things as black power and religion. In the low-similarity condition, the questionnaire responses indicated that the other person had attitudes that were highly dissimilar to those of the subjects. The experimenter had, in fact, individually tailored each questionnaire so as to produce the desired degree of similarity or dissimilarity. Subjects were told that the experiment was concerned with people's ability to make judgments about others on the basis of limited information. After examining the picture and attitude questionnaire, the subjects made a series of judgments regarding the other person's intelligence, knowledge of current events, and so forth. These judgments were followed by some measurements of various aspects of interpersonal attraction. The most direct of these questions concerned liking for the other person. The exact wording of this question and the seven alternative responses are contained in Table 1.1. Numbers are assigned to the seven alternatives so that 7 indicates the highest degree of liking, and 1, the lowest degree of liking.

In a situation such as the present one, common sense has contradictory implications. Thus, common sense indicates that "birds of a feather flock together," and also that "opposites attract"; i.e., that greater attraction may result from high similarity or from low similarity. In view of their interest in either result, Stroebe et al., thus, formulated the alternative hypothesis:

$$H_1: \mu_1 \neq \mu_2$$

that the population mean liking in the high-similarity condition, μ_1, is different from the population mean liking in the low-similarity condition, μ_2. The null hypothesis is the logical contradictory of H_1:

$$H_0: \mu_1 = \mu_2$$

namely, that the two means are equal. Stroebe et al. specified α at .05, which set the critical ts at values that cut off .025 of the t distribution in each tail. Since each condition contained 60 subjects, the df for a t test of differences between means was:

$$df = n_1 + n_2 - 2$$
$$= 60 + 60 - 2$$
$$= 118$$

and the critical values (Table II, Appendix C) were approximately $+1.98$ and -1.98.

The liking data for the high-similarity condition are contained in Table 3.1; and the liking data for the low-similarity condition, in Table 3.2. These tables also indicate that the means for high similarity and low similarity were 5.70 and 3.27, respectively; i.e., $\bar{X}_1 = 5.70$, and $\bar{X}_2 = 3.27$. The sample variances for high and low similarity were 1.26 and 1.86, respectively; i.e., $\hat{s}_1^2 = 1.26$ and $\hat{s}_2^2 = 1.86$. The calculation of $\hat{s}_1^2$ is contained in Table 4.5, and the calculation of $\hat{s}_2^2$ was asked for in Exercise 10 of Chapter 4. Assuming H_0 to be true, the t was computed using formula 9–13:

$$t = \frac{(\bar{X}_1 - \bar{X}_2) - (\mu_1 - \mu_2)}{\sqrt{\frac{(n_1 - 1)\hat{s}_1^2 + (n_2 - 1)\hat{s}_2^2}{n_1 + n_2 - 2} \left(\frac{1}{n_1} + \frac{1}{n_2}\right)}}$$

$$= \frac{5.70 - 3.27 - 0}{\sqrt{\frac{(60 - 1)1.26 + (60 - 1)1.86}{60 + 60 - 2} \left(\frac{1}{60} + \frac{1}{60}\right)}}$$

$$= 10.66$$

Since the obtained t of 10.66 fell in the critical region beyond 1.98, H_0 was rejected in favor of H_1. Further, the most reasonable assumption, given the direction of the outcome, is that high similarity leads to greater liking than does low similarity.

Effects of Violating Assumptions

As previously mentioned, certain assumptions are routinely made prior to using a t test for differences between means. The researcher is rarely confident, however, that these assumptions are true. It is possible to question, for example, whether the population distributions are well approximated by theoretical normal distributions or whether samples are drawn from populations having equal variances. What effect does the violation of these assumptions have on the outcome of hypothesis testing? The results of t tests are not greatly affected by deviations of population distributions from normality. Provided sample sizes are moderately large (say n_1 and $n_2 = 10$) and deviations from normality are not extreme (e.g., markedly skewed, bimodal, etc.) conclusions reached by assuming normality and using the theoretical t distribution will usually be correct. The more extreme the deviations from normality, the larger the sample sizes should be.

The t test is also little affected by deviations from homogeneity of population variances. Populations from which samples are drawn may have variances differing by a ratio as high as 4:1, and yet the t distribution will yield quite accurate probability estimates for t values in extreme intervals, provided the sample sizes are of at least moderate size (8 to 10) and equal ($n_1 = n_2$). If the investigator suspects that the two population variances are different, he may use the t test but should use samples of moderate and equal size.

> The assumption regarding independent random sampling is really two assumptions: random selection of subjects from a population of subjects and random assignment of subjects to treatment conditions. It is the random assignment of subjects to conditions that assures independence of the two conditions. Usually, there is no problem in randomly assigning subjects to conditions. However, there typically is a problem in randomly selecting subjects from a specified population. In fact, such random assignment typically is not done. For example, although an investigator may randomly assign rats to conditions, the rats are not randomly selected from the population of all rats, or even the population of rats of the same breed. A similar situation exists with human research. Although human subjects may be randomly assigned to conditions, subjects typically are not randomly selected from the population of concern. Random selection from a population may occur if the population of concern is narrowly defined (for example, the population of sixth grade students in a given school district). In most theoretically oriented research, however, the population of concern consists of all people—or at least of all "normal" adults. A moment's thought will make it apparent that random sampling

from such a population is impossible. It is very difficult to draw a random sample of just U.S. adults, much less a random sample of adults from all countries. And assuming this could be done, how does one sample from those who are dead or as yet unborn?

What is the effect of violating the assumption of random sampling? Violating this assumption undoubtedly has effects ranging all the way from trivial to extreme. It is reasonable to suppose that the exact effect of non-random sampling will depend partially upon the extent of variation in the properties of subjects that influence the magnitude of the difference between treatment conditions. Beyond this, however, little is known. There is reason to believe that, if the problem of induction can be solved, the solution will revolve around the development of generalizations regarding the amount of bias introduced through the use of non-random samples. Finally, it should be pointed out that the more developed sciences have progressed without the use of random samples. Thus, a given result in one laboratory may be checked at a later time, in a different laboratory, with a different non-random sample. Clearly, it is possible to build a science with non-random samples.

Turning to the matter of independence between treatment conditions, we find that in some situations an investigator may, for reasons of increased experimental control, introduce dependence between observations. He or she could observe the same subject under both treatment conditions, or select pairs of subjects matched on a variable that is thought to be related to the dependent variable and then assign one subject from each pair to each treatment group. The advantage of repeated observations of the same subjects or matching subjects is the reduction of variability in scores due to sources other than the experimental conditions. When either of these procedures is used, observations in the two groups are expected to be positively correlated. Thus, the assumption of independence among observations is violated.

Testing Hypotheses Concerning Differences Between Dependent Sample Means

When dependence is systematically created by repeated measures either of the same subjects or of matched pairs of subjects, the t test for differences between independent sample means should not be used. However, provided subjects or subject pairs have been randomly selected, a t test for a single mean can be employed to test the $H_0: \mu_1 = \mu_2$. This is accomplished by calculating difference scores—each score being the difference between a given subject's score in condition one and the score of the matched or same subject in condition two. Thus, the two-sample case for dependent samples is reduced to a one-sample case in which the scores of interest are differences. Aside from the fact that we are working with difference scores, the procedures are in all respects similar to the one-sample case described earlier.

t FORMULA. Formula 9-1 for the one-sample *t*:

$$t = \frac{\bar{X} - \mu}{\text{est. } \sigma_{\bar{x}}}$$

can be used for the two-sample case with dependent observations if the various operations in the formula are performed on difference scores, *d*s, rather than upon *X* scores. In order not to confuse *d*s and *X*s, however, it is preferable to rewrite formula 9-1 with appropriate subscripts:

$$t = \frac{\bar{X}_d - \mu_d}{\text{est. } \sigma_{\bar{x}_d}} \quad (9\text{-}14)$$

Degrees of freedom for this *t* are the number of differences (n_d) minus one. In formula 9-14, $\bar{X}_d$ is the mean of the sample of difference scores, μ_d is the hypothesized mean of the population of difference scores, and est. $\sigma_{\bar{x}_d}$ is the estimated standard error of the mean of differences. The estimated standard error of the mean of differences can be calculated by formula 9-15:

$$\text{est. } \sigma_{\bar{x}_d} = \frac{\hat{s}_d}{\sqrt{n_d}} \quad (9\text{-}15)$$

The estimate of the standard deviation of difference scores ($\hat{s}_d$) in the numerator of formula 9-15 can be calculated by formula 9-16:

$$\hat{s}_d = \sqrt{\frac{n_d \sum X_d^2 - (\sum X_d)^2}{n_d(n_d - 1)}} \quad (9\text{-}16)$$

Formula 9-15 is analogous to formula 9-2 for *X* scores, and formula 9-16 is analogous to formula 4-9 for *X* scores.

AN EXAMPLE PROBLEM. Suppose, once again, that an experimenter is interested in determining the effect of two or zero hours food deprivation upon food intake during a subsequent test period. Further, suppose that the experimenter uses a form of dependent sampling in which every animal in one condition has a litter mate in the other condition.

The alternative hypothesis that the true (or population) mean of the differences is greater than zero is stated:

$$H_1: \mu_d = (\mu_1 - \mu_2) > 0$$

and the logical contradiction of the (one-tailed) alternative hypothesis, the null hypothesis, is stated:

$$H_0: \mu_d = (\mu_1 - \mu_2) \leq 0$$

Finally the α level is set at .01 and the critical region located in the upper tail. Why is the upper, or right tail chosen? If two hours deprivation produces more

TABLE 9.6. *Data for Dependent Samples.*

TWO HOURS DEPRIVATION X_1	ZERO HOURS DEPRIVATION X_2	X_d	X_d^2
2.0	0.3	1.7	2.89
3.1	0.9	2.2	4.84
2.6	0.8	1.8	3.24
2.8	1.1	1.7	2.89
4.0	1.7	2.3	5.29
3.3	1.2	2.1	4.41
2.3	0.5	1.8	3.24
1.9	0.6	1.3	1.69
3.5	1.9	1.6	2.56
3.2	1.3	1.9	3.61
		$\Sigma = 18.40$	$\Sigma = 34.66$

$$\bar{X}_d = \Sigma X_d / n_d = 18.40 / 10 = 1.84$$
$$(\Sigma X_d)^2 = (18.40)^2 = 338.56$$

food intake than zero hours deprivation, the result of subtracting the scores in the zero hours condition from the scores in the two hours condition would be a set of difference scores whose mean, $\bar{X}_d$, is positive. Further, if the mean of the population of difference scores, μ_d, is set equal to zero, $\bar{X}_d - \mu_d$, the numerator of the t formula (formula 9–14) would be positive; and therefore, the calculated t would also be positive.

The data for the experiment are contained in Table 9.6 and the t is calculated in Table 9.7. The obtained t of 20.44 exceeds the critical value of t (for $n_d - 1$ or 9 df) in Appendix C, Table II, 2.821. Thus, the experimenter rejects H_0 and accepts H_1.

Why would an experimenter ever choose dependent rather than independent sampling? The rationale for either matching or repeated testing of the same subjects lies in the reduction of variability between the two conditions due to factors other than the presence or absence of a treatment. For example, in the above experiment the use of litter mates in the two conditions reduces the possibility of the animals in one condition just happening to be bigger eaters than the animals in the other condition. In general, the better the matching procedure utilized, the greater the reduction in chance differences between conditions.

The correlation between scores in the two sample sets is a direct index of the success of the matching procedure. The effect of a correlation between sample sets is a smaller standard error of the difference between sample means. The variability of the difference between pairs of sample means across repeated selection of two samples decreases as the correlation between sample means increases. Pairs of sample means are correlated to the extent that corresponding paired sample sets of scores are correlated. Formula 9–17 gives the standard error of the difference between correlated means:

$$\sigma_{\bar{x}_1 - \bar{x}_2} = \sqrt{\sigma_{\bar{x}_1}^2 + \sigma_{\bar{x}_2}^2 - 2\sigma_{\bar{x}_1}\sigma_{\bar{x}_2} r_{\bar{x}_1 \bar{x}_2}} \qquad (9\text{--}17)$$

The larger $r_{\bar{x}_1 \bar{x}_2}$, the smaller $\sigma_{\bar{x}_1 - \bar{x}_2}$. Thus, matching or repeated measurement procedures, to the extent that they result in a correlation between sample sets of scores and thus sample means, decrease the variability of sample mean differences. This effect has the direct result of decreasing the variability of t, for the difference between sample means is the variable in the numerator of t:

$$t = \frac{(\bar{X}_1 - \bar{X}_2) - (\mu_1 - \mu_2)}{\text{est. } \sigma_{\bar{x}_d}}$$

If the difference $\bar{X}_1 - \bar{X}_2$ varies less across repeated selection of sample pairs, then t will vary less across repeated selections. And a decrease in the standard error of t produces an increase in the power of the t test to detect a true alternative hypothesis.

TABLE 9.7. *Calculation of t for Correlated Samples.*

$$\hat{s}_d^2 = \frac{n_d \sum X_d^2 - (\sum X_d)^2}{n_d(n_d - 1)}$$

$$= \frac{10(34.66) - 338.56}{10(9)}$$

$$= \frac{346.60 - 338.56}{90}$$

$$= \frac{8.04}{90}$$

$$= .09$$

$$\hat{s}_d = .3$$

$$\text{est. } \sigma_{\bar{x}_d} = \frac{\hat{s}_d}{\sqrt{n_d}}$$

$$= \frac{.3}{\sqrt{10}}$$

$$= \frac{.3}{3.16}$$

$$= .09$$

$$t = \frac{\bar{X}_d - \mu_d}{\text{est. } \sigma_{\bar{x}_d}}$$

$$= \frac{1.84 - 0}{.09}$$

$$= \frac{1.84}{.09}$$

$$= 20.44$$

Despite the above advantages, matching and repeated measures procedures are not a panacea. For example, repeated measures of the same organisms under different experimental conditions is practical only if the effects of one experimental treatment do not alter response of the organism to the next treatment. Also, matching procedures increase control and the power of a test only if increased correlation between sample sets actually results. If little or no correlation results then power is lost. This is due to a reduction in degrees of freedom by one half, from $(n_1 + n_2 - 2)$ to $(n_d - 1)$, where $n_d = n_1 = n_2$. A reduction in degrees of freedom increases the standard error of t, thus decreasing power of the t test. For matching to be an effective procedure, the correlation between sample means must sufficiently increase the power of the test to more than compensate for the loss of power resulting from the reduced degrees of freedom.

RECAPITULATION: TAILS AND TESTS

At this point it may be helpful to review a matter that some students find confusing. Typically, students find the calculation of t fairly straightforward, but are sometimes confused when determining whether or not the calculated t will enable rejection of the null hypothesis.

Suppose we use the conventional .05 level of significance and, for purposes of illustration, have 60 degrees of freedom. Consider, first, a two-tailed test. A two-tailed .05 means that one-half of the critical interval (.025) is placed in the upper, or right, tail and one-half of the critical interval (.025) is placed in the lower, or left, tail. Appendix C, Table II for the theoretical t distribution gives only the upper tail. According to Table II, the critical value that cuts off .025 of the upper tail is 2.00 (with 60 df). Since the t distribution is symmetrical, the critical value that cuts off .025 of the lower tail is -2.00. Therefore, in order to reject a two-tailed null hypothesis with 60 df, the calculated t must either be greater than $+2.00$ or less than -2.00. In general, *in order to reject a two-tailed null hypothesis, the absolute value of the calculated t must be greater than the absolute value of the positive and negative critical values.*

Now consider a one-tailed test. Again, suppose that we use the .05 level of significance and have 60 degrees of freedom. With a one-tailed test the entire interval is placed in either the left tail or the right tail. According to Table II, the critical value that cuts off .05 of the upper tail is 1.671 (with 60 df). Since the t distribution is symmetrical, the critical value that cuts off .05 of the lower tail is -1.671. Therefore, in order to reject a one-tailed null hypothesis with 60 df, the calculated t must be greater than $+1.671$ if a positive t is expected, or less than -1.671 if a negative t is expected. In general, *in order to reject a one-tailed null hypothesis, the calculated t must be greater than a positive critical value if a positive t is expected, or less than a negative critical value if a negative t is expected.*

This, then, leaves only the matter of determining whether a positive t

or a negative *t* is expected. In general, *a positive t is expected if a confirmed alternative hypothesis would produce a positive numerator in the appropriate t formula; and a negative t is expected if a confirmed alternative hypothesis would produce a negative numerator in the appropriate t formula.* The calculated *t* will be positive if the numerator is positive and negative if the numerator is negative.

REVIEW OF FORMULAS

Table 9.8 presents a summary of formulas used thus far to test hypotheses about population means and differences between means. The two right squares in the upper row are empty because a discussion of testing hypotheses about the difference between population means when population standard deviations are known has not been included. Although such discussion would be quite straightforward, it has been excluded because population standard deviations are seldom known.

TESTING EQUALITY OF TWO POPULATION VARIANCES

Two types of questions give rise to the need for a test of the null hypothesis stating that two population distributions have equal variances. The first question was raised in the previous section.

TABLE 9.8. *Review of Formulas.*

	ONE-SAMPLE CASE	TWO-SAMPLE CASE	
		DEPENDENT	INDEPENDENT
Pop. S.D. σ Known	$z = \dfrac{\bar{X} - \mu}{\sigma / \sqrt{n}}$		
σ Not Known	$t = \dfrac{\bar{X} - \mu}{\hat{s} / \sqrt{n}}$ $df = n - 1$	$t = \dfrac{\bar{X}_d - \mu_d}{\hat{s}_d / \sqrt{n_d}}$ $df = n_d - 1$	$t = \dfrac{\bar{X}_1 - \bar{X}_2}{\sqrt{\hat{s}_p^2 \left(\dfrac{1}{n_1} + \dfrac{1}{n_2}\right)}}$ $\hat{s}_p^2 = \dfrac{(n_1 - 1)\hat{s}_1^2 + (n_2 - 1)\hat{s}_2^2}{n_1 + n_2 - 2}$ $df = n_1 + n_2 - 2$

1. In the process of using a t distribution to test $H_0: \mu_1 = \mu_2$, it is assumed that the population variances for the two conditions are equal:

$$\sigma_1^2 = \sigma_2^2$$

Provided the proper test is available, this assumption can be stated as a null hypothesis:

$$H_0: \sigma_1^2 = \sigma_2^2$$

which can be tested.

2. Just as knowledge about the effect of experimental treatments on a population mean may be desired, knowledge about the effects of such treatments on a population variance may also be desired. For example, in the food deprivation experiment, it might be hypothesized that two processes—one inhibitory in nature and the other excitatory—are activated during food deprivation. If the inhibitory process dominated, food intake would decrease following deprivation, while if the excitatory process dominated, food intake would increase following deprivation. Provided animals vary as to which process dominates, a larger variance of food intake after deprivation, rather than before, would be expected. Or, consider the problem of humans setting their level of aspiration on a task. It might be predicted that physically handicapped persons would either tend to overestimate grossly their abilities (i.e., set their aspirations high above their performance level) or to underestimate grossly their abilities; while physically normal persons would tend to overestimate just slightly their abilities, or to be quite accurate. This prediction is based on the assumption that physically handicapped persons tend to feel inferior (in spite of being able to perform as well as normal persons on the task under consideration) and react to this feeling by either expecting too little of themselves or denying the feeling and projecting themselves as superior to their actual ability.

The F Distribution

To test a null hypothesis regarding the equality of two population variances, a ratio of estimates of the two population variances, an F statistic, is computed; and then, a theoretical probability distribution called the F distribution (see Table III, Appendix C) is consulted. Just as was the case for t, the sampling distribution of the F statistic follows a theoretical F distribution, provided certain assumptions (specified below) are met. The F statistic is given by formula 9–18:

$$F = \frac{\hat{s}_1^2}{\hat{s}_2^2} \tag{9-18}$$

where $\hat{s}_1^2$ and $\hat{s}_2^2$ are estimates of the same population variance (σ^2).

In subsequent discussion we will refer to a sampling distribution of F as an F distribution, and assume that a theoretical F distribution (Appendix C, Table III) is an accurate model for a particular sampling distribution of F.

The shape of an F distribution depends on the sample sizes from which the two estimates are derived (n_1 and n_2). More accurately, the shape of an F distribution depends on the degrees of freedom associated with each of the two variance estimates. It was previously stated that the degrees of freedom associated with a single sample estimate of a population variance equal the sample size "minus" 1 ($n - 1$). Thus, the particular combination of ($n_1 - 1$) and ($n_2 - 1$) dictates the specific shape of an F distribution.

ASSUMPTIONS. Provided certain assumptions are true, the sampling distribution of F will be closely approximated by a theoretical F distribution. Theoretical F distributions are used as models just as the theoretical normal and t distributions have been used. We refer to the theoretical model, Table III in Appendix C, to obtain information concerning the probability of obtained Fs, and this information allows us to test hypotheses about the equality of two population variances.

If a theoretical F distribution is to be an accurate model, observations should be obtained according to the following assumptions:

1. The sample sets are selected by independent random sampling.
2. A normal distribution accurately describes the population distributions.
3. The populations have equal variances, σ_1^2 equals σ_2^2.

This latter assumption becomes the null hypothesis to be tested.

F TABLE. Values of F that cut off .05 and .01, and also, .025 and .005 of the upper tail of an F distribution, are given in Table III, Appendix C. Since the distribution of F is different for each combination of df, different F values are given for each combination. The appropriate Fs are determined by finding (in Table III) the cell corresponding to the intersection of df for the numerator of F, represented along the top of Table III; and the df for the denominator, represented along the left side of Table III. For example, if df for the numerator is 7, and df for the denominator is 9; then, an F of 3.29 will cut off .05 of the distribution and an F of 5.62 will cut .01 of the distribution.

Determining the critical values for a two-tailed F test is not as straightforward as determining critical values of t or z. F is neither distributed symmetrically nor centered on zero. The critical value of F in the upper tail is not the same distance from the mean of F as the critical value of F in the lower tail. Yet, the values of F in Table III are only values for the upper tail of F. Actually, *the value in the lower tail of F that cuts off a certain percentage of the F distribution in the lower tail is the reciprocal of the value of F that cuts off the same percentage in the upper tail with the two dfs reversed.* (The reciprocal of F is $1 / F$.) Thus, to find the critical values for a two-tailed test at $\alpha = .05$, follow three steps: (1) Determine the dfs in the numerator and denominator of F. (2) Find the F value that cuts off .025 of the upper tail of F. (3) Reverse the dfs for the numerator and denominator and take the reciprocal of the F value that cuts off .025 in the upper tail. For example, if the dfs associated with the numerator and denominator of F are, respectively, 7 and 10, the critical value of F at the upper tail would be some value that cuts off .025 at dfs 7 and 10 (3.95 according to Table III). The critical value of F at the lower tail would be the reciprocal of that value cutting off .025 at the upper tail where the dfs are 10 and 7, respectively. With 10 and 7 dfs, the F that cuts off .025 of the upper tail is 4.76. Taking the reciprocal gives the F for the lower tail:

$$\frac{1}{F_{(10,7)}} = \frac{1}{4.76}$$
$$= .21$$

With a two-tailed test, one should randomly assign one variance estimate to the numerator, and one variance estimate to the denominator. The null hypothesis is rejected if the obtained F is smaller than the critical F in the lower tail (.21 in the example) or larger than the critical F in the upper tail (3.95 in the example).

Example Problem: Two-Tailed Test

We can now use the F distribution to test a null hypothesis regarding two population variances:

$$H_0: \sigma_1^2 = \sigma_2^2$$

The alternative hypothesis in this case would be stated as:

$$H_1: \sigma_1^2 \neq \sigma_2^2$$

Three assumptions were stated above for the use of the theoretical F distribution. The third of these assumptions is the null hypothesis. If the first two assumptions concerning independent random sampling from a normal population hold, and the third assumption is not true; i.e., $\sigma_1^2 \neq \sigma_2^2$, we should expect that either large or small F values will be more probable than expected from the theoretical F distribution. A test of H_0 is, thus, a questioning of the third assumption.

As an example test for $H_0: \sigma_1^2 = \sigma_2^2$, we will use the deprivation versus no deprivation experiment discussed above. The hypothesis is that the population variance of food intake is greater after two hours food deprivation than after no deprivation. However, since it is desirable that the test be sensitive to the possibility of food deprivation actually decreasing the population variance of food intake, the non-directional hypothesis, $H_1: \sigma_1^2 \neq \sigma_2^2$, is chosen as the alternative hypothesis. This selection dictates a two-tailed test. Thus, after the α level is set, say at .05, the critical region is split equally between the two tails of the F

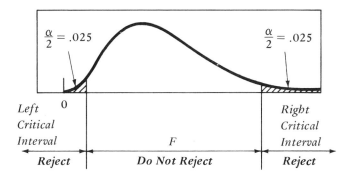

FIGURE 9-3. *Critical intervals for two-tailed F test where* $\alpha = .05$.

distribution, as illustrated in Fig. 9-3. In the present example, the variance estimate for the population under deprivation was randomly assigned to the numerator so that:

$$F = \frac{\text{est. } \sigma^2_{\text{deprivation}}}{\text{est. } \sigma^2_{\text{no deprivation}}} = \frac{s^2_{\text{deprivation}}}{s^2_{\text{no deprivation}}}$$

Since sample sizes for both the deprivation and no deprivation conditions are 10, the df for both numerator and denominator of F is 9 (10 − 1). The critical value of F for the upper tail is obtained by locating the F value (in the cell representing df concurrence 9, 9) that cuts off .025 of the upper tail. This value is 4.03. The critical value for the lower tail is then computed by taking the reciprocal of 4.03, since reversal of dfs also gives 9 for the numerator and 9 for the denominator:

$$F_{\text{lower .025}} = \frac{1}{4.03} = .25$$

Table 9.9 illustrates the computation of F using the variance estimates in Table 9.4. The obtained F of 1.37 does not fall within the critical intervals. It is neither greater than 4.03 nor less than .25. Thus, the experimenter fails to reject the null hypothesis that $\sigma^2_1 = \sigma^2_2$.

Example Problem: One-Tailed Test

If the investigator had been certain that the effects of deprivation, if any, would be to increase and not decrease the variance of food intake, then a one-tailed rather than two-tailed test would have been appropriate. The alternative and null hypotheses in this case would be:

$$H_1: \sigma^2_1 > \sigma^2_2$$
$$H_0: \sigma^2_1 \leq \sigma^2_2$$

TABLE. 9.9. Computation of F for the Two-tailed Example Problem.

$$F = \frac{\text{est. } \sigma^2_{\text{deprivation}}}{\text{est. } \sigma^2_{\text{no deprivation}}}$$

$$= \frac{\hat{s}^2_1}{\hat{s}^2_2}$$

$$= \frac{.458}{.333}$$

$$= 1.37$$

where σ^2_1 represents the variance of food intake under the deprivation condition, and σ^2_2 the variance after no deprivation.

For a one-tailed test, the variance estimate for the treatment predicted (in H_1) to effect the larger variance should be placed in the numerator. In this case the upper tail of the F distribution is the only concern when performing a one-tailed test. Selecting α as .05, the critical value of F with 9 and 9 df is 3.18. Since the obtained F in this problem, 1.37, does not exceed 3.18, the null hypothesis is not rejected with a one-tailed test.

A FURTHER CONTROVERSIAL ISSUE: MAGNITUDE OF EFFECT

Chapter 8 discussed two controversial issues: the arbitrariness of the significance level and one- versus two-tailed tests. After reading the contents of the present chapter, the student is in a better position to appreciate the importance of still a third controversial issue. This issue relates to the importance of the magnitude of an observed effect, or observed difference.

Consider the Chapter 8 example of the dean who over a ten-year period found a significant increase in freshman IQ—an increase from 110 to 113.25. In this instance the size of the effect or, more exactly, the magnitude of the obtained difference between means was 3.25 IQ points. Does a difference this small make a difference? If it does, what about a difference as small as 2 IQ points, or 1 IQ point?

One circumstance having an effect on the meaningfulness of the magnitude of an observed effect, or the variance of an observed effect, is the amount of total observed variability. Thus, a mean difference of 1 IQ point would be more meaningful in a situation in which IQ varied by 3 points than in a situation in which IQ varied by 50 points. It is for this reason that the measures of effect magnitude, to be presented in Chapter 12 (point biserial correlation and correlation ratio), express the variance of an effect relative to the total observed variance.

Historically, the tendency has been for psychologists to assume that a significant effect is necessarily an effect of meaningfully large magnitude. It is the case, however, that exceedingly small differences become significant if the sample size is increased sufficiently. For example, with a sample size of up to 2,000, our hypothetical dean would have found that a difference of only one-half of one (0.50) IQ point would have been significant at the .05 level. It has, in fact, been argued (cf. Bakan, 1966) that the null hypothesis is generally false. If measurement is carried out to enough decimal points, some difference, however small, generally appears; and with a sufficiently large sample, small differences become significant. In actual practice, of course, sample size typically does not exceed several hundred so that the problem may not be as acute as first thought. Nonetheless there is an important theoretical issue. The issue concerns the magnitude of an effect that is required in order to make an effect meaningful. Or, as Bakan (1966, p. 435) stated it: "How much of a difference makes a difference?" As Bakan indicates, there is no one answer to this question. It appears that for applied research in which the interest is in a large change or improvement, the larger the effect, the greater is its importance. This is particularly true if the research is conducted in the field setting for which the new technique or procedure is designed. On the other hand in theoretically oriented research, effects of small magnitude may be of extreme importance. An illustrative example is Madame Curie's discovery of radium through the observation of an exceedingly small effect. In the more developed sciences, such as physics, small effects are frequently of great theoretical importance. However, in less developed research areas there seems to be some basis for regarding effects of large magnitude as more important than effects of small magnitude—although there is reason to avoid being too doctrinaire about this.

One of the reasons for being cautious regarding the greater importance of large rather than small effects relates to the fact that the observed magnitude of an effect is partially dependent on the nature of the particular research situation. With precise measurement and a high degree of control over the various possible sources of error, the total observed variability declines, and the relative amount of variability due to the independent property increases. Also, with a more extreme manipulation of the independent property, a large effect is more likely to occur. For example, relative to a low-shock control, an electric shock of large magnitude is more likely to produce a larger difference in reported pain than is an electric shock of moderate magnitude.

In view of such complexities, the student should readily appreciate the opportunity for controversy and uncertainty. Statistics is a growing, changing field, and change and controversy appear to be sociologically yoked. Chapter 8 discussed two controversial issues. The present chapter has added a third issue. Still other controversial issues could be considered. One of these relating to distribution-free statistical tests as opposed to the classical t and F tests will be discussed in Chapter 13. Another controversial issue will only be alluded to in passing. This relates to the relative merits of the classical hypothesis testing procedure (presented in this book) and the newer Bayesian hypothesis testing procedure. The interested student can find an example of the Bayesian procedure applied to an ESP experiment in an article by Anthony Greenwald in the *Journal of Experimental Social Psychology*, 1975, pp. 180–191. The data (ESP scores) are analyzed with a traditional analysis of variance (as will be discussed in Chapter 11), but the hypothesis testing procedure is non-traditional.

HISTORICAL PERSPECTIVE ON THE t TEST

William Sealy Gosset—"Student"

The first and most important name associated with the t test is that of William Sealy Gosset (1876–1937). Gosset studied mathematics and chemistry and took his degree in natural science at Oxford in 1899. In that same year, he went to work for Arthur Guinness and Sons, Ltd., the brewers, in Dublin, Ireland. At the brewery he confronted an important practical problem. The problem concerned the size of sample needed to make the probability sufficiently large that the sample mean lay within defined limits. (One can readily appreciate the importance of taking small samples in a brewery!) Gosset urged the company to seek mathematical advice; and in 1906, he was sent to study with Karl Pearson at University College, London. In the next few years Gosset made his most important contributions to statistical theory. He, however, remained with Arthur Guinness, Ltd. throughout his life. All of Gossett's theoretical work was directed toward the solution of practical problems arising at the brewery.

Gosset's most famous theoretical paper, "The Probable Error of a Mean," was published in 1908 under the pen name "Student." The problem was to devise a test for means that was feasible with small samples. As Gosset or Student pointed out:

> The usual method of determining the probability that the mean of the population lies within a given distance of the mean is to assume a normal distribution about the mean of the sample with a standard deviation equal to $s/\sqrt{n}$, where s is the standard deviation of the sample, and to use the tables of the probability integral (1908, p. 1).

In other words, at that time the accepted procedure was to use formula 8–4 for z, and Appendix C, Table I for the standard normal distribution. This procedure, which had been worked out a century earlier in the time of Gauss, requires knowledge of σ, the standard deviation of the population. With a sufficiently large sample, the standard deviation of the population could be accurately estimated with s, the standard deviation of the sample. Gosset, however, was concerned with practical situations in which the taking of large samples was not economically feasible. Thus he continues:

> Again, although it is well known that the method of using the normal curve is only trustworthy when the sample is "large," no one has yet told us very clearly where the limit between "large" and "small" samples is to be drawn.
>
> The aim of the present paper is to determine the point at which we may use the tables of the probability integral in judging of the significance of the mean of a series of experiments, and to furnish alternative tables for use when the number of experiments is too few (1908, p. 2).

Gosset then proceeded to do just that.

E. S. Pearson (1938) pointed out that Karl Pearson failed to appreciate the importance of Gosset's work. Karl Pearson was suspicious of all research based on small samples. E. S. Pearson goes on to note, however, that Gosset and Pearson did have mutual respect for each other, and, in fact, got along rather well. Karl Pearson's attitude was one of "humorous protest" in which he chastised "naughty brewers" who took samples with n too small.

Ronal Aylmer Fisher
As previously indicated, R. A. Fisher (1890–1962) succeeded Karl Pearson in the Galton chair of eugenics at University College, London. Fisher developed analysis of variance, the topic treated in Chapter 10.

Unlike Karl Pearson, R. A. Fisher recognized the importance of Gosset's work on small samples. In 1925, Fisher published a paper, "Application of 'Student's' Distribution," in which the t test, as we now know it, was developed.

EXERCISES

1. Define the following terms:
(a) test statistic
(b) Student's t distribution
(c) one-sample case
(d) two-sample case
(e) homogeneity of variance assumption
(f) pooled estimate of the common population variance
(g) dependent sampling

2. Using the t table, find the approximate probabilities of the following outcomes:
(a) $df = 11$, $2.201 \leq t \leq \infty$
(b) $df = 27$, $1.703 \leq t \leq \infty$
(c) $df = 6$, $t \leq -1.440$
(d) $df = 4$, $t \geq 1.919$
(e) $df = 24$, $t \leq 2.313$
(f) $df = 18$, $-\infty \leq t \leq -1.330$

3. Using the t table, find the critical value of t for the following:
(a) $df = 14$, $\alpha = .05$, two-tailed
(b) $df = 40$, $\alpha = .01$, one-tailed, positive t expected
(c) $df = 5$, $\alpha = .001$, one-tailed, positive t expected
(d) $df = 8$, $\alpha = .01$, two-tailed
(e) $df = 13$, $\alpha = .05$, one-tailed, negative t expected
(f) $df = 40$, $a = .02$, two-tailed
(g) $df = 3$, $\alpha = .01$, one-tailed, negative t expected
(h) $df = 20$, $\alpha = .001$, two-tailed

4. Test the following null hypotheses. Assume independent samples for (d), (e), (f), (j), and (k).
(a) $H_0: \mu = 10$
$H_1: \mu \neq 10$
$\bar{X} = 8$, $\hat{s} = 2$, $n = 20$, $\alpha = .05$
(b) $H_0: \mu \leq 40$
$H_1: \mu > 40$
$\bar{X} = 44$, $\hat{s}^2 = 9$, $n = 15$, $\alpha = .01$
(c) $H_0: \mu \geq 5$
$H_1: \mu < 5$
$\bar{X} = 4$, $\hat{s}^2 = 4$, $n = 25$, $\alpha = .05$
(d) $H_0: \mu_1 = \mu_2$
$H_1: \mu_1 \neq \mu_2$
$\bar{X}_1 = 12$, $\bar{X}_2 = 15$, $\hat{s}_1 = 5$, $\hat{s}_2 = 3$, $n_1 = 4$, $n_2 = 6$, $\alpha = .01$
(e) $H_0: \mu_1 \geq \mu_2$
$H_1: \mu_1 < \mu_2$
$\bar{X}_1 = 12$, $\bar{X}_2 = 48$, $\hat{s}_1 = 2$, $\hat{s}_2 = 6$, $n_1 = 10$, $n_2 = 10$, $\alpha = .05$
(f) $H_0: \mu_1 = \mu_2$
$H_1: \mu_1 \neq \mu_2$
$\bar{X}_1 = 3$, $\bar{X}_2 = 2$, $\hat{s}_1^2 = 4$, $\hat{s}_2^2 = 2$, $n_1 = 25$, $n_2 = 36$, $\alpha = .02$
(g) $H_0: \mu_d \geq 0$
$H_1: \mu_d < 0$
$\bar{X}_d = -10$, $\hat{s}_d = 5$, $n_d = 6$, $\alpha = .05$
(h) $H_0: \mu \geq 6$
$H_1: \mu < 6$
$\bar{X} = 5.2$, $\hat{s} = 4$, $n = 16$, $\alpha = .05$
(i) $H_0: \mu = 100$
$H_1: \mu \neq 100$
$\bar{X} = 125$, $\hat{s} = 50$, $n = 10$, $\alpha = .01$
(j) $H_0: \mu_1 = \mu_2$
$H_1: \mu_1 \neq \mu_2$
$\bar{X}_1 = 20$, $\bar{X}_2 = 17$, $\hat{s}_1 = 4$, $\hat{s}_2 = 6$, $n_1 = 10$, $n_2 = 10$, $\alpha = .02$
(k) $H_0: \mu_1 \leq \mu_2$
$H_1: \mu_1 > \mu_2$
$\bar{X}_1 = 4$, $\bar{X}_2 = 2$, $\hat{s}_1 = 1$, $\hat{s}_2 = 5$, $n_1 = 6$, $n_2 = 2$, $\alpha = .01$

(l) $H_0: \mu_d = 0$
$H_1: \mu_d \neq 0$
$\bar{X}_d = 4, \hat{s}_d = 2, n_d = 16, \alpha = .01$

(m) From prior experiments an investigator knows that the mean time to reach agreement in a bargaining task is 45 seconds. He wishes to know whether or not his current sample of bargaining time scores came from a population whose mean is 45 seconds. His current sample has a variance estimate of 25 and a mean of 55, and contains 8 scores. Test the hypothesis that $\mu = 45$ seconds using $\alpha = .05$.

5. Test the following hypotheses:

(a) $H_0: \sigma_1^2 = \sigma_2^2$
$H_1: \sigma_1^2 \neq \sigma_2^2$
$\hat{s}_1^2 = 10, \hat{s}_2^2 = 5, n_1 = 30, n_2 = 20, \alpha = .05$

(b) $H_0: \sigma_1^2 \leq \sigma_2^2$
$H_1: \sigma_1^2 > \sigma_2^2$
$\hat{s}_1^2 = 4, \hat{s}_2^2 = 2, n_1 = 5, n_2 = 5, \alpha = .01$

(c) $H_0: \sigma_1^2 = \sigma_2^2$
$H_1: \sigma_1^2 \neq \sigma_2^2$
$\hat{s}_1^2 = 6, \hat{s}_2^2 = 4, n_1 = 20, n_2 = 5, \alpha = .01$

(d) $H_0: \sigma_1^2 \leq \sigma_2^2$
$H_1: \sigma_1^2 > \sigma_2^2$
$\hat{s}_1^2 = 6, \hat{s}_2^2 = 4, n_1 = 20, n_2 = 5, \alpha = .01$

(e) $H_0: \sigma_1^2 \leq \sigma_2^2$
$H_1: \sigma_1^2 > \sigma_2^2$
$\hat{s}_1^2 = 6, \hat{s}_2^2 = 3, n_1 = 31, n_2 = 25, \alpha = .05$

6. Test $H_0: \mu_1 = \mu_2$ and $H_0: \sigma_1^2 = \sigma_2^2$ ($\alpha = .05$), given the following two independent samples:

T_1	T_2
6	2
8	2
4	3
10	4

7. Under what circumstances is t preferable to z? Discuss the relationship of t to z.

8. What are the assumptions underlying the use of the t test?

9. What are the effects of violating the assumptions of a t test?

10. What are the assumptions underlying the use of the F test?

11. Discuss the differences among the following: one-sample case, two-sample case for dependent samples, two-sample case for independent samples.

12. What determines the probability of making a Type I error?

13. What are the factors affecting the probability of making a Type II error with a t test?

14. What is the sampling distribution of differences between independent sample means?

15. Consider the following formulas:

$$t = \frac{\bar{X} - \mu}{\text{est. } \sigma_{\bar{x}}}, \quad t = \frac{(\bar{X}_1 - \bar{X}_2) - (\mu_1 - \mu_2)}{\text{est. } \sigma_{\bar{x}_1 - \bar{x}_2}},$$

and

$$t = \frac{\bar{X}_d - \mu_d}{\text{est. } \sigma_{\bar{x}_d}}$$

How can all three expressions be equated to t?

Prologue to Chapter 10

In a very real sense Chapter 10 is the culmination of, and pay off for, all of the preceding discussion of statistical inference. After Chapter 10, "its all downhill." You have made it this far; do not give up now.

The purpose of Chapter 10 is to describe a statistical procedure known as analysis of variance. Analysis of variance is a technique for analyzing data obtained from simple or complex experimental designs. Because of this, it is perhaps the most important statistical tool that experimental psychologists have at their disposal. Much of the primary research literature in psychology cannot be understood without some comprehension of analysis of variance.

10

Analysis of Variance

In this chapter, procedures designed to test the effects of two or more experimental treatments will be discussed. Analysis of variance, which incorporates the F statistic introduced in Chapter 9, provides a technique for testing the significance of such multiple effects.

There is an obvious need for a statistical test applicable to the effects of more than two experimental treatments. For example, the deprivation experiment discussed in the last chapter might be expanded to include not only an experimental and control group (i.e., groups receiving no deprivation and two hours deprivation) but also groups receiving four and six hours food deprivation. Other examples of multi-group comparisons are those among several types of psychotherapy or several education techniques. Questions posed by psychologists are seldom limited to the examination of only two treatments or two population sets, or the classical comparison of an experimental group receiving a treatment and a control group receiving no treatment.

One possible procedure for making multiple comparisons involves using t tests for differences between all possible combinations of two means. For example, in a three-treatment experiment, t test comparisons of treatment one with two, one with three, and two with three could be computed.

While such a procedure is not infrequent in psychological literature, there are strong reasons for avoiding it. One reason is that not all of the pair comparisons are independent. Suppose sample means for three treatments were ordered $\bar{X}_1 < \bar{X}_2 < \bar{X}_3$. If t tests showed $\bar{X}_1$ to be significantly less than $\bar{X}_2$, and $\bar{X}_2$ to be significantly less than $\bar{X}_3$, it is almost certain that $\bar{X}_1$ will be significantly less than $\bar{X}_3$. The third comparison is not independent of the first two; in fact, the outcomes of the first two allow one to predict with near certainty the outcome of the third. The probabilities normally used to evaluate ts do not apply to the third comparison.

Because of problems with such multiple t tests, a single overall test of significance is needed. Analysis of variance provides such a test. After a brief overview, we will consider one-factor and then two-factor analysis of variance. One-factor analysis of variance is used for an experiment with one manipulated property; two-factor analysis of variance is used for an experiment with two manipulated properties.

OVERVIEW OF ANALYSIS OF VARIANCE

Before becoming "bogged down" in all of the mathematical details, it may be helpful to give a brief overview of analysis of variance. The basic idea behind analysis of variance is both reasonably simple and intuitively compelling. Because of this, it is unfortunate if the basic idea becomes lost in a string of mathematical formulas.

As an example, consider again the Stroebe et al. experiment. Recall that subjects were assigned to high- and low-similarity conditions prior to the assessment of liking for the anonymous other person. Although this involves only two treatment conditions (groups), analysis of variance is general enough to be applied to any number of treatment conditions. A frequency polygon for the high-similarity condition is pictured in Fig. 2-3, and a frequency polygon for the low-similarity condition is pictured in Fig. 2-4. In the high-similarity condition, the scores range from 3 to 7. The "peak" (mode) of the distribution is at 6. In the low-similarity condition, the scores range from 1 to 7. In this case, the "peak," or mode, is at 3. Examination of the two frequency polygons makes it obvious that the two sample distributions differ.

Although it is apparent that the liking scores in the high-similarity condition tend to be higher than the liking scores in the low-similarity condition, the main concern of Stroebe et al. was not with description of the samples but with making an inference about the population from which the samples were drawn. Stated differently, Stroebe et al. were mainly concerned with drawing a conclusion, that in pre-acquaintance situations high-similarity leads to more liking than does low-similarity.

What are the considerations that affect one's confidence in the validity of the above generalization? Undoubtedly, there are a large number. Analysis of variance, however, focuses upon two. One of these is the variability between conditions or groups, and the other is the variability within conditions or groups.

Without attempting to be overly precise at this point, we can regard between groups variability as a reflection of the extent to which the two groups differ. As pointed out above, examination of Figs. 2-3 and 2-4 makes it apparent that there is a tendency for the two groups to differ. What is it that contributes to between group variability? Why is it that the two groups differ? Since subjects were randomly assigned to conditions, it is possible that the scores differ because of chance. Stated differently, it is possible that the difference between groups arises because of unknown and unspecified causes. Such chance variability is referred to as error. Beyond this, however, it is also possible that the scores differ because the experimental manipulation of similarity had a genuine effect.

Within groups variability is a reflection of the variability of the scores within each of the two groups. Examination of Figs. 2-3 and 2-4 reveals that there is variation of the liking scores within each of the two groups. In the high-similarity condition, the liking scores range from 3 to 7; and in the low-similarity condition, the liking scores range from 1 to 7. The analysis of variance procedure is based on the assumption that within groups variability is solely due to chance, or error. The greater is the variability of scores within each of the two groups, the greater the contribution of chance effects, or the greater the amount of error.

It is apparent that, other things being equal, an increase in error will lead to a decrease in confidence regarding the validity of the above generalization regarding the similarity treatment. As the within groups variability increases, there is a decrease in confidence that a given between groups difference reflects a genuine treatment effect.

Briefly, let us review the argument to this point. First, it has been stated that between groups variability is a reflection of both treatment effects and error effects. Second, it has been stated and/or assumed that within groups variability is a reflection of only error effects. Third, and finally, it has been stated that error decreases confidence in the generalization that similarity has a genuine effect on liking.

Since between groups variability reflects both treatment and error, and since within groups variability reflects only error, observe now that it is reasonable to compare directly these two variabilities, or mathematical indices of these two variabilities. One way of comparing these mathematical indices would be to subtract the mathematical index of within groups variability from the mathematical index of between groups variability; i.e., to compute (*mathematical index of between groups variability*) − (*mathematical index of within group variability*). For certain technical reasons, however, this is not done. Rather, the comparison is made by taking the ratio of the mathematical index of between groups variability to the mathematical index of within groups variability; i.e., by computing (*mathematical index of between groups variability*) / (*mathematical index of within groups variability*).

It is compelling intuitively that the ratio of the mathematical index of between groups variability to the mathematical index of within groups variability bears directly upon one's confidence regarding the genuineness of the treatment effect. The larger this ratio is, the greater the potential confidence. This is the central idea upon which analysis of variance is based.

It is explained below that, within the context of analysis of variance, mathematical indices of variability are referred to as *mean squares*; and that the ratio of the between groups mean square to the within groups mean square results in F, the test statistic introduced in the previous chapter as the ratio of two variance estimates. It should be intuitively apparent that the larger the F, the greater the potential confidence that a given treatment effect is genuine.

ONE-FACTOR ANALYSIS OF VARIANCE

An Example Experiment

Consider an experiment employing four experimental treatments (T_1, T_2, T_3, and T_4), each of which has a different effect on the population mean. More specifically, assume that the treatments result in the following order of population means: $\mu_1 < \mu_2 < \mu_3 < \mu_4$. Figure 10–1 displays the four population distributions of scores that might result from applying four such treatments to a population of organisms. In this case both normality and homogeneity of variance are assumed.

The corresponding sampling distribution of means are displayed in Fig. 10–2. The means of these sampling distributions are identical to their respective population means and the variances are less by a fraction of $1/n_i$; that is:

$$\sigma^2_{\bar{x}_2} = \frac{\sigma^2_2}{n_2}$$

Suppose that four samples of size n were randomly and independently chosen from a population of organisms and one of the four experimental treatments was

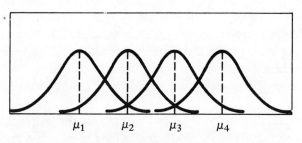

FIGURE 10–1. Four population distributions.

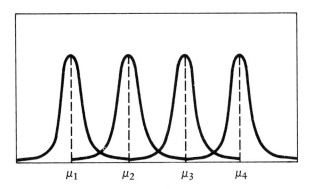

FIGURE 10–2. *Four sampling distributions.*

applied to each sample. This would be equivalent to sampling from four different populations corresponding to the four treatments. Given the ordering of population means, a likely ordering of sample means would be $\bar{X}_1 < \bar{X}_2 < \bar{X}_3 < \bar{X}_4$. Since the four sampling distributions in Fig. 10-2 do overlap, this ordering is not a necessary outcome (it would be possible to obtain $\bar{X}_2 < \bar{X}_1 < \bar{X}_4 < \bar{X}_3$); however, it is a likely outcome.

The four treatments could, for example, represent four levels of food deprivation in rats. Table 10.1 lists food intake scores that might be obtained

TABLE 10.1. *Food Intake Scores After Four Levels of Food Deprivation.*

	DEPRIVATION IN HOURS				
SUBJECTS	0	2	4	6	
1	1.3	2.0	5.5	5.8	
2	0.9	3.1	4.3	6.5	
3	2.3	2.6	6.1	6.7	
4	1.5	2.8	5.7	6.1	
5	1.7	4.0	4.8	7.2	
6	0.7	3.3	5.1	5.9	
7	2.5	2.3	6.0	7.5	
8	1.6	1.9	4.5	6.3	
9	1.4	3.5	5.2	7.1	
10	2.1	3.2	5.6	6.9	
$\sum X$	16.0	28.7	52.8	66.0	163.50
$\sum X^2$	28.60	86.49	282.14	438.60	835.83

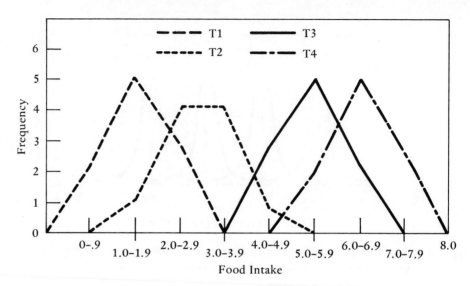

FIGURE 10–3. Grouped sample distributions for deprivation experiment.

from each sample of rats ($n_i = 10$) after application of each of the four treatments. From Table 10.1, it can be seen that the sum of all scores across all four levels or groups is $16.0 + 28.7 + 52.8 + 66.0 = 163.50$, and that the sum of all squared scores is $28.60 + 86.49 + 282.14 + 438.60 = 835.83$. Figure 10–3 displays frequency polygons for each of the four sample distributions.

Sums of Squares

TOTAL SUM OF SQUARES. A sum of squares ($\sum x^2$) is a sum of squared deviations of scores in a particular set from the arithmetic mean of the set:

$$\sum x^2 = \sum (X - \bar{X})^2$$

The expression $\sum x^2$ is also the numerator of $\hat{s}^2$ and s^2. In the context of analysis of variance, a sum of squares is symbolized "SS" rather than "$\sum x^2$." If a sum of squares is calculated for all of the scores in Table 10.1, ignoring the classification of scores into groups, we have the total sum of squares (SS_t). The total sum of squares is defined as:

$$SS_t = \sum (X - \bar{X}_t)^2 \qquad (10\text{–}1)$$

where $\bar{X}_t$ is the *grand mean* or mean for all of the scores.

Formula 10–1 is the definitional formula for the total sum of squares. The computational formula is:

$$SS_t = \sum X^2 - \frac{(\sum X)^2}{n} \qquad (10\text{–}2)$$

where $n = n_1 + n_2 + \cdots + n_k$, or the total number of observations (where k is the number of groups).

For the data in Table 10.1, we have:

$$SS_t = \sum X^2 - \frac{(\sum X)^2}{n}$$

$$= (1.3)^2 + (.9)^2 + \cdots + (6.9)^2 - \frac{(1.3 + .9 + \cdots + 6.9)^2}{40}$$

$$= 835.83 - \frac{(163.5)^2}{40}$$

$$= 167.52$$

The total sum of squares is 167.52.

The total sum of squares is a positive function of the total variability of all the scores. The greater the variability among all of the scores, the greater the total sum of squares.

The total sum of squares is comprised of two components, the *within groups sum of squares* (SS_{WG}) and the *between groups sum of squares* (SS_{BG})*:

$$SS_t = SS_{WG} + SS_{BG} \qquad (10\text{–}3)$$

We will consider these component sums of squares in order.

WITHIN GROUPS SUM OF SQUARES. If the quantity $\sum x^2$ is computed within each group and the results summed across groups, we have the within groups sum of squares:

$$SS_{WG} = SS_1 + SS_2 + \cdots + SS_k \qquad (10\text{–}4)$$

The subscripts, $1, 2, \ldots, k$, refer to each of the k treatment groups.

---

* To be grammatically correct, when there are more than two groups, we should refer to *among* groups rather than *between* groups sum of squares. Statisticians, however, prefer to use "between." We will follow precedent.

One-Factor Analysis of Variance

The within groups sum of squares is computed by calculating:

$$SS_1 = \sum X_1^2 - \frac{(\sum X_1)^2}{n_1}$$

$$SS_2 = \sum X_2^2 - \frac{(\sum X_2)^2}{n_2}$$

$$\vdots \qquad \vdots \qquad \vdots$$

$$SS_k = \sum X_k^2 - \frac{(\sum X_k)^2}{n_k}$$

and summing across the k groups. For the data in Table 10.1:

$$SS_1 = 28.60 - \frac{(16.0)^2}{10} = 3.00$$

$$SS_2 = 86.49 - \frac{(28.7)^2}{10} = 4.12$$

$$SS_3 = 282.14 - \frac{(52.8)^2}{10} = 3.36$$

$$SS_4 = 438.60 - \frac{(66.0)^2}{10} = 3.00$$

Summing these four sums of squares gives 13.48, the within groups sum of squares.

As noted above, the within groups sum of squares is one of the components of the total sum of squares. The other component is the between groups sum of squares (formula 10–6).

It is possible to look at the total sum of squares in still a different way. According to the alternative conception, one contributor to the total sum of squares is the differential effects of the experimental treatments. An additional contributor to the total sum of squares is those uncontrolled variables that produce effects unrelated to the experimental treatments. These variables may include such things as subject type, temperature, time of day, humidity, mood of experimenter, etc. The variability among scores attributed to these uncontrolled and typically unspecified variables is collectively referred to as *error variance*.

The within groups sum of squares is a positive function of error variance and, in some instances, experimental treatments. The greater the amount of error variance, the greater the within groups sum of squares. Since the within groups sum of squares concerns the variation among subjects receiving the same experimental treatment, this sum of squares is usually assumed not to be affected by experimental treatments. If experimental treatments affect the within groups sum of squares, it will be because the treatments differ in their effects on population variability. This would be the case, for example, if six hours deprivation produced more variability in food intake than did two hours deprivation. Analysis of variance procedures assume homogeneity of variance across treatment populations, i.e., that treatments do not differentially affect population variability.

BETWEEN GROUPS SUM OF SQUARES. As the sample distributions in Fig. 10–3 show, scores not only vary around their respective sample means, but also around the central tendency of all sample groups. Part of the total variability is variability between group means.

The *between means sum of squares* ($SS_{\bar{X}}$) is an index of variability among sample means. The sum of squares for sample means is the sum of squared deviations of sample means from the grand mean:

$$SS_{\bar{X}} = (\bar{X}_1 - \bar{X}_t)^2 + (\bar{X}_2 - \bar{X}_t)^2 + \cdots + (\bar{X}_k - \bar{X}_t)^2 \qquad (10\text{–}5)$$

The between means sum of squares is a positive function of the differential effects of experimental treatments and error variance. If experimental treatments do produce changes in a population mean, the group means are expected to vary among themselves and thus from the grand mean. Error variance may also contribute to the between means sum of squares. To the extent that there is error, the group means will vary among themselves and thus vary from the grand mean.

A slight transformation of the between means sum of squares ($SS_{\bar{X}}$) yields a quantity referred to as the *between groups sum of squares* (SS_{BG}). The transformation involves the multiplication of each squared deviation of a sample mean from the grand mean by n_i, the number of subjects comprising each group:

$$SS_{BG} = n_1(\bar{X}_1 - \bar{X}_t)^2 + n_2(\bar{X}_2 - \bar{X}_t)^2 + \cdots + n_k(\bar{X}_k - \bar{X}_t)^2$$
$$= \sum_{i=1}^{i=k} n_i(\bar{X}_i - \bar{X}_t)^2 \qquad (10\text{–}6)$$

Since $\sum_{i=1}^{i=k}$ is somewhat awkward to write, from now on we will simply write $\sum_i$ for the summation across treatment groups. The between groups sum of squares (SS_{BG}) is also a function of the effects of experimental treatments and error variance.

> The rationale for transforming the between means sum of squares to the between groups sum of squares is not at all intuitively obvious. The basic reason for the transformation is that the between groups sum of squares allows one to obtain an estimate of the variance of the population of error effects (when H_0 is true), and the between means sum of squares does not. The mathematical proof of this statement is somewhat complex and will not be presented here.

Formula 10–6 is the definitional formula for the between groups sum of squares. The computational formula is:

$$SS_{BG} = \frac{(\sum X_1)^2}{n_1} + \frac{(\sum X_2)^2}{n_2} + \cdots + \frac{(\sum X_k)^2}{n_k} - \frac{(\sum X)^2}{n}$$

$$= \sum_i \frac{(\sum X_i)^2}{n_i} - \frac{(\sum X)^2}{n} \tag{10-7}$$

In formula 10–7, $(\sum X)^2$ without a subscript refers to the squared sum of scores across all treatment groups, and n without a subscript refers to the total number of scores across all treatment groups.

The algebraic proof of the equivalence of formulas 10–6 and 10–7 follows. Beginning with formula 10–6 and squaring the terms in parentheses:

$$\sum_i n_i(\bar{X}_i - \bar{X}_t)^2 = \sum_i n_i(\bar{X}_i^2 - 2\bar{X}_i\bar{X}_t + \bar{X}_t^2)$$

$$= \sum_i n_i\bar{X}_i^2 - \sum_i n_i 2\bar{X}_i\bar{X}_t + \sum_i n_i\bar{X}_t^2$$

Moving the constants $\bar{X}_t$ and 2 to the other side of $\sum_i$:

$$\sum_i n_i(\bar{X}_i - \bar{X}_t)^2 = \sum_i n_i\bar{X}_i^2 - 2\bar{X}_t \sum_i n_i\bar{X}_i + \bar{X}_t^2 \sum_i n_i$$

Since $\bar{X}_i$ equals $\sum X_i / n_i$ and $\sum n_i$ equals n:

$$\sum_i n_i(\bar{X}_i - \bar{X}_t)^2 = \sum_i n_i \frac{(\sum X_i)^2}{n_i^2} - 2\bar{X}_t \sum_i n_i \frac{(\sum X_i)}{n_i} + n\bar{X}_t^2$$

$$= \sum_i \frac{(\sum X_i)^2}{n_i} - 2\bar{X}_t \sum_i (\sum X_i) + n\bar{X}_t^2$$

Multiplying and dividing the middle term by n and noting that $\sum_i (\sum X_i)$ is the total sum of scores across all groups, $\sum X$:

$$\sum_i n_i(\bar{X}_i - \bar{X}_t)^2 = \sum_i \frac{(\sum X_i)^2}{n_i} - 2n\bar{X}_t \frac{\sum X}{n} + n\bar{X}_t^2$$

and since $\sum X / n$ equals $\bar{X}_t$:

$$\sum_i n_i(\bar{X}_i - \bar{X}_t)^2 = \sum_i \frac{(\sum X_i)^2}{n_i} - 2n\bar{X}_t^2 + n\bar{X}_t^2$$

$$= \sum_i \frac{(\sum X_i)^2}{n_i} - n\bar{X}_t^2$$

$$= \sum_i \frac{(\sum X_i)^2}{n_i} - n\frac{(\sum X)^2}{n^2}$$

$$= \sum_i \frac{(\sum X_i)^2}{n_i} - \frac{(\sum X)^2}{n}$$

For the data in Table 10.1:

$$SS_{BG} = \frac{(16.0)^2}{10} + \frac{(28.7)^2}{10} + \frac{(52.8)^2}{10} + \frac{(66.0)^2}{10} - \frac{(163.5)^2}{40}$$

$$= 822.34 - \frac{(163.5)^2}{40}$$

$$= 154.04$$

The between groups sum of squares is 154.04.

It is possible to use formula 10–3, $SS_t = SS_{WG} + SS_{BG}$, to check our calculations. The total sum of squares, 167.52, should equal the within groups sum of squares, 13.48, plus the between groups sum of squares, 154.04. This, in fact, is the case; 167.52 = 13.48 + 154.04.

Degrees of Freedom and Mean Squares

In Chapter 4, it was pointed out that division of the sum of squares for a sample by the appropriate degrees of freedom resulted in a variance estimate. In the language of analysis of variance, a variance estimate is referred to as a *mean square*. When the between groups sum of squares is divided by its degrees of freedom, the *between groups mean square* is the result. When the within groups sum of squares is divided by its degrees of freedom, the *within groups mean square* is the result. Analysis of variance uses the F statistic, which is a ratio of two variance estimates. Thus it is necessary to transform sums of squares to mean squares.

BETWEEN GROUPS MEAN SQUARE. The between groups sum of squares (SS_{BG}) has $df = k - 1$, where k is the number of groups. Thus, the between groups mean square (MS_{BG}) is defined as:

$$MS_{BG} = \frac{SS_{BG}}{k - 1} \tag{10–8}$$

Table 10.2 summarizes the calculations for the data in Table 10.1. The columns are labeled *Source of Variance, Sum of Squares, df, Mean Square,* and *F*. The

TABLE 10.2. Summary Table for the Deprivation Experiment.

SOURCE OF VARIANCE	SUM OF SQUARES	df	MEAN SQUARE	F
Between Groups	154.04	3	51.3467	137.14
Within Groups	13.48	36	0.3744	
Total	167.52	39		

previously calculated between groups sum of squares and within groups sum of squares are listed in the second column.

The sum of squares between groups has:

$$df = k - 1$$
$$= 4 - 1$$
$$= 3$$

The 3 is shown in the third column of Table 10.2. According to formula 10–8, the between groups mean square is obtained by dividing the between groups sum of squares by the appropriate degrees of freedom ($154.04 \div 3 = 51.3467$), as shown in the fourth column of Table 10.2.

Suppose that the experimental treatments produce no differential effects. This is the same as assuming that the null hypothesis $H_0: \mu_1 = \mu_2 = \cdots = \mu_k$ is true. In this case the mean square between groups reflects only error variance. If we conceive of the population of effects produced by error, then when H_0 is true the between groups mean square is an estimate of population error variance ($\hat{s}_e^2$):

$$MS_{BG} = \frac{SS_{BG}}{k - 1} = \hat{s}_e^2 \qquad (10\text{–}9)$$

WITHIN GROUPS MEAN SQUARE. The within groups sum of squares (SS_{WG}) has $df = n_1 + n_2 + \cdots + n_k - k$, where k is the number of groups. Since one degree of freedom is lost within each group across all groups, degrees of freedom is equal to the total number of subjects minus the number of groups. In the case in which all groups have an equal number of observations, $df = k(n_i - 1)$.

The within groups mean square (MS_{WG}) is defined as:

$$MS_{WG} = \frac{SS_{WG}}{n_1 + n_2 + \cdots + n_k - k} \qquad (10\text{–}10)$$

For the data in Table 10.1:

$$df = n_1 + n_2 + n_3 + n_4 - k$$
$$= 10 + 10 + 10 + 10 - 4$$
$$= 36$$

Since all n_is are equal, it is also the case that:

$$df = k(n_i - 1)$$
$$= 4(10 - 1)$$
$$= 36$$

According to formula 10–10, the within groups mean square is obtained by dividing SS_{WG}, 13.48, by the appropriate df, 36. The result, .3744, is shown in Table 10.2.

If we conceive of the population of effects produced by error, the within groups mean square is an estimate of the variance of that population, $(\hat{s}_e^2)$:

$$MS_{WG} = \frac{SS_{WG}}{n_1 + n_2 + \cdots + n_k - k} = \hat{s}_e^2 \qquad (10\text{–}11)$$

Thus, when H_0 is true, both the between groups mean square and the within groups mean square are estimates of the variance of the population of error effects.

Ratio of MS_{BG} to MS_{WG}

The primary goal of analysis of variance is to determine whether or not a set of experimental treatments produced differential effects. If the between groups mean square (MS_{BG}) were solely a function of the differential effects of experimental treatments, the matter would be simple. In that case, assuming that the absence of experimental treatments would produce no difference between groups, the simultaneous presence of experimental treatments and differences between groups, would directly indicate a treatment effect. Unfortunately, however, the situation is more complicated. The complication is produced by error variance—the between groups mean square is a positive function of both treatment effects and error effects. When H_0 is true (i.e., there are no treatment effects), the between groups mean square is, in fact, an estimate of the variance of the population of error effects.

The between groups mean square has to be transformed in order to yield information regarding treatment effects. The essential ingredient in this transformation is the within groups mean square (MS_{WG}). The between groups mean square is transformed by dividing it by the within groups mean square:

$$\frac{MS_{BG}}{MS_{WG}}$$

The numerator of this ratio is a positive function of treatment effects and error; the denominator is a positive function of error. Thus, the larger the ratio, the greater the probability that treatments did produce a differential effect.

The three graphs in Fig. 10–4 illustrate experimental results that would generate three differing mean square between group to mean square within group ratios. In graph (1), the ratio will be sizeable. Relative to the variability within groups, the separation between groups is quite large. In graph (2), the variability within groups is the same as in graph (1), but the variability between groups is markedly reduced. Thus, the ratio is smaller for the data in graph (2) than for

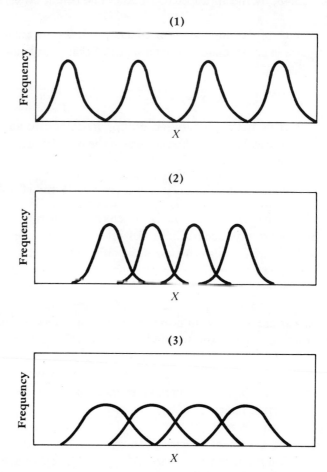

FIGURE 10–4. *Frequency polygons picturing three possible patterns of results for a four-treatment experiment.*

the data in graph (1). The data in graph (3) will generate an even smaller ratio. The variability among groups is the same as in graph (2), but the variability within groups is markedly larger.

Assumptions and F

In Chapter 9, it was stated that, if certain assumptions are made, the ratio of two estimates of the same population variance follows a theoretical F distribution. The between groups mean square and the within groups mean square are variance estimates. Thus, if a number of assumptions are made, the ratio of the two variance estimates, MS_{BG} and MS_{WG}, has an F distribution:

$$F = \frac{MS_{BG}}{MS_{WG}} \qquad (10\text{–}12)$$

The practice is always followed of placing MS_{BG} in the numerator and MS_{WG} in the denominator. The degrees of freedom that determine the particular F distribution are the df associated with MS_{BG} ($k - 1$) and MS_{WG} ($n_1 + n_2 + \cdots + n_k - k$).

The following assumptions are necessary prerequisites for the F computed by formula 10–12 to be considered a value from a theoretical distribution of F for which critical values are given in Table III (Appendix C):

ASSUMP-
TIONS

1. The k sample sets are selected by independent random sampling.
2. A normal distribution accurately describes the populations.
3. The populations have equal means, $\mu_1 = \mu_2 = \ldots = \mu_k$ (i.e., H_0 is true).
4. The k sample variances estimate the same population variance (i.e., homogeneity of variance exists).

In order for the F defined in formula 10–12 to be a value from the theoretical distribution of F represented in Table III, it is necessary that MS_{BG} and MS_{WG} be estimates of the same population variance. The within groups mean square is an estimate of the variance of the population of error effects. When H_0 is true this is also the case for the between groups mean square. Thus, the prerequisite that the two mean squares be estimates of the same population variance requires the third assumption—that $H_0: \mu_1 = \mu_2 = \cdots = \mu_k$ is true.

The fourth, or homogeneity of variance assumption, is necessary if the mean square within groups is to be a function only of error and not of treatment effects.

Hypothesis Testing

We are now in a position to complete the analysis of the deprivation experiment. The alternative hypothesis, which is the logical contradictory of the null hypothesis, is very general:

$$H_1: \mu_1 \neq \mu_2 \quad \text{and/or} \quad \mu_1 \neq \mu_3 \quad \text{and/or} \quad \mu_2 \neq \mu_3, \text{ etc.}$$

Expressed verbally, the alternative hypothesis is simply that *the population mean is not the same under every experimental treatment*. At least one of the treatments must have an effect on the mean that is different from the other treatments. This hypothesis obviously does not specify the direction of the difference between means.

The previously stated null hypothesis is:

$$H_0: \mu_1 = \mu_2 = \cdots = \mu_k$$

Expressed verbally, the null hypothesis is that *all the population means are equal*.

Setting α at .01, the critical interval is placed in the right tail of the F distribution for H_0. In view of the fact that the alternative hypothesis does not specify the direction of the difference between means, why is the critical region placed in just one tail? The reason lies in the fact that F is not sensitive to the direction of differences between treatment means. The greater the plus or minus differences between means, the greater the between groups mean square. Since the between groups mean square is the numerator of the F ratio (formula 10–12), the greater the plus or minus differences between means, the greater is F. In view of F's insensitivity to the direction of differences between means, it is pointless for an investigator to formulate a directional alternative hypothesis. This does not mean, however, that after rejecting the null hypothesis and accepting the logically implied, non-directional alternative hypothesis, the experimenter is not justified in noting which mean is largest and which is smallest.

As shown in Table 10.2, the MS_{BG} has $df = 3$, and the MS_{WG} has $df = 36$. Table III in Appendix C contains the F values cutting off .01 of the upper tail, if H_0 is true. For 3 and 36 df, the F cutting off .01 of the upper tail is 4.38. Computing F by formula 10–12:

$$F = \frac{MS_{BG}}{MS_{WG}}$$

$$= \frac{51.3467}{0.3744}$$

$$= 137.14$$

The F of 137.14, which is recorded in the last column of Table 10.2, is larger than the critical F of 4.38. Thus, the null hypothesis is rejected and the alternative hypothesis is accepted. Hours of food deprivation did affect the mean amount of food intake in the population. From Table 10.1, the investigator may further note that the greatest amount of food intake occurred with six hours of deprivation and the least amount with zero hours of deprivation.

A Further Example of One-Factor Analysis of Variance

Since the previous example of one-factor analysis of variance was interspersed with discussion of the rationale for the various steps, it will be helpful to give a second example in which the calculational procedures are all gathered together. For this example we can return again to the Stroebe *et al.* experiment. Up to this point the Stroebe *et al.* experiment has been described as a study of the effect of high and low similarity upon attraction. It was explained that subjects received pictures of other people and questionnaires supposedly filled out by these people. The questionnaires were, in fact, individually constructed for each subject so as to create either high or low similarity. For present purposes, however, the similarity manipulation can be ignored, and a previously undescribed manipulation of physical attractiveness is focused upon. The pictures used in the experiment had been selected from a large pool of yearbook pictures on the basis of judges' ratings. Such ratings enabled the selection of pictures that were either high, average, or low in physical attractiveness. For male subjects the pictures were of females, and for female subjects the pictures were of males. For present purposes the factor to be considered is, thus, physical attractiveness. The experiment was set up so that at each of the three levels of the physical attractiveness factor there were an equal number of high- and low-similarity questionnaires.

Although Stroebe et al. examined the effect of physical attractiveness on all three of the assessments of attraction in Table 1.1 (liking, dating preference, marriage preference), we will use the dating preference assessment as the dependent property in the present example. The dating scores are contained in Table 10.3. Each of the three conditions (or levels of physical attractiveness) has 40 scores. The total n is, thus, 120. As indicated in Table 10.3, the sum of all scores across the three groups is $199 + 159 + 97 = 455$; and the sum of all squared scores across the three groups is $1,131 + 745 + 311 = 2,187$. The null hypothesis was that the mean dating preference was the same in each of the three conditions. This hypothesis was tested with a significance level of $\alpha = .05$.

First, the total sum of squares, SS, was computed:

$$SS_t = \sum X^2 - \frac{(\sum X)^2}{n}$$

$$= (7)^2 + (4)^2 + \cdots + (2)^2 - \frac{(7 + 4 + \cdots + 2)^2}{120}$$

$$= 2187 - \frac{(455)^2}{120}$$

$$= 461.79$$

The within groups and between groups sums of squares were then calculated. Using data from Table 10.3:

$$SS_1 = 1131 - \frac{(199)^2}{40} = 140.98$$

$$SS_2 = 745 - \frac{(159)^2}{40} = 112.98$$

$$SS_3 = 311 - \frac{(97)^2}{40} = 75.78$$

TABLE 10.3. *Dating Preference Scores As a Function of Physical Attractiveness.*

	PHYSICAL ATTRACTIVENESS		
	1 HIGH	2 AVERAGE	3 LOW
	7	3	2
	4	4	2
	7	5	4
	6	2	1
	7	5	5
	7	5	3
	7	7	1
	5	6	4
	7	6	1
	7	5	3
	5	4	1
	5	3	3
	1	3	1
	2	3	1
	3	3	1
	7	7	1
	5	2	1
	6	3	1
	4	5	1
	7	4	1
	7	5	4
	7	5	5
	5	6	3
	7	6	3
	4	3	2
	5	7	4
	6	5	4
	7	4	5
	6	5	3
	7	6	4
	2	2	2
	3	4	5
	2	3	1
	3	1	3
	4	1	1
	3	2	2
	3	3	1
	4	2	3
	3	3	2
	2	1	2
$\sum X$	199	159	97	455
$\sum X^2$	1,131	745	311	2,187

Therefore:
$$SS_{WG} = 140.98 + 112.98 + 75.78 = 329.74$$
and:
$$SS_{BG} = \frac{(199)^2}{40} + \frac{(159)^2}{40} + \frac{(97)^2}{40} - \frac{(455)^2}{120}$$
$$= 132.07$$

Using formula 10–3 to check the calculations:
$$SS_t = SS_{WG} + SS_{BG}$$
$$SS_t = 329.74 + 132.07$$
$$SS_t = 461.81$$

The value for SS_t obtained from formula 10–3, 461.81, is within rounding error of the value obtained above by directly calculating SS_t, 461.79. The degrees of freedom associated with SS_{WG} is given by:
$$df = 40 + 40 + 40 - 3 = 117$$

One degree of freedom is lost within each of the three groups. Since all n_is are equal, it is also the case that:
$$df = k(n_i - 1)$$
$$= 3(40 - 1)$$
$$= 117$$

The degrees of freedom associated with the SS_{BG} is given by:
$$df = 3 - 1 = 2$$

Thus:
$$MS_{WG} = \frac{329.74}{117} = 2.82$$
and:
$$MS_{BG} = \frac{132.07}{2} = 66.04$$

Referring to Table III (Appendix C), we find that for $\alpha = .05$ and $df = 2$ and 117, the critical value of F is approximately 3.08. Calculating F:
$$F = \frac{MS_{BG}}{MS_{WG}}$$
$$= \frac{66.04}{2.82}$$
$$= 23.42$$

One-Factor Analysis of Variance

TABLE 10.4. *Summary Table for Analysis of Variance of the Effect of Physical Attractiveness on Dating Preference.*

SOURCE OF VARIANCE	SUM OF SQUARES	df	MEAN SQUARE	F
Between Groups	132.07	2	66.04	23.42
Within Groups	329.74	117	2.82	
Total	461.81	119		

Since the calculated F is larger than the tabled entry, the null hypothesis that all three levels of physical attractiveness result in the same dating preference is rejected. A summary of this analysis is given in Table 10.4. High physical attractiveness resulted in greater dating preference than did low physical attractiveness.

> The significant F of 23.42 indicates that physical attractiveness had a significant effect upon dating preference. The F, however, does not tell us exactly which level of physical attractiveness differs from which level. Although it is reasonable to conclude that the two extreme levels differ from each other, we do not know if the intermediate (or average) level differs from either of the extremes (high or low). If a factor has more than three levels, the potential uncertainty regarding the number of levels that differ from each other is, of course, even greater. In order to gain information regarding the significance of differences between or among levels of a factor, any of a number of so-called multiple range tests may be used (for example, a Newman-Keuls test). Discussion of such tests is subject matter for a more advanced course in statistics.

TWO-FACTOR ANALYSIS OF VARIANCE

There are many circumstances in which the psychologist finds it valuable to manipulate two different properties or factors in the same experiment. How can the data from such an experiment be analyzed? It is possible, of course, to conduct two separate one-factor analysis-of-variance designs and draw separate conclusions about each factor. However, two-factor analysis of variance provides a more sophisticated method of data analysis. This technique provides information not only about the main effects of each treatment or factor, but also about the possible interaction between the two factors.

The Concept of Interaction

Recall that in the Stroebe *et al.* experiment male subjects received information about female others, and female subjects received information about male others. Thus, if sex of subject is considered a factor, along with the above physical attractiveness factor, we have a two-factor design. In this design there are a total of six cells or conditions: male-subjects-low-physical-attractiveness, male-subjects-average-physical-attractiveness, male-subjects-high-physical-attractiveness, female-subjects-low-physical-attractiveness, female-subjects-average-physical-attractiveness, female-subjects-high-physical-attractiveness. In general, the number of cells can always be obtained by multiplying the number of levels of the factors; thus 3 (levels of physical attractiveness) times 2 (levels of sex) equals 6 (cells).

Figure 10-5 shows the effects of physical attractiveness and sex of subject on dating preference. A two-factor analysis of variance resulted in a physical attractiveness main effect and a physical attractiveness by sex interaction. The physical attractiveness main effect indicates a general tendency for dating preference to increase with an increase in physical attractiveness. This is, of course, what was found above when physical attractiveness was the only factor. This main effect is indicated in Fig. 10-5 by a general tendency for the two lines (one

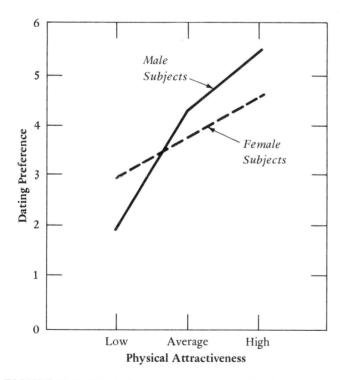

FIGURE 10-5. Physical attractiveness by sex of subject interaction for dating preference in the Stroebe et al. experiment.

for male subjects and one for female subjects) to rise as physical attractiveness increases. Ignoring the difference between males and females, the general tendency is for dating preference to increase as physical attractiveness increases. The physical attractiveness by sex interaction, however, indicates that the difference between male and female subjects should not be ignored. The effect of physical attractiveness on dating preference was more marked for males than for females. In Fig. 10–5, it can be seen that the two lines are not parallel. The line for male subjects is steeper than the line for female subjects.

A two-factor interaction is always indicated by a *departure from parallelness* in a plot of the two functions. The two functions in Fig. 10–5 are not parallel. When there is no interaction, the two functions are parallel. This is illustrated in Fig. 10–6, which has been specifically constructed to illustrate results that might have been obtained if there were no interaction.

Another way of describing a two-factor interaction is as *a qualification of one factor by a second factor*. Thus, in the Stroebe et al. experiment the effect of physical attractiveness was qualified by the sex of the subject.

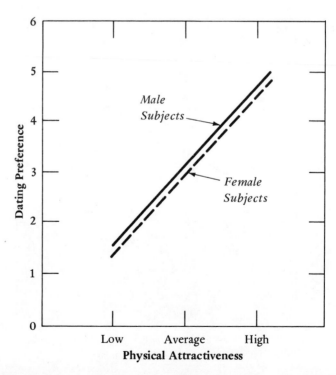

FIGURE 10–6. *Hypothetical results: main effect for physical attractiveness.*

Qualifications, or interactions, may take any of a number of forms. In Fig. 10–5, which represents the actual Stroebe *et al.* results, the line for female subjects has a non-zero slope. When Stroebe *et al.* did a one-factor analysis of variance on just the data for female subjects, it was found that physical attractiveness still had a significant effect. Thus, although the effect of physical attractiveness is qualified by the sex of subject, the qualification is by no means complete. Even though physical attractiveness had a greater effect for males than for females, physical attractiveness still had some effect for females.

Of course, it could have been the case that physical attractiveness had no effect for females. Such a hypothetical situation is pictured in Fig. 10–7. Here, the departure from parallelness is even more marked. The function for females has zero slope. In this case the qualification is complete.

Still another possible interaction is pictured in Fig. 10–8. Here, physical attractiveness has a positive relation with dating preference for males and a negative relation with females. For males, the greater the physical attractiveness is,

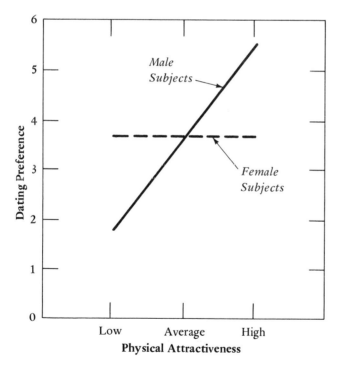

FIGURE 10–7. Hypothetical results: physical attractiveness by sex of subject interaction in which physical attractiveness has no effect for females.

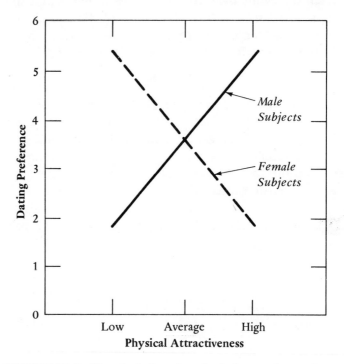

FIGURE 10–8. *Hypothetical results: physical attractiveness by sex of subject interaction in which physical attractiveness has a negative (or reverse) effect for females.*

the greater the dating preference; for females, the greater the physical attractiveness is, the less the dating preference. With these hypothetical results the departure from parallelness is as extreme as it can become.

An Example Experiment

The deprivation experiment discussed earlier will be expanded to illustrate computation of a two-factor analysis of variance. *The analysis of variance procedure discussed below assumes that all sample groups have equal ns.* No such assumption was made in the above one-factor analysis of variance. Procedures for analyzing two or more factor experiments with unequal *n*s are available, but will not be discussed here.

Suppose an experimenter is interested in studying not only the effects of hours deprivation on subsequent food intake but also the effects of different foods on amount of food intake. In the prior one-factor analysis of variance

TABLE 10.5. *Treatment Groups for Two-Way Analysis of Variance.*

FOOD TYPE	DEPRIVATION			
	0	2	4	6
1	1,0	1,2	1,4	1,6
2	2,0	2,2	2,4	2,6
3	3,0	3,2	3,4	3,6

problem, the effects of hours deprivation on food intake was examined only for a food of known low preference. It was found that increased hours deprivation resulted in decisive increases in food intake. The question arises as to whether intake of more prefered foods would increase similarly with increasing food deprivation.

To answer this question an experiment was designed allowing a two-factor analysis of variance. Four levels of deprivation, 0, 2, 4, and 6 hours, were to be studied in conjunction with each of three types of food. Table 10.5 displays the various treatment groups needed for such a two-factor design. Notice that at each level of each variable a separate group of subjects is designated to receive each of the levels of the other variable. The experiment thus requires twelve different sample sets of ten rats each. Table 10.6 lists by sample sets the food intake scores obtained in the experiment. The four sets under food type 1 are identical to those in the prior one-factor experiment. The two-factor analysis of variance design described in the remainder of this section requires that all n_is be equal.

Sums of Squares

TOTAL SUM OF SQUARES. The total, within groups and between groups sums of squares can be computed for a two-factor experiment just as they can for a one-factor experiment. Formula 10–2 is the computational formula for the total sum of squares:

$$SS_t = \sum X^2 - \frac{(\sum X)^2}{n}$$

For the data in Table 10.6:

$$SS_t = (1.3)^2 + (.9)^2 + \cdots + (10.7)^2 - \frac{(1.3 + .9 + \cdots + 10.7)^2}{120}$$

$$= 3397.91 - \frac{(588.5)^2}{120}$$

$$= 511.81$$

TABLE 10.6. *Food Intake Data for Food Type and Deprivation Experiment.*

		\multicolumn{4}{c}{HOURS OF DEPRIVATION}			
		0	2	4	6
FOOD 1		1.3	2.0	5.5	5.8
		0.9	3.1	4.3	6.5
		2.3	2.6	6.1	6.7
		1.5	2.8	5.7	6.1
		1.7	4.0	4.8	7.2
		0.7	3.3	5.1	5.9
		2.5	2.3	6.0	7.5
		1.6	1.9	4.5	6.3
		1.4	3.5	5.2	7.1
		2.1	3.2	5.6	6.9
	$\sum X$	16.0	28.7	52.8	66.0
	$\sum X^2$	28.60	86.49	282.14	438.60
FOOD 2		2.8	4.2	4.5	6.7
		1.9	4.1	5.2	7.1
		1.8	3.2	7.2	5.9
		3.2	1.8	6.6	6.6
		2.2	2.7	7.5	8.0
		2.9	3.9	5.8	7.3
		2.8	2.1	6.1	6.9
		3.4	3.6	6.4	6.5
		1.7	3.7	7.3	6.2
		2.3	2.7	8.4	6.8
	$\sum X$	25.0	32.0	65.0	68.0
	$\sum X^2$	65.76	108.78	434.60	465.50
FOOD 3		6.0	4.2	6.3	8.1
		4.9	4.4	6.1	4.8
		4.9	4.1	7.2	7.5
		6.1	6.0	6.9	7.8
		3.8	5.8	7.5	7.6
		4.3	4.7	6.9	4.9
		4.7	5.1	6.8	6.6
		6.2	3.9	7.5	8.8
		4.1	4.2	7.8	5.2
		3.0	3.6	6.0	10.7
	$\sum X$	48.0	46.0	69.0	72.0
	$\sum X^2$	240.50	217.36	479.54	550.04

WITHIN GROUPS SUM OF SQUARES. As in the case of a one-factor analysis of variance, the within groups sum of squares is computed by calculating:

$$SS_1 = \sum X_1^2 - \frac{\sum (X_1)^2}{n_1}$$

$$SS_2 = \sum X_2^2 - \frac{\sum (X_2)^2}{n_2}$$

$$\vdots \qquad \vdots \qquad \vdots$$

$$SS_k = \sum X_k^2 - \frac{\sum (X_k)^2}{n_k}$$

and summing across the k groups. For the data in Table 10.6:

$SS_1 = 28.60 - \frac{(16.0)^2}{10} = 3.00 \qquad SS_7 = 434.60 - \frac{(65.0)^2}{10} = 12.10$

$SS_2 = 86.49 - \frac{(28.7)^2}{10} = 4.12 \qquad SS_8 = 465.90 - \frac{(68.0)^2}{10} = 3.10$

$SS_3 = 282.14 - \frac{(52.8)^2}{10} = 3.36 \qquad SS_9 = 240.50 - \frac{(48.0)^2}{10} = 10.10$

$SS_4 = 438.60 - \frac{(66.0)^2}{10} = 3.00 \qquad SS_{10} = 217.36 - \frac{(46.0)^2}{10} = 5.76$

$SS_5 = 65.76 - \frac{(25.0)^2}{10} = 3.26 \qquad SS_{11} = 479.54 - \frac{(69.0)^2}{10} = 3.44$

$SS_6 = 108.78 - \frac{(32.0)^2}{10} = 6.38 \qquad SS_{12} = 550.04 - \frac{(72.0)^2}{10} = 31.64$

Summing these twelve sums of squares gives 89.26, the within groups sum of squares.

BETWEEN GROUPS SUM OF SQUARES. Formula 10–7 for the between groups sum of squares is:

$$SS_{BG} = \frac{\sum X_1^2}{n_1} + \frac{\sum X_2^2}{n_2} + \cdots + \frac{\sum X_k^2}{n_k} - \frac{\sum X^2}{n}$$

Since all n_i are equal:

$$SS_{BG} = \frac{\sum X_1^2 + \sum X_2^2 + \cdots + \sum X_k^2}{n_i} - \frac{\sum X^2}{n}$$

Two-Factor Analysis of Variance

For the data in Table 10.6:

$$SS_{BG} = \frac{(16.0)^2 + (28.7)^2 + \cdots + (72.0)^2}{10} - \frac{(1.3 + .9 + \cdots + 10.7)^2}{120}$$

$$= 3308.65 - \frac{(588.5)^2}{120}$$

$$= 422.55$$

Formula 10–3 can be used to check our calculations to this point:

$$SS_t = SS_{WG} + SS_{BG}$$
$$511.81 = 89.26 + 422.55$$
$$511.81 = 511.81$$

The total sum of squares equals the within groups sum of squares plus the between groups sum of squares.

In a two-factor design, the between groups sum of squares can be broken down into three component parts: factor A sum of squares (SS_A), factor B sum of squares (SS_B), and the sum of squares of the interaction of A and B (SS_{AB}):

$$SS_{BG} = SS_A + SS_B + SS_{AB} \qquad (10\text{--}13)$$

Formula 10–13 makes the rationale for the term, "analysis of variance," more apparent. Analysis of variance is a variance analyzing or partitioning technique. According to formula 10–13, the numerator of the between groups variance (SS_{BG}) for the groups in Table 10.6 can be divided into three components. One of these reflects variations among the four columns or among the four levels of deprivation. Another reflects variation among the three rows or among the three types of food. And, the third and final component reflects variation due to the interaction of deprivation and food type.

Each of the three component sums of squares is a positive function of treatments and error, just as is the total between groups sum of squares.

FACTOR A SUM OF SQUARES. In order to facilitate calculation of the component sums of squares for the between groups sum of squares, it is helpful to arrange the group sums as in Table 10.7. Each of the table entries is a sum for a particular group or sample set. The marginals of the table contain the column and row sums. The lower right-hand corner contains the total sum. This total sum may be calculated by either summing the column sums or summing the row sums.

Hours of deprivation can be designated arbitrarily as factor A, and food type as factor B. The deprivation factor is a column factor in Table 10.7. The concern is with the variation among the column sums (89.0, 106.7, 186.8, and 206.0), each of which is the sum of the 30 scores for a given deprivation period.

TABLE 10.7. *Group Sums for the Food Type and Deprivation Experiment.*

	HOURS OF DEPRIVATION				
	0	2	4	6	Σ
FOOD 1	16.0	28.7	52.8	66.0	163.5
FOOD 2	25.0	32.0	65.0	68.0	190.0
FOOD 3	48.0	46.0	69.0	72.0	235.0
Σ	89.0	106.7	186.8	206.0	588.5

Altogether there are 120 scores, (4)(30) = 120. The calculation of SS_A is as follows:

$$SS_A = \frac{(89.0)^2 + (106.7)^2 + (186.8)^2 + (206.0)^2}{30} - \frac{(588.5)^2}{120}$$

$$= 335.10$$

The deprivation or factor A sum of squares is 335.10.

FACTOR B SUM OF SQUARES. The factor B, or food type, sum of squares is a row factor in Table 10.7. The concern, in this instance, is with the variation among the row sums (163.5, 190.0, and 235.0), each of which is the sum of the 40 scores for one of the food types. The calculation of SS_B is as follows:

$$SS_B = \frac{(163.5)^2 + (190.0)^2 + (235.0)^2}{40} - \frac{(588.5)^2}{120}$$

$$= 65.33$$

The food type or factor B sum of squares is 65.33.

SUM OF SQUARES FOR THE INTERACTION OF A AND B. The sum of squares for the interaction of A and B is, for the present experiment, a sum of squares for the interaction of hours of deprivation and food type. A variant of formula 10–13 is used to calculate the interaction. Formula 10–13 is: $SS_{BG} = SS_A + SS_B + SS_{AB}$. Solving for SS_{AB} gives: $SS_{AB} = SS_{BG} - SS_A - SS_B$.

If the previously calculated values for the between groups, deprivation, and food type sums of squares are inserted, then:

$$SS_{AB} = 422.55 - 335.10 - 65.33$$

$$= 22.12$$

The sum of squares for the interaction of deprivation and food type is 22.12.

Two-Factor Analysis of Variance

Mean Squares

Just as in the case of one-factor analysis of variance, a sum of squares divided by appropriate degrees of freedom yields a mean square. Two-factor analysis of variance differs in that there are more sums of squares with which to deal.

FACTOR A MEAN SQUARE. Degrees of freedom for the factor A (deprivation) mean square is equal to the number of treatment levels minus one. Since there are four levels of deprivation, $df = 4 - 1 = 3$.

Dividing the deprivation sum of squares, 335.10, by the df, 3, gives the deprivation mean square, 111.70. Table 10.8 summarizes the sum of squares, df, and mean square for the various sources of variance.

FACTOR B MEAN SQUARE. Degrees of freedom for the factor B mean square is also equal to the number of treatment levels minus one. Since factor B is food type and there are three types of food, $df = 3 - 1 = 2$. Dividing the food type sum of squares, 65.33, by its df, 2, gives the food type mean square, 32.67. These values are entered in Table 10.8.

MEAN SQUARE FOR THE INTERACTION OF A AND B. Just as the between groups sum of squares can be divided into three components (formula 10–13): $SS_{BG} = SS_A + SS_B + SS_{AB}$, the between groups df can be also divided into three components: $df_{BG} = df_A + df_B + df_{AB}$. The between groups df equals the df for A, plus the df for B, plus the df for the interaction of A and B. Solving for df_{AB}:

$$df_{AB} = df_{BG} - df_A - df_B$$

The between groups df equals the number of groups minus one. Since there are twelve groups, $df_{BG} = 12 - 1 = 11$. Further, since $df_A = 3$ and $df_B = 2$:

$$df_{AB} = 11 - 3 - 2$$
$$= 11 - 5$$
$$= 6$$

TABLE 10.8. *Summary Table for the Food Type and Deprivation Experiment.*

SOURCE OF VARIANCE	SUM OF SQUARES	df	MEAN SQUARE	F
Deprivation	335.10	3	111.70	134.58
Food Type	65.33	2	32.67	39.36
Deprivation and Food Type Interaction	22.12	6	3.69	4.44
Within Groups	89.26	108	0.83	

An alternative procedure for calculating an interaction df is to multiply the degrees of freedom for the interacting variables:

$$df_{AB} = (df_A)(df_B)$$
$$= (3)(2)$$
$$= 6$$

Both procedures give 6 as the df for the deprivation by food interaction.

As is indicated in Table 10.8, the mean square for the interaction of deprivation and food type is 3.69. This value is obtained by dividing the sum of squares for the interaction, 22.12, by the appropriate df, 6.

WITHIN GROUPS MEAN SQUARE. One degree of freedom is lost within each group. Thus, the within groups df equals $n_i - 1$ times the number of groups. In the present case there are twelve groups of ten observations each. Thus $df_{WG} = 12(10 - 1) = 108$. Dividing the within groups sum of squares, 89.26, by the appropriate df, 108, gives the within groups mean square, .83. These values are entered in Table 10.8.

The previously calculated df should sum to the total df:

$$df_t = df_A + df_B + df_{AB} + df_{WG}$$

The df_t is one less than the total number of observations, or $120 - 1 = 119$; thus:

$$119 = 3 + 2 + 6 + 108$$
$$119 = 119$$

This provides a check on our calculations of the df for the component sums of squares.

Assumptions and F

For the previous one-factor design, F was the ratio of the MS_{BG} to the MS_{WG}. For the present two-factor design, however, there are three Fs:

$$F_A = \frac{MS_A}{MS_{WG}} \qquad (10\text{--}14)$$

$$F_B = \frac{MS_B}{MS_{WG}} \qquad (10\text{--}15)$$

$$F_{AB} = \frac{MS_{AB}}{MS_{WG}} \qquad (10\text{--}16)$$

In every case the MS_{WG} is in the denominator and the MS for the treatment effect being tested is in the numerator.

The rationale behind these three Fs is the same as for the single F in the one-factor design. The denominators are a function of error. In fact, the MS_{WG} is an estimate of variance of the population of error effects. The numerators for each of the ratios are a function of some experimental effect and error. When H_0 is true (i.e., there is no experimental effect), the numerators are estimates of the variance of the population of error effects. The larger the F ratios, the greater the likelihood that there was in fact an experimental effect. By virtue of the same assumptions made earlier, the calculated Fs can be considered values from the theoretical sampling distribution of F.

Hypothesis Testing

For the present two-factor analysis of variance there are three alternative hypotheses and three null hypotheses to be stated. For deprivation, an alternative hypothesis is:

$$H_1: \mu_0 \neq \mu_2 \quad \text{and/or} \quad \mu_0 \neq \mu_4 \quad \text{and/or} \quad \mu_2 \neq \mu_4, \text{etc.}$$

According to this hypothesis the population mean is not the same under every deprivation treatment. The null hypothesis is:

$$H_0: \mu_0 = \mu_2 = \mu_4 = \mu_6$$

The negation of this null hypothesis is, of course, the above alternative hypothesis.

For the food factor:

$$H_1: \mu_1 \neq \mu_2 \quad \text{and/or} \quad \mu_1 \neq \mu_3 \quad \text{and/or} \quad \mu_2 \neq \mu_3$$

and:

$$H_0: \mu_1 = \mu_2 = \mu_3$$

Since statement of the null and alternative hypotheses for the AB (hours deprivation and food type) interaction is very complex, we shall simply state that if H_1 is true and there is an interaction effect,

$$\frac{MS_{AB}}{MS_{WG}}$$

is expected to be greater than one. If H_0 is true and thus both MS_{AB} and MS_{WG} are estimates of the error variance of the *same* population of error effects, the ratio,

$$\frac{MS_{AB}}{MS_{WG}}$$

is expected to be approximately one.

Rejection of a given null hypothesis logically implies an alternative hypothesis. An experimenter, however, may not expect certain alternative hypotheses to be supported. For example, an experimenter may expect the alternative hypothesis for deprivation and food type, but not the one for deprivation and food interaction, to be supported. This means, of course, that an experimenter's research hypotheses may or may not be the same as the alternative hypotheses.

Even in the case in which an experimenter expects a given alternative hypothesis to be supported, his research hypothesis may be more specific. For example, an experimenter may expect that the population mean for food intake will increase with increasing deprivation. The alternative hypothesis, however, is not that specific. The alternative hypothesis simply states that the population mean for food intake is not the same under every deprivation treatment.

Setting α at .01, the critical F values from Table III (Appendix C) are obtained for the appropriate df. For deprivation, df equals 3 and 108, and the critical F approximately equals 3.98. For food type, df equals 2 and 108, and the critical F approximately equals 4.82. For the deprivation and food type interaction, df equals 6 and 108, and the critical F approximately equals 2.99.

Using formulas 10–14, 10–15, and 10–16:

$$F_A = \frac{MS_A}{MS_{WG}} = \frac{111.70}{.83} = 134.58$$

$$F_B = \frac{MS_B}{MS_{WG}} = \frac{32.67}{.83} = 39.36$$

$$F_{AB} = \frac{MS_{AB}}{MS_{WG}} = \frac{3.69}{.83} = 4.44$$

In every case the calculated Fs exceed the critical values. Thus the three null hypotheses are rejected and the three alternative hypotheses are accepted.

Figure 10–9 graphically presents the results. The deprivation effect is indicated by the increasing heights of the three curves across deprivation levels. The food type effect is indicated by the differential heights of the three curves. At all levels of deprivation, food 3 is most preferred and food 1 least preferred. Finally, the interaction is indicated by the departure from parallelness in the three curves. The differential heights of the three curves are greatest at the low levels of deprivation and least at the high levels of deprivation. When the animals are not very hungry they are "finicky" about what they will eat. As hunger increases, however, they show less differential preference.

A Further Example of Two-Factor Analysis of Variance

For a final example of two-factor analysis of variance, we once again can use the dating preference scores from the Stroebe *et al.* experiment. The physical attractiveness factor was discussed above—as was also the sex of subject factor.

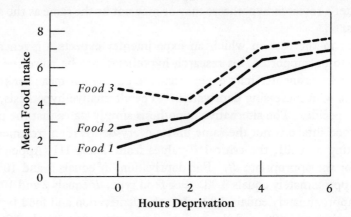

FIGURE 10-9. *Mean food intake results for the deprivation and food type experiment.*

For this present example, we will use similarity (high or low) rather than sex of subject as one factor and physical attractiveness as the other factor. The 20 scores in each of the 6 cells of the 3 by 2 design are presented in Table 10.9. Since each score is from a different subject, the total is 120 subjects ($n = 120$). Half of the 20 scores in each cell are from male subjects, and half from female subjects.

First, the total sum of squares, SS_t, is computed across all 120 scores:

$$SS_t = \sum X^2 - \frac{(\sum X)^2}{n}$$

$$= (7)^2 + (4)^2 + \cdots + (2)^2 - \frac{(7 + 4 + \cdots + 2)^2}{120}$$

$$= 2187 - \frac{(455)^2}{120}$$

$$= 461.79$$

This is the same answer as was obtained above, when the only concern was physical attractiveness. Next, the SS_{WG} is calculated by first finding and then summing the sums of squares for each of the six different groups:

$$SS_{1HS} = 803 - \frac{(125)^2}{20} = 21.75$$

$$SS_{2HS} = 532 - \frac{(100)^2}{20} = 32.00$$

$$SS_{3HS} = 231 - \frac{(63)^2}{20} = 32.55$$

$$SS_{1LS} = 328 - \frac{(74)^2}{20} = 54.20$$

$$SS_{2LS} = 213 - \frac{(59)^2}{20} = 38.95$$

$$SS_{3LS} = 80 - \frac{(34)^2}{20} = 22.20$$

Thus, $SS_{WG} = 21.75 + 32.00 + 32.55 + 54.20 + 38.95 + 22.20 = 201.65$.

TABLE 10.9. *Dating Preference Scores As a Function of Physical Attractiveness and Similarity.*

	PHYSICAL ATTRACTIVENESS		
	1 HIGH	2 AVERAGE	3 LOW
HIGH SIMILARITY (HS)	7 4 7 6 7 7 7 5 7 7 7 7 5 7 4 5 6 7 6 7	3 4 5 2 5 5 7 6 6 5 5 5 6 6 3 7 5 4 5 6	2 2 4 1 5 3 1 4 1 3 4 5 3 3 2 4 4 5 3 4
$\sum X$	125	100	63
$\sum X^2$	803	532	231

(cont.)

TABLE 10.9 (continued)

	PHYICAL ATTRACTIVENESS		
	1 HIGH	2 AVERAGE	3 LOW
	5	4	1
	5	3	3
	1	3	1
	2	3	1
	3	3	1
	7	7	1
	5	2	1
	6	3	1
LOW	4	5	1
SIMILARITY	7	4	1
(LS)	2	2	2
	3	4	5
	2	3	1
	3	1	3
	4	1	1
	3	2	2
	3	3	1
	4	2	3
	3	3	2
	2	1	2
$\sum X$	74	59	34
$\sum X^2$	328	213	80

The sums of squares between all six groups is given by:

$$SS_{BG} = \frac{(125)^2 + (100)^2 + \cdots + (34)^2}{20} - \frac{(7 + 4 + \cdots + 2)^2}{120}$$

$$= 1985.35 - \frac{(455)^2}{120}$$

$$= 260.14$$

Using formula 10-3 to check the calculations to this point:

$$SS_t = SS_{WG} + SS_{BG}$$
$$461.79 = 201.65 + 260.14$$
$$461.79 = 461.79$$

In order to facilitate calculation of the component sums of squares for the between groups sum of squares, it is helpful to arrange the group sums as in

TABLE 10.10. *Group Sums for Dating Preference Scores in Table 10.9.*

	PHYSICAL ATTRACTIVENESS			
	HIGH	AVERAGE	LOW	Σ
HIGH SIMILARITY (HS)	125	100	63	288
LOW SIMILARITY (LS)	74	59	34	167
Σ	199	159	97	455

Table 10.10. Each of the table entries is a sum for a particular group or sample set in Table 10.9.

The sum of squares between groups can be partitioned into sum of squares for attractiveness, SS_A, sum of squares for similarity, SS_S, and sum of squares for the attractiveness by similarity interaction SS_{SA}. In the case of SS_A, the concern is with variation among the column sums (199, 159, and 97), each of which is the sum of 40 scores:

$$SS_A = \frac{(199)^2 + (159)^2 + (97)^2}{40} - \frac{(455)^2}{120}$$
$$= 132.06$$

In the case of SS_S, the concern is with variation among the row sums (288, 167), each of which is the sum of 60 scores:

$$SS_S = \frac{(288)^2 + (167)^2}{60} - \frac{(455)^2}{120}$$
$$= 122.01$$

Sum of squares for the interaction, SS_{SA}, can be found by subtracting SS_A and SS_S from SS_{BG}:

$$SS_{SA} = 260.14 - 132.06 - 122.01$$
$$= 6.07$$

Next, mean squares are computed by dividing each sum of squares by its degrees of freedom:

$$MS_{WG} = \frac{SS_{WG}}{k(n_i - 1)} \quad MS_A = \frac{SS_A}{df_A} \quad MS_S = \frac{SS_S}{df_S}$$

$$= \frac{201.65}{114} \quad\quad = \frac{132.06}{2} \quad\quad = \frac{122.01}{1}$$

$$= 1.77 \quad\quad\quad = 66.03 \quad\quad\quad = 122.01$$

Two-Factor Analysis of Variance

and:

$$MS_{SA} = \frac{SS_{SA}}{df_{SA}}$$

$$= \frac{6.07}{2}$$

$$= 3.04$$

Number of degrees of freedom for the A and S sum of squares is equal to the number of treatment levels minus 1. Since there are three levels of physical attractiveness, $df_A = 2$. Also, since there are two levels of similarity, $df_B = 1$. Number of degrees of freedom for the SA interaction can be determined as follows:

$$df_{SA} = (df_S)(df_A)$$
$$= (1)(2)$$
$$= 2$$

Finally, F statistics are computed for factors A, S, and the SA interaction:

$$F_A = \frac{MS_A}{MS_{WG}} \qquad F_S = \frac{MS_S}{MS_{WG}}$$

$$= \frac{66.03}{1.77} \qquad = \frac{122.01}{1.77}$$

$$= 37.30 \qquad = 68.93$$

and:

$$F_{SA} = \frac{3.04}{1.77}$$

$$= 1.72$$

A summary of this analysis is contained in Table 10.11.

TABLE 10.11. *Analysis of Variance Summary for Effect of Physical Attractiveness and Similarity on Dating Preference.*

SOURCE OF VARIANCE	SUM OF SQUARES	df	MEAN SQUARE	F
PHYSICAL ATTRACTIVENESS (A)	132.06	2	66.03	37.30
SIMILARITY (S)	122.01	1	122.01	68.93
INTERACTION (SA)	6.07	2	3.04	1.73
WITHIN GROUPS	201.65	114	1.77	

Assuming a significance level of $\alpha = .05$, and checking Table III (Appendix C), we find that the critical value of F for the physical attractiveness factor, df equaling 2 and 114, is approximately 3.08. Thus, the obtained F of 37.30 falls in the region of rejection; and the null hypothesis that all three levels of attraction lead to the same mean dating preference is rejected. Likewise, the null hypothesis that high and low similarity lead to the same dating preference is rejected. The F of 68.93 exceeds the critical F, approximately 3.93, at $\alpha = .05$ and df equalling 1 and 114.

The interaction of physical attractiveness and similarity, however, is not significant. The obtained F of 1.72 does not exceed the critical F, approximately 3.08, at $\alpha = .05$ and df equaling 2 and 114. There is no evidence that the effect of similarity on dating preference is qualified by the three levels of physical attractiveness.

THREE-OR-MORE FACTOR ANALYSIS OF VARIANCE

In the immediately preceding section, dating preference scores from the Stroebe et al. experiment were used to provide an example of two-factor analysis of variance. The factors were similarity (high or low), and physical attractiveness (high, average, or low). The analysis of variance revealed that both of these so-called main effects were significant but that the interaction was not. Earlier in this chapter, we discussed a different two-factor analysis of the Stroebe et al. dating preference scores. These two factors were physical attractiveness and sex of subject. Altogether, there were a total of three factors: similarity (high or low), physical attractiveness (high, average, or low), and sex of subject. The three factors generate 12 cells ($3 \times 2 \times 2 = 12$), which are shown in Table 10.12. If all three of the factors are considered to be of importance, the most appropriate analysis of the data from such an experimental design is not a series of two-factor analyses of variance, rather, a three-factor analysis of variance.

TABLE 10.12. *Three-Factor Design for Stroebe et al. Experiment.*

		SIMILARITY			
		HIGH		LOW	
		MALES	FEMALES	MALES	FEMALES
PHYSICAL ATTRACTIVENESS	HIGH				
	AVERAGE				
	LOW				

In this introductory book, it is not our intent to describe the calculational prodecures for three-factor analysis of variance. However, we do wish to acquaint the student with three-factor analysis of variance so that he or she will be in a better position to read and understand the psychological literature.

With a three-factor analysis of variance, there are a total of seven effects; three main effects, three double interactions, and a triple interaction. Let us consider the main effects, double interactions, and triple interaction in that order. First, the three main effects; in the Stroebe *et al.* experiment, these include: the main effect for similarity, the main effect for physical attractiveness, and the main effect for sex of subject. The similarity main effect, for example, indicates whether or not there is any overall difference between high and low similarity—an overall difference that ignores the level of physical attractiveness and the sex of the subject. In like manner, the physical attractiveness and sex main effects indicate the existence of such overall effects.

Stroebe *et al.* had a number of different dependent properties. The questions assessing the three most important of these (liking, dating preference, and marital preference) are contained in Table 1.1. Three-factor analyses of variance of these dependent properties resulted in significant main effects for similarity and physical attractiveness, and non-significant main effects for sex for all three dependent properties. Liking, dating preference, and marital preference increased with an increase in similarity and an increase in physical attractiveness.

Second, a three-factor analysis of variance tests three double, or two-factor, interactions. For the Stroebe *et al.* design, these double interactions are: similarity by physical attractiveness, similarity by sex, and physical attractiveness by sex. As indicated above, two-factor, or double, interactions indicate the extent to which the effect of one factor is qualified by the effect of a second factor. Consider the physical attractiveness by sex interaction that was previously described for dating preference. The interaction indicates that the effect of physical attractiveness on dating preference is greater for males than for females; i.e., that the effect of physical attractiveness is qualified by the sex of subject. This same interaction is also significant for marital preference; it is not, however, significant for liking. Males do not differ from females in the extent to which physical attractiveness affects liking, but they do differ in the extent to which physical attractiveness affects dating and marital preference. Such results are intriguing. Superficially, it might have been thought that the reason males are more affected than females by physical attractiveness is simply that, for males, physical attractiveness has a relatively greater effect on liking. This, however, is not the case. Perhaps, considerations of status are involved. It may be that males gain more status by associating with physically attractive females than females do by associating with physically attractive males.

The similarity by physical attractiveness interaction is not significant for any of the three dependent properties, but the similarity by sex interaction is significant for liking only. This interaction is pictured in Fig. 10-10. The form of this interaction is interesting—similarity has a greater effect on liking for females than it does for males. Notice the departure from parallelness in the two lines in Fig. 10-10. The greater effect of similarity on liking for females than males is intriguing in view of the previously described greater effect of physical attractiveness on dating and marital preference for males than females.

The third and final effect in a three-factor analysis of variance is the triple interaction. A triple interaction can be defined as: *the qualification of a double interaction by a third factor*. In the Stroebe *et al.* experiment, the triple interaction is not significant for any of the three dependent properties. Thus, the above double interactions are not qualified by a third factor. It is not necessary, for example, to qualify the interaction of physical attractiveness and sex by reference to the level of similarity. It is possible, of course, that such an interaction would have occurred. For example, it could have been that the greater effect of physical attractiveness for males than for females was more apparent when similarity

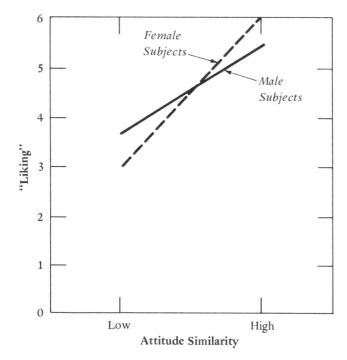

FIGURE 10-10. Attitude similarity by sex of subject interaction for liking scores in the Stroebe et al. experiment.

was low than when similarity was high. In many experiments such complex triple interactions do occur.

Finally, the student should be aware that it is possible to continue adding factors. An investigator may have four factors, five factors, or any number of factors. The only problem is that the number and complexity of the interactions continue to grow so that the investigator's ability to interpret such effects is increasingly taxed. In many instances, however, an adequate understanding of some psychological phenomenon may depend on the conceptualization of a complex interaction. It would be nice if the psychological world were made up of main effects. Unfortunately, this is not always the case.

Analysis of variance is an extremely powerful research tool. In fact, R. A. Fisher, who developed analysis of variance, succeeded in revolutionizing research design; and thus, he greatly advanced our ability to analyze and understand psychological phenomena.

R. A. FISHER AND ANALYSIS OF VARIANCE

R. A. Fisher (1890–1962) graduated in 1912 from Cambridge, where he studied mathematics and theoretical physics. After graduation he wandered from job to job—working for an investment house, doing farm chores in Canada, and teaching high school. All of this changed, however, in 1919 when he went to an agricultural research organization, the Rothamsted Experimental Station. He was employed as a statistician to analyze a 66 year accumulation of data on manurial field trials and weather records. During the 14 years in which Fisher remained at Rothamsted, he established a reputation as one of the leading statisticians in the world. While at Rothamsted, he published a book, *Statistical Methods for Research Workers*, which is considered a landmark in the history of statistics. This book was written especially for researchers. They, nonetheless, found it difficult. Mathematicians also found it difficult, however; all of the proofs were omitted.

In 1933, Fisher left Rothamsted to become Galton Professor of Eugenics at University College, London. Unfortunately, Fisher's relationship with Karl Pearson deteriorated. Jerzy Newman, a contemporary statistician, writes that in 1934: "There was a sharp feud raging between Karl Pearson and R. A. Fisher with, seemingly, the population of the earth divided into two categories, K. P.'s school and R. A. F.'s school. Both E. S. Pearson and I did not like the situation and did our best to avoid becoming involved" (1967, p. 1456). Newman goes on to recount that later on he and Fisher fell out on other grounds. Fisher possessed a strong personality. He was generous and supportive of his followers, and hostile and aggressive toward his dissenters.

In 1943, Fisher left the University of London to become Balfour Professor of Genetics at Cambridge. In 1959, he was knighted. Throughout his life he remained a tireless and dedicated worker. Over a period of half a century, he published an average of one paper every two months.

Newman characterizes Fisher as an "ideological descendent" of Karl Pearson. Newman states that, although Fisher would object to this characterization, it is nonetheless true since Fisher really took up various statistical problems where Karl Pearson left off. This is particularly evident with regard to correlation. One of Fisher's first contributions was the solution to the previously elusive problem of establishing the exact sampling distribution for r. Fisher also worked out the sampling distributions for regression coefficients, partial rs and multiple rs.

Unlike Pearson, Fisher realized the importance of Gosset's work with small samples. After developing the familiar t test from Gosset's work, Fisher went on to develop analysis of variance and a theory of hypothesis testing. His theory makes use of the now familiar null hypothesis and the use of exact cutoff points such as the .05 and .01.

Fisher's paramount contribution was undoubtedly the development of analysis of variance. When Fisher came to Rothamsted, the standard research procedure was to vary one factor at a time. Fisher realized that the simultaneous variation of factors was necessary for the testing of interactions, and that the carrying out of one multi-factor experiment did not necessarily involve more time, effort, and money than the carrying out of a series of single-factor experiments. Fisher actually succeeded in revolutionizing research design, and thus may be considered one of the major historical figures in the history of induction.

One of Fisher's better known points of view had to do with a belief in the importance of randomization in experimental design. He was very skeptical about the possibility of inferring cause and effect from correlations between non-experimentally manipulated properties. Thus, for example, he argued that the correlation between smoking and lung cancer did not prove that smoking causes lung cancer. Because of the fact that subjects had not been randomly assigned to possible smoking and non-smoking conditions, alternative causal explanations were possible. It could be, for example, that cancer prone individuals are, for some reason or other, more likely to smoke. Fisher argued that such alternative explanations can be eliminated through random assignment of subjects to conditions.

The whole issue of inferring causation from correlation is a complex one. There are some instances in which we do not wish to infer causation from correlation. This is true of the positive correlation between the number of saloons and the number of churches in U.S. cities. Despite the fact that the more churches there are the more saloons there are, common sense indicates that the causal inference that churches cause saloons is absurd. There are other instances, however, in which we do wish to infer causation from correlation. This is the case for the correlation between the distances to the sun and moon and the height of the tides. There are currently some interesting methodological developments in what has come to be called path analysis. Path analysis is directly concerned with the testing of causal models with correlational data (cf. Blalock, 1961). Fisher's emphasis upon randomization stands as the orthodox position against which such new developments are opposed.

Whatever the outcome of future developments in methodology and statistics, Fisher's place in the history of both these traditions is secure. Fisher stands out in a strong British tradition of statistics. Most of the important early contributions to statistics came from the continent. With regard to later developments, however, men like Galton, Pearson, Gosset, and Fisher have assured Britain of a preeminent position.

EXERCISES

1. Define the following terms:

(a) one-factor analysis of variance
(b) sum of squares
(c) total sum of squares
(d) within groups sum of squares
(e) between means sum of squares
(f) between groups sum of squares
(g) mean square
(h) between groups mean square
(i) within groups mean square
(j) two-factor analysis of variance
(k) two-factor interaction

2. Test $H_0: \mu_1 = \mu_2 = \mu_3$ against the alternative $H_1: \mu_1 \neq \mu_2$ and/or $\mu_1 \neq \mu_3$, etc., for the following sets of scores. Choose an α level.

T_1	T_2	T_3
6	2	10
5	1	9
4	7	4
7	6	10
3	5	11
3	5	9

3. Test H_0: $\mu_1 = \mu_2 = \mu_3 = \mu_4$, given the following sample set of scores. Choose an α level.

T_1	T_2	T_3	T_4
1	4	3	14
5	6	9	8
6	6	9	15
8	9	9	13

4. Listed below are the conformity scores obtained under four different social conditions, T_1, T_2, T_3, and T_4. The scores were obtained from independent samples of human subjects. Test the null hypothesis that the population mean is the same under all treatments. Choose an α level.

T_1	T_2	T_3	T_4
4	5	4	5
5	3	5	4
4	3	4	5
2	2	1	3
4	4	4	5
2	2	2	3
2	2	2	3
3	5	4	3
5	4	5	6
2	2	2	3

5. Three different management styles were tested with different work groups, comparable in type of work and randomly selected from work groups in a given company. In all, eighteen work groups were studied, six under each style. After six months, performance ratings of each group were made.

Given the performance scores below, test the null hypothesis that all three management styles effect the same mean performance level. Choose an α level.

Management Styles

1	2	3
15	4	10
16	5	14
8	12	9
9	6	20
11	8	13
12	6	12

6. Test the null hypothesis for factor A, factor B, and the A-by-B interaction with the data from the following two-factor design. Choose an α level.

		FACTOR A	
		A_1	A_2
FACTOR B	B_1	10	8
		11	4
		5	4
		8	10
	B_2	4	11
		2	11
		6	8
		6	12

7. The data below give the number of seconds taken to reach agreement concerning the division of a monetary reward. Conditions A_1 and A_2 represent, respectively, instructions to bargain competitively and instructions to bargain cooperatively. Conditions B_1 and B_2 represent, respectively, two types of mixed-sex dyads, married and stranger. Test the null hypothesis for factor A, factor B, and the A-by-B interaction. Choose an α level.

		FACTOR A	
		A_1	A_2
FACTOR B	B_1	20	15
		15	10
		23	8
		18	20
		30	11
	B_2	25	13
		18	20
		32	16
		35	25
		19	21

8. A_1, A_2, and A_3 represent three different psychotherapeutic techniques, and B_1 and B_2 represent two different types of patients in a mental hospital. Use a two-factor analysis of variance to test null hypotheses about the two main factors and their interaction. The data below represent ratings of improvement after one year of psychotherapy. A rating of 1 represents no improvement, and 10 represents great improvement. Choose an α level.

FACTOR A

		A_1	A_2	A_3
FACTOR B	B_1	1	6	9
		1	2	8
		5	4	9
		6	5	4
		2	2	5
	B_2	10	5	4
		8	9	5
		9	8	3
		6	7	6
		7	8	4

9. The following data are ratings (on a 12-point scale) of persons sampled from four different socio-economic levels (B_1, B_2, B_3, and B_4) and administered three different treatments (A_1, A_2, and A_3). Test the null hypotheses for factor A, factor B, and the A-by-B interaction. Choose an α level.

FACTOR A

		A_1	A_2	A_3
FACTOR B	B_1	2	8	3
		2	9	2
		3	4	4
		6	5	4
	B_2	4	4	3
		6	4	7
		7	3	6
		5	8	5
	B_3	1	5	10
		1	6	9
		2	4	5
		3	8	8
	B_4	9	5	2
		11	8	3
		8	3	1
		6	4	3

10. Graph the group means in Exercise 9.

11. What effects contribute to the within groups sum of squares and the between groups sum of squares?

12. What is the rationale for dividing the between groups mean square by the within groups mean square in a one-factor analysis of variance?

13. Why is the homogeneity of variance assumption necessary for analysis of variance?

14. Suppose that 10 different subjects are tested in each of the 4 cells of a 2-by-2 design. One two-level factor can be labeled "A", and the other "B." Suppose, further, that there is no variance within any of the cells so that any obtained effect is significant. What effect or effects are significant for each of the following patterns of results? The numbers in the cells are arithmetic means of the 10 scores.

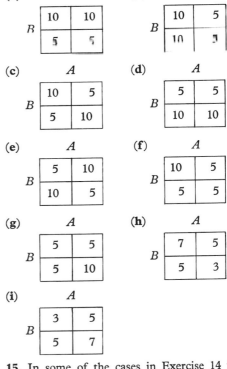

15. In some of the cases in Exercise 14 there were interactions, and in some there were not. For each case examine the means of the two means in each diagonal and then formulate a simple rule that differentiates the cases that do, and do not, show interactions.

16. List and describe the possible main effects and interactions in a three-factor (A, B, and C) design.

Prologue to Chapter 11

Compared to Chapter 10, Chapter 11 is fairly easy. Chapter 11 describes one of the procedures for analyzing frequency data. The chi square statistic, which is introduced in this chapter, is one of the oldest and most widely used test statistics in the entire field of statistics. Thus, anyone who reads widely in any number of disciplines, whether they be psychology, sociology, anthropology, public health, economics, biology, or genetics, will sooner or later encounter the chi square statistic.

11

An Introduction to Chi Square

This chapter introduces *chi square*, χ^2, a test statistic that can be applied to data expressed as frequencies. Psychological data often take the form of frequencies with which measurement classes occur. This is particularly true where the measurement classes represent only qualitative distinctions. The frequencies of true and false answers to a questionnaire item or the frequencies of voters registered Republican, Democrat, and Independent are examples of qualitative classification resulting in frequency data.

Frequency data are in no sense limited to the results of qualitative measurement, however. Examples of frequency distributions for quantified properties, IQ for example, were given in Chapter 2.

CHI SQUARE STATISTIC

The chi square statistic is defined by:

$$\chi^2 = \sum_{i=1}^{k} \frac{(f_{O_i} - f_{E_i})^2}{f_{E_i}} \qquad (11\text{-}1)$$

The subscript i refers to a particular measurement class and k to the number of measurement classes. The symbol f_{O_i} stands for the *obtained frequency* of a particular measurement class and f_{E_i} for the *expected frequency* of that class given a particular null hypothesis. To state formula 11-1 verbally, the χ^2 statistic equals the sum across k measurement classes of the ratio of the squared difference between *obtained* (f_{O_i}) and *expected* (f_{E_i}) frequencies to the *expected* frequency (f_{E_i}).

Provided certain conditions (discussed below) can be met, the sampling distribution of the χ^2 statistic defined in formula 11-1 follows the theoretical continuous chi square distribution for which certain values and probabilities are given in Table IV, Appendix C. The values of χ^2 in Table IV cut off .20, .10, .05, .02, .01, and .001 of the upper tail of a theoretical chi square distribution. The hypothesis testing discussed in this chapter makes use of only the upper tail of the chi square distribution. Also note that (as with t and F) there is a different chi square distribution for different degrees of freedom (df). The determination of df is discussed below.

Certain conditions must be met or assumptions made in order for the theoretical chi square distribution (Table IV) to be an adequate model for the sampling distribution of the χ^2 statistic defined in formula 11–1. These conditions are:

1. Simple random sampling exists.
2. The classified observations are independent of each other.
3. No expected frequency (f_{E_i}) is smaller than 5.
4. The sum of obtained frequencies (f_{O_i}) and the sum of expected frequencies (f_{E_i}) are equal.

The first condition guarantees that each unit in the population has an equal chance of being selected. The second condition requires that the selection and classification of one observation in no way depends on the selection and classification of any other observation. For example, if the data of interest are the frequencies of true and false responses to a given test item by students in a given class, the condition of independence means that each student's answer (true or false) to the item is completely independent of (in no way depends on) any other student's answer. The condition of independence would be violated if one or more students were copying from other students or if certain students had studied together and thus increased the probability that they would give the same answers. In such cases certain students' answers to the item are not independent of every other student's answer.

The third condition, that no expected frequency (f_{E_i}) should be smaller than 5, is particularly important when the number of measurement classes is small (4 or less). While the specific cutoff at 5 is somewhat arbitrary, it does reflect the fact that as the expected frequencies decrease in size, the theoretical chi square distribution becomes a poorer model of the sampling distribution of the χ^2 statistic.

The fourth condition regarding the equality of $\sum f_{O_i}$ and $\sum f_{E_i}$ guarantees that the obtained χ^2 will reflect differences in the distributions of obtained and expected frequencies across measurement classes rather than a simple difference between the totals of observed and expected frequencies.

Related to this fourth condition is the requirement that when a researcher's interest is focused on the *occurrence* of a particular value of some property (e.g., the occurrence of a head to the toss of a coin or the occurrence of a male after sampling from a mixed sex group) he or she must also count *non-occurrences* (tail and female for the cited examples), if the χ^2 statistic is to be used. The researcher cannot simply compare the obtained and expected frequencies of the value that is focused upon, but must also compare obtained and expected frequencies of non-occurrences of that value. For example, consider a situation in which a coin is tossed 100 times and the frequency of heads counted. The expected frequency of heads is 50. Suppose, though, that the obtained frequency of heads is 70. In this case, since the obtained and expected frequencies differ, the above-stated fourth assumption has been violated. The problem can be avoided, however, by attending to the obtained frequency of *both* heads and tails (70 + 30 = 100), and the expected frequency of *both* heads and tails (50 + 50 = 100). In this case the sums of obtained and expected frequencies are equal, and the fourth assumption is met.

Chi square may be applied to frequency data obtained from a single sample or to frequency data obtained from two or more samples. We will consider these cases in order.

CHI SQUARE APPLIED TO A SINGLE SAMPLE: TEST OF GOODNESS OF FIT

One of the uses of χ^2 is to determine whether a single sample of data was selected from a population having a certain probability distribution. This use of χ^2 involves comparing the frequency distribution for a single sample with a frequency distribution expected if a hypothesized population probability distribution exists. This single sample case is frequently referred to as a test of *goodness of fit*. The concern is with the goodness of fit between obtained frequencies and frequencies expected on the basis of a hypothesized population distribution. This use of χ^2 is analogous to the use of t to test hypotheses concerning a population mean. In both instances there is only one sample.

Obtained and Expected Frequencies

Suppose that an experimenter selects ninety subjects from a population and has each subject designate which of three kinds of orange juice, A, B, or C, he prefers. The results from such a hypothetical survey are presented in Table 11.1.

The second column, labeled f_{O_i}, presents *the obtained frequency of preference* for each of the three types of orange juice. In this case we have three measurement classes—three types of orange juice. The third column, labeled

TABLE 11.1. *Distribution of Preferences for Three Kinds of Orange Juice and Calculation of χ^2.*

(1) JUICE	(2) f_{O_i}	(3) f_{E_i}	(4) $f_{O_i} - f_{E_i}$	(5) $(f_{O_i} - f_{E_i})^2$	(6) $(f_{O_i} - f_{E_i})^2 / f_{E_i}$
A	25	30	−5	25	0.83
B	25	30	−5	25	0.83
C	40	30	10	100	3.33
Σ	90	90			$\chi^2 = 4.99$

f_{E_i}, *presents expected frequencies that follow from a hypothesis* about the population probability distribution of preferences. In this case the hypothesis is that each juice is equally preferred, i.e., the probability of sampling a person who prefers one juice is equal to the probability of sampling a person who prefers one of the other juices. When simple random sampling exists the probability of selecting an element belonging to a particular subset equals the proportion of elements in the population that belong to that subset. Thus, in this case, the hypothesis of equal preference can be though of as equivalent to the hypothesis that one third of the population prefers each type of orange juice.

Expected frequencies are derived directly from the probabilities of each measurement class as designated by the hypothesis the researcher plans to test, i.e., the null hypothesis. In the example situation the null hypothesis is designated as:

$$H_0: p(A) = p(B) = p(C) = 1/3$$

where $p(A)$, $p(B)$, and $p(C)$ refer to the probabilities of preference for juices A, B, and C, respectively. The expected frequency for a given measurement class is equal to the total sample size multiplied by the probability designated in the null hypothesis:

$$f_{E_A} = p(A)(n) = (1/3)(90) = 30$$
$$f_{E_B} = p(B)(n) = (1/3)(90) = 30$$
$$f_{E_C} = p(C)(n) = (1/3)(90) = 30$$

Hypothesis Testing

The χ^2 statistic can be used to test H_0 if, as in the present case, the four stated conditions are met:

1. Simple random sampling is used; i.e., each subject in the population has an equal chance of being selected.
2. Subjects are selected and express preferences independently of each other.

3. All of the f_Es are greater than 5 (in fact all are greater than 29).
4. The sum of observed frequencies equals the sum of expected frequencies:

$$\sum f_{O_i} = \sum f_{E_i} = 90$$

The next step is to state the alternative hypothesis H_1; in this case:

$$H_1: p(A) \neq p(B) \quad \text{and/or} \quad p(A) \neq p(C) \quad \text{and/or} \quad p(B) \neq p(C)$$

Stated verbally the alternative hypothesis is that all three juices are not equally preferred. This alternative hypothesis is a general statement of inequality rather than a specific directional hypothesis such as $H_1 : p(A) > p(B) > p(C)$. As stated above, the null hypothesis is:

$$H_0 : p(A) = p(B) = p(C)$$

Next, the α level is set (say at .05), and the critical region is designated. For the type of χ^2 test discussed here, the critical region for χ^2 is never split between the two tails, but is always placed in the upper tail. The reason for this lies in the fact that the χ^2 formula is insensitive to the direction of differences between obtained and expected frequencies. In Table 11.1, the obtained frequencies for juices A, B, and C are 25, 25, and 40. As far as the computed χ^2 value is concerned, however, the obtained frequencies for A, B, and C could have been 25, 40, and 25, or 40, 25, and 25. Since the null hypothesis is equal preference ($E_A = E_B = E_C = 30$), formula 11-1 would yield identical χ^2s for each of these sets of data. The t statistic varies from negative to positive depending upon the direction of the difference between means. Chi square, on the other hand, varies from zero to positive depending upon the absolute differences between obtained and expected frequencies. The larger the absolute differences between obtained and expected frequencies the larger is χ^2. For that reason the entire critical region is placed in the right tail of the theoretical probability distribution.

Chi square's insensitivity to the direction of the differences between obtained and expected frequencies makes it pointless for the experimenter to formulate an alternative hypothesis that specifies the direction of difference. The most straightforward procedure is to formulate the non-directional alternative hypothesis and then, after rejecting the null hypothesis and accepting the non-directional alternative hypothesis, take note of the direction of differences.

In order to enter Table IV to obtain the critical value of χ^2, degrees of freedom (df) must be specified. When a hypothesis about a population probability distribution is being tested, the df associated with the test is equal to the number of measurement classes minus 1 ($k - 1$). For the present example, df equals 2 since there are three classes:

$$\begin{aligned} df &= k - 1 \\ &= 3 - 1 \\ &= 2 \end{aligned}$$

Given the total frequency of 90 (see Table 11.1), the frequencies of any two classes are completely free; the third is determined. The third class is determined by the restriction that the total frequency for all three classes must sum to 90. With an α level of .05 and 2 df, the critical value of χ^2 is 5.99. Thus, the critical region is from 5.99 to infinity.

Calculation of χ^2 for the juice preference experiment is illustrated in columns 4, 5, and 6 of Table 11.1. The obtained χ^2 of 4.99 does not lie in the critical region (is not larger than 5.99) and thus the experimenter fails to reject the null hypothesis that the three juices are equally preferred.

A Further Example

Over the past five years, in a given school, the proportions (probabilities) of third graders scoring pass, marginal, and fail on an achievement test were respectively .50, .20, and .30. During the present year the third graders participated in an experimental curriculum designed to improve performance. At the end of the year, out of thirty students, twenty-four scored pass, five marginal, and one fail. Given this data, the χ^2 statistic can be used to test the null hypothesis that the experimental curriculum had no effect, that is, that the current students were selected from a population with the same probability distribution as that of the past five years.

First, expected frequencies are computed:

$$f_{E_{pass}} = (.50)(30) = 15$$
$$f_{E_{marginal}} = (.20)(30) = 6$$
$$f_{E_{fail}} = (.30)(30) = 9$$

Then the χ^2 statistic is calculated using formula 11–1:

$$\chi^2 = \sum_{i=1}^{k} \frac{(f_{O_i} - f_{E_i})^2}{f_{E_i}}$$

$$= \frac{(24 - 15)^2}{15} + \frac{(5 - 6)^2}{6} + \frac{(1 - 9)^2}{9}$$

$$= 12.68$$

If the significance level has been set at $\alpha = .01$, then, referring to Table IV, Appendix C, the critical χ^2 value is 9.210 at 2 df. Since the obtained χ^2 of 12.68 exceeds 9.210, the null hypothesis of no effect is rejected. We conclude by examination of the data that the experimental curriculum improved performance.

CHI SQUARE APPLIED TO TWO OR MORE SAMPLES: TEST OF INDEPENDENCE

A second major use of χ^2 is to determine whether or not the probability distributions of two or more populations differ. Suppose the juice preference experiment had been completed on two samples, one of children and one of adults. In this case the concern would be with determining whether or not children and adults differ in their preferences for the three types of juice. This situation is analogous to a t test applied to two samples in order to test a hypothesis concerning the equality of two population means. The χ^2 test can, however, be applied to more than two samples. This would be the case, for instance, if we compared the juice preferences of children, teenagers, and adults. This latter situation would be analogous to using F in order to test a hypothesis concerning the equality of three population means.

Chi square applied to two or more samples is frequently referred to as a test of independence. In Chapter 6, it was stated that the outcomes comprising an intersection of outcomes are said to be independent if the probability of each outcome is in no way altered by or dependent on the occurrence of the other outcome(s). In terms of the above problem, the question relates to the independence of the outcomes, child and adult, and preference between any two of the three types of juice (A and B, A and C, B and C). For example, does the knowledge that one is concerned with a child alter the probability that juice A will be preferred to juice B? Altogether there are six outcome intersections: child and A versus B, child and A versus C, child and B versus C, adult and A versus B, adult and A versus C, and adult and B versus C.

In the present situation the multiplicity of outcome intersections makes it convenient to speak of the independence of variables. *Two variables, X and Y (representing measures of two properties), are independent if each outcome intersection, arising from a pairing of a value of X with a value of Y, is comprised of independent outcomes.* Stated differently, variables X and Y are independent if the certain occurrence of any value of X in no way alters the probability of occurrence of any value of Y (and vice versa). Two variables are statistically related or associated (not independent), if specifying a value of one variable provides information regarding the probability of occurrence of different values for the other variable. If the probabilities of values of Y change depending on which value of X has occurred, or is certain to occur, then X and Y are statistically related, dependent to some degree. In terms of the juice preference experiment, the X variable (child or adult) would be independent of the Y variable (juice type) if specifying that a given individual was either a child or an adult provided no information regarding juice preference. Such independence would exist if children and adults did not differ in their juice preferences. Dependence among

X (child or adult) and Y (juice type) would exist if probabilities of preferences for the different juices changed depending on whether the individual was a child or an adult.

Obtained and Expected Frequencies

Suppose an investigator is interested in determining whether or not in a given community the size of family income is related to owning more than one car. In order to study this he arranges for household heads from a random sample of high income families (income greater than $10,000) and household heads from a random sample of low income families (income less than $10,000) to be interviewed. Hypothetical data derived from "yes" and "no" answers to the question, "Do you own more than one car?" are presented in Table 11.2. The investigator wishes to test a null hypothesis designating the independence of variable X, representing two levels of income, and variable Y, representing "yes" and "no" replies.

Following from the concept of independent variables discussed above such a null hypothesis states that the probability of a family having more than one car is identical for high and low income families and thus, by implication, that the probability of a family having no more than one car is identical for high and low income families. In other words, the certain occurrence of any value of X (high or low income) in no way alters the probability of any value of Y (yes or no responses). This null hypothesis can be symbolized as:

$$H_0: p(\text{yes} \mid \text{high}) = p(\text{yes} \mid \text{low})$$

and,

$$p(\text{no} \mid \text{high}) = p(\text{no} \mid \text{low})$$

This null hypothesis should be read: The probability of a "yes" response given that the family has a high income, equals the probability of a "yes"

TABLE 11.2. *Distribution of "Yes" and "No" Responses for High and Low Income Groups.*

		Y		
		YES	NO	Σ
X	HIGH INCOME	60	40	100
	LOW INCOME	20	80	100
	Σ	80	120	200

response, given that the family has a low income; and the probability of a "no" response, given that the family has a high income, equals the probability of a "no" response, given that the family has a low income.

If independence exists, it is also true that the certain occurrence of any value of variable Y (yes or no) does not alter the probabilities of occurrence of any value of variable X (high or low income). That is, the null hypothesis could be stated:

$$H_0: p(\text{high} \mid \text{yes}) = p(\text{high} \mid \text{no})$$

and,

$$p(\text{low} \mid \text{yes}) = p(\text{low} \mid \text{no})$$

The first statement of H_0 is preferred because the condition of high or low income is known prior to asking the question about number of cars owned and because if there is dependence between the two variables, it seems more likely that size of income influences the number of cars owned than vice versa. The first statement of H_0 is in terms of "yes" or "no" responses given the condition of high or low income.

To compute a χ^2 value and test the null hypothesis of independence, it is necessary to obtain expected frequencies of outcome intersections when independence exists. In the present example, four different expected frequencies must be determined, corresponding to the four outcome intersections: yes ∩ high, yes ∩ low, no ∩ high, no ∩ low. As with the prior example, expected frequency of an outcome (measurement class) is determined by multiplying total sample size times the probability of the outcome of the truth of H_0. Thus, for the present example the four expected frequencies can be obtained by multiplying n ($n_{\text{high}} + n_{\text{low}} = 100 + 100 = 200$) times the probability of each outcome intersection, if variables X and Y are independent (H_0).

From Chapter 6, we know that if outcomes comprising an intersection of outcomes are independent, the probability given the outcome intersection equals the product of probabilities of the individual outcomes. This is a statement of the multiplication rule. In the present case, if H_0 is true, the four joint outcomes are comprised of independent outcomes; and thus, their probabilities can be determined by simple multiplication:

$$p(\text{yes} \cap \text{high}) = p(\text{yes}) \cdot p(\text{high})$$
$$p(\text{yes} \cap \text{low}) = p(\text{yes}) \cdot p(\text{low})$$
$$p(\text{no} \cap \text{high}) = p(\text{no}) \cdot p(\text{high})$$
$$p(\text{no} \cap \text{low}) = p(\text{no}) \cdot p(\text{low})$$

Since a null hypothesis of independence makes no specification of probabilities of values of the X and Y variables, i.e., no specification of $p(\text{yes})$, $p(\text{no})$, $p(\text{high})$,

$p(\text{low})$, these probabilities must be estimated before expected frequencies can be calculated. The estimation of the population probability for a particular value of a variable is based on the relative frequency of that value in the total sample. For example, the best estimate of the probability of a "yes" response in the population is the relative frequency of occurrence of a "yes" response in the total sample:

$$\text{est. } p(\text{yes}) = \frac{f(\text{yes})}{n}$$

$$= \frac{80}{200}$$

$$= .4$$

In a similar manner:

$$\text{est. } p(\text{no}) = \frac{f(\text{no})}{n}$$

$$= \frac{120}{200}$$

$$= .6$$

$$\text{est. } p(\text{high}) = \frac{f(\text{high})}{n}$$

$$= \frac{100}{200}$$

$$= .5$$

and,

$$\text{est. } p(\text{low}) = \frac{f(\text{low})}{n}$$

$$= \frac{100}{200}$$

$$= .5$$

Having these estimates, we can then estimate the probabilities of the four outcome intersections using the multiplication rule:

$$\text{est. } p(\text{yes} \cap \text{high}) = \text{est. } p(\text{yes}) \cdot \text{est. } p(\text{high})$$

$$= \frac{f(\text{yes})}{n} \cdot \frac{f(\text{high})}{n}$$

$$= (.4)(.5)$$

$$= .2$$

$$\text{est. } p(\text{yes} \cap \text{low}) = \text{est. } p(\text{yes}) \cdot \text{est. } p(\text{low})$$
$$= \frac{f(\text{yes})}{n} \cdot \frac{f(\text{low})}{n}$$
$$= (.4)(.5)$$
$$= .2$$

$$\text{est. } p(\text{no} \cap \text{high}) = \text{est. } p(\text{no}) \cdot \text{est. } p(\text{high})$$
$$= \frac{f(\text{no})}{n} \cdot \frac{f(\text{high})}{n}$$
$$= (.6)(.5)$$
$$= .3$$

and,

$$\text{est. } p(\text{no} \cap \text{low}) = \text{est. } p(\text{no}) \cdot \text{est. } p(\text{low})$$
$$= \frac{f(\text{no})}{n} \cdot \frac{f(\text{low})}{n}$$
$$= (.6)(.5)$$
$$= .3$$

Expected frequencies for outcome intersections are then determined by simply multiplying joint probability estimates "times" the total sample size n, in this case 200. Expected frequencies for a test of independence in the present example are thus:

$$f_E(\text{yes} \cap \text{high}) = \text{est. } p(\text{yes} \cap \text{high}) \cdot n$$
$$= (.2)(200)$$
$$= 40$$

$$f_E(\text{yes} \cap \text{low}) = \text{est. } p(\text{yes} \cap \text{low}) \cdot n$$
$$= (.2)(200)$$
$$= 40$$

$$f_E(\text{no} \cap \text{high}) = \text{est. } p(\text{no} \cap \text{high}) \cdot n$$
$$= (.3)(200)$$
$$= 60$$

TABLE 11.3. *Expected Frequencies of "Yes" and "No" Responses for High and Low Income Groups.*

		Y		
		YES	NO	Σ
X	HIGH INCOME	40	60	100
	LOW INCOME	40	60	100
	Σ	80	120	200

and,

$$f_E(\text{no} \cap \text{low}) = \text{est. } p(\text{no} \cap \text{low}) \cdot n$$
$$= (.3)(200)$$
$$= 60$$

Table 11.3 summarizes these expected frequencies.

Notice that the row and column totals are the same for Table 11.2 as for Table 11.3. This fact allows one to calculate readily the three remaining expected frequencies after the first one has been obtained as above. For example, the low-income-yes frequency must be 40 since the high-income-yes frequency is 40 and the column total is 80.

Hypothesis Testing

Assuming simple random sampling and independent observations, and noting that the smallest f_{E_i} is greater than 5 and that $\sum f_{O_i} = \sum f_{E_i} = 200$, the χ^2 statistic can be used to test H_0.

The alternative hypothesis, H_1, is a statement that variables X and Y are not independent:

$$H_1: p(\text{yes} \mid \text{high}) \neq p(\text{yes} \mid \text{low})$$

and,

$$p(\text{no} \mid \text{high}) \neq p(\text{no} \mid \text{low})$$

This hypothesis states that the probabilities of "yes" and "no" replies depend on whether income level is high or low.

The null hypothesis, which is the logical contradictory of the alternative hypothesis, was stated above as:

$$H_0: p(\text{yes} \mid \text{high}) = p(\text{yes} \mid \text{low})$$

and,
$$p(\text{no} \mid \text{high}) = p(\text{no} \mid \text{low})$$

Next, an α level is set (say at .01) and the critical region is designated. As with the chi square test for a population probability distribution, the critical region for a test of independence is always one-tailed and always in the upper tail. As noted previously, the chi square statistic is not sensitive to the direction of differences between obtained and expected frequencies but only to absolute size of the differences.

If l represents the number of classes for X, and m the number of classes for Y, df for a chi square test of independence is given by: $df = (l - 1)(m - 1)$. For the present example, where X represents two classes of income and Y represents two classes of replies, "yes" and "no":

$$\begin{aligned} df &= (l - 1)(m - 1) \\ &= (2 - 1)(2 - 1) \\ &= 1 \end{aligned}$$

Given the row and column totals in Table 11.3, only one of the cell frequencies is free to vary. After one of the cell frequencies is determined, the others are fixed by the totals. For 1 df and an α of .01, the critical value is 6.64, according to Table IV, Appendix C.

The use of formula 11–1 to calculate χ^2 for the obtained frequencies in Table 11.2 is illustrated in Table 11.4. The obtained χ^2 of 33.34 exceeds the critical value of 6.64. Thus the null hypothesis of independence is rejected. In the examined community, family income is related to number of cars owned.

The above test indicates that there is a difference between high and low income families in the probabilities of families owning two or more cars. It is obvious that this difference has to be in one of two directions—either among high income families there is a greater probability of owning two or more cars than among low income families, or vice versa. Since the obtained frequencies in Table 11.1 are consistent with the former difference, the experimenter may

TABLE 11.4. *Calculation of χ^2 for the Data in Table 11.2.*

X	Y	f_{O_i}	f_{E_i}	$f_{O_i} - f_{E_i}$	$(f_{O_i} - f_{E_i})^2$	$(f_{O_i} - f_{E_i})^2 / f_{E_i}$
HIGH INCOME	Yes	60	40	20	400	10
	No	40	60	−20	400	6.67
LOW INCOME	Yes	20	40	−20	400	10
	No	80	60	20	400	6.67
	Σ	200	200			$\chi^2 = 33.34$

conclude that his data support the directional hypothesis of a greater probability of high income families owning two or more cars than low income families.

Correction for Continuity

As noted above, the chi square distribution referred to is a theoretical distribution. The statistic obtained from formula 11–1 has a sampling distribution that is well approximated by a theoretical chi square distribution. It is also possible to calculate the exact probabilities for certain obtained frequencies. When df for the χ^2 statistic equals 1, the approximation to these exact probabilities can be improved through a correction for continuity. This correction is made by subtracting .5 from each of the absolute discrepancies, $f_{O_i} - f_{E_i}$, before squaring them:

$$\chi^2 = \sum_{i=1}^{i=k} \left[\frac{(|f_{O_i} - f_{E_i}| - .5)^2}{f_{E_i}} \right] \tag{11–2}$$

The correction for continuity reduces the size of the obtained χ^2. The correction is most important for samples in which *any expected frequency is less than 10*. The correction is used only when df equals 1.

The use of formula 11–2 to calculate χ^2 is illustrated in Table 11.5. Here, variable X has two measurement classes, as does variable Y, so that df equals 1. The correction for continuity is necessary because df is 1 and one of the expected frequencies is less than 10.

Computing χ^2 for More than Two Samples

Table 11.6 contains frequency data for more than two samples, and Table 11.7 illustrates computation of χ^2 for the data in Table 11.6. Here, we have three samples, representing, respectively, adults, teenagers, and children; and preferences for three juices, A, B, and C. The computed χ^2 would be used to test a null hypothesis stating the equivalence of juice preferences in the three age

TABLE 11.5. *Calculation of χ^2 with the Correction for Continuity.*

		f_{O_i}	f_{E_i}	$f_{O_i} - f_{E_i}$	$\|f_{O_i}-f_{E_i}\| - .5$	$(\|f_{O_i}-f_{E_i}\|-.5)^2$	$\dfrac{(\|f_{O_i}-f_{E_i}\|-.5)^2}{f_{E_i}}$
X_1	Y_1	20	13.33	6.67	6.17	38.07	2.86
	Y_2	0	6.67	−6.67	6.17	38.07	5.71
X_2	Y_1	20	26.67	−6.67	6.17	38.07	1.43
	Y_2	20	13.33	6.67	6.17	38.07	2.86
	Σ	60	60.00				$\chi^2 = 12.86$

TABLE 11.6. *Frequency of Juice Preferences for Three Samples: Adults, Teenagers, and Children.*

		Y JUICES			
		A	B	C	Σ
X	ADULTS	25	25	40	90
	TEENAGERS	20	35	35	90
	CHILDREN	40	25	25	90
	Σ	85	85	100	270

groups, i.e., the independence of variable X, representing age groups, and variable Y, representing the different juices.

The f_Es were calculated as follows. First, probabilities for outcome intersections (assuming independence) were estimated:

$$\text{est. } p(\text{adult} \cap A) = \text{est. } p(\text{adult}) \cdot \text{est. } p(A)$$
$$= \frac{(25 + 25 + 40)}{270} \cdot \frac{(25 + 20 + 40)}{270}$$
$$= \frac{(90)(85)}{(270)^2}$$

$$\text{est. } p(\text{adult} \cap B) = \text{est. } p(\text{adult}) \cdot \text{est. } p(B)$$
$$= \frac{(25 + 25 + 40)}{270} \cdot \frac{(25 + 35 + 25)}{270}$$
$$= \frac{(90)(85)}{(270)^2}$$

TABLE 11.7. *Computation of χ^2 for Data (in Table 11.6) from Three Samples.*

X	Y	f_{O_i}	f_{E_i}	$f_{O_i} - f_{E_i}$	$(f_{O_i} - f_{E_i})^2$	$(f_{O_i} - f_{E_i})^2 / f_{E_i}$
	A	25	28.33	−3.33	11.09	0.39
ADULTS	B	25	28.33	−3.33	11.09	0.39
	C	40	33.33	6.67	44.49	1.33
	A	20	28.33	−8.33	69.39	2.45
TEENAGERS	B	35	28.33	6.67	44.49	1.57
	C	35	33.33	1.67	2.79	0.08
	A	40	28.33	11.67	136.19	4.81
CHILDREN	B	25	28.33	−3.33	11.09	0.39
	C	25	33.33	−8.33	69.39	2.08
	Σ	270	~270			$\chi^2 = 13.49$

$$\text{est. } p(\text{adult} \cap C) = \text{est. } p(\text{adult}) \cdot \text{est. } p(C)$$
$$= \frac{(25 + 25 + 40)}{270} \cdot \frac{(40 + 35 + 25)}{270}$$
$$= \frac{(90)(100)}{(270)^2}$$

$$\text{est. } p(\text{teenager} \cap A) = \text{est. } p(\text{teenager}) \cdot \text{est. } p(A)$$
$$= \frac{(20 + 35 + 35)}{270} \cdot \frac{(25 + 20 + 40)}{270}$$
$$= \frac{(90)(85)}{(270)^2}$$

$$\text{est. } p(\text{teenager} \cap B) = \text{est. } p(\text{teenager}) \cdot \text{est. } p(B)$$
$$= \frac{(20 + 35 + 35)}{270} \cdot \frac{(25 + 35 + 25)}{270}$$
$$= \frac{(90)(85)}{(270)^2}$$

$$\text{est. } p(\text{teenager} \cap C) = \text{est. } p(\text{teenager}) \cdot \text{est. } p(C)$$
$$= \frac{(20 + 35 + 35)}{270} \cdot \frac{(40 + 35 + 25)}{270}$$
$$= \frac{(90)(100)}{(270)^2}$$

$$\text{est. } p(\text{child} \cap A) = \text{est. } p(\text{child}) \cdot \text{est. } p(A)$$
$$= \frac{(40 + 25 + 25)}{270} \cdot \frac{(25 + 20 + 40)}{270}$$
$$= \frac{(90)(85)}{(270)^2}$$

$$\text{est. } p(\text{child} \cap B) = \text{est. } p(\text{child}) \cdot \text{est. } p(B)$$
$$= \frac{(40 + 25 + 25)}{270} \cdot \frac{(25 + 35 + 25)}{270}$$
$$= \frac{(90)(85)}{(270)^2}$$

$$\text{est. } p(\text{child} \cap C) = \text{est. } p(\text{child}) \cdot \text{est. } p(C)$$
$$= \frac{(40 + 25 + 25)}{270} \cdot \frac{(40 + 35 + 25)}{270}$$
$$= \frac{(90)(100)}{(270)^2}$$

Then, f_Es were computed:

$$f_E(\text{adult} \cap A) = \text{est. } p(\text{adult} \cap A) \cdot n$$
$$= \frac{(90)(85)}{(270)^2}(270) = 28.33$$

$$f_E(\text{adult} \cap B) = \text{est. } p(\text{adult} \cap B) \cdot n$$
$$= \frac{(90)(85)}{(270)^2}(270) = 28.33$$

$$f_E(\text{adult} \cap C) = \text{est. } p(\text{adult} \cap C) \cdot n$$
$$= \frac{(90)(100)}{(270)^2}(270) = 33.33$$

$$f_E(\text{teenager} \cap A) = \text{est. } p(\text{teenager} \cap A) \cdot n$$
$$= \frac{(90)(85)}{(270)^2}(270) = 28.33$$

$$f_E(\text{teenager} \cap B) = \text{est. } p(\text{teenager} \cap B) \cdot n$$
$$= \frac{(90)(85)}{(270)^2}(270) = 28.33$$

$$f_E(\text{teenager} \cap C) = \text{est. } p(\text{teenager} \cap C) \cdot n$$
$$= \frac{(90)(100)}{(270)^2}(270) = 33.33$$

$$f_E(\text{child} \cap A) = \text{est. } p(\text{child} \cap A) \cdot n$$
$$= \frac{(90)(85)}{(270)^2}(270) = 28.33$$

$$f_E(\text{child} \cap B) = \text{est. } p(\text{child} \cap B) \cdot n$$
$$= \frac{(90)(85)}{(270)^2}(270) = 28.33$$

$$f_E(\text{child} \cap C) = \text{est. } p(\text{child} \cap C) \cdot n$$
$$= \frac{(90)(100)}{(270)^2}(270) = 33.33$$

Notice that the square in the denominator can always be cancelled. This allows for the formulation of a simple rule for determination of the f_E for any one cell. The rule is to multiply the row and column sums for the cell in question and then divide by the total sum. For the adult ∩ A cell in Table 11.6:

$$f_E(\text{adult} \cap A) = \frac{(90)(85)}{270}$$

$$= 28.33$$

This is, of course, the same answer as was obtained above. This rule also can be used for the example cited earlier involving only two samples (Tables 11.2 and 11.3).

In the present example, after four of the f_Es have been calculated by the rule, the remaining four can be more simply obtained through subtraction from the marginal (row or column) totals. Thus since $f_E(\text{adult} \cap A) = 25$ and $f_E(\text{adult} \cap B) = 25$, $f_E(\text{adult} \cap C) = 40$ because the row total is 90 (90 − 25 − 25 = 40). In general, the rule for calculating f_Es need only be used for as many cells as there are degrees of freedom. The remaining f_Es are determined by the marginal totals.

Finally, the χ^2 value of 13.49 was calculated as shown in Table 11.7. In this situation there are 4 degrees of freedom:

$$df = (l - 1)(m - 1)$$
$$= (3 - 1)(3 - 1)$$
$$= 4$$

Using the .05 level of significance, the critical χ^2 value is 9.49. Since the calculated χ^2, 13.49, exceeds this critical value, the null hypothesis may be rejected. Children, teenagers, and adults differ in their relative preferences for the three types of juice.

EXERCISES

1. Define the following terms:
   (a) Chi square statistic
   (b) obtained frequency
   (c) expected frequency
   (d) test for goodness of fit
   (e) test of independence
   (f) df for χ^2 test of independence
   (g) correction for continuity

2. Test the null hypothesis that a sample containing 10 freshmen, 20 sophomores, 30 juniors, and 35 seniors was selected randomly from a high school population where the proportions of freshmen, sophomores, juniors, and seniors are, respectively: .40, .30, .20, and .10. Choose an α level.

3. In a given community the proportions of voters registered Democrat, Republican, and Independent were, respectively: .60, .30, and .10. A researcher was interested in whether voting preferences for a current election followed the same proportions. Therefore, he picked a random sample of 100 voters and asked which

candidate they preferred, the Democrat, Republican, or Independent. Test H_0: $p(D) = .60$; $p(R) = .30$; and $p(I) = .10$. Choose an α level. State the assumptions and conclusions of your test.

	DEMOCRAT	REPUBLICAN	INDEPENDENT
f_{0_i}	40	40	20

4. A district of a city is composed primarily of three cultural groups: Italian, Polish, and Black. The data below represent the number of votes cast for each of two candidates in a city council election. Test the null hypothesis that voting pattern is independent of cultural group. Choose an α level.

	ITALIAN	BLACK	POLISH
CANDIDATE A	100	125	150
CANDIDATE B	125	100	90

5. Married dyads and stranger mixed-sex dyads performed a bargaining task. The data below give the frequencies of dyads of each type performing the task competitively or cooperatively. Test the null hypothesis that the married–stranger variable is independent of the competitive–cooperative variable. Choose an α level.

	MARRIED	STRANGER
COMPETITIVE	20	26
COOPERATIVE	30	24

6. The letters A and B represent two different environments in which a measure of anxiety was made. The results were coded as high, medium, or low anxiety. Test the null hypothesis that anxiety level is independent of environment. Choose an α level.

	A	B
HIGH	20	10
MEDIUM	20	15
LOW	15	30

7. Two classes of first grade students (total $n = 50$) were grouped according to their scores on a reading achievement test. The groupings were in three categories: pass, marginal, and fail. A researcher was interested in determining whether these 50 students could be considered a random sample from the group on which the test was standardized. In the standardization group, the proportions falling into the score intervals labeled pass, marginal, and fail were, respectively: .70, .20, and .10. Test H_0: $p(\text{Pass}) = .70$, $p(\text{Marginal}) = .20$, and $p(\text{Fail}) = .10$. With an α of .01, can H_0 be rejected?

	PASS	MARGINAL	FAIL
f_{0_i}	25	15	10

8. Two hundred persons tasted five different rosé wines and selected the one they preferred most. Test the null hypothesis that, in the population from which the 200 persons were randomly chosen, there is an equal preference for the five wines. Choose an α level and state a conclusion.

	WINES				
	1	2	3	4	5
f_{0_i}	50	30	30	35	55

9. Two sections ($n_1 = 39$, $n_2 = 31$) of a personality course took the same final examination. Choose an α level and test the null hypothesis that sections are independent of grades. What conclusions can be drawn?

	GRADES			
	A	B	C	D
SECTION 1	4	7	10	18
SECTION 2	10	7	10	4

10. The following data are frequencies of true and false answers to an item in a personality inventory by a sample of males ($n_{\text{male}} = 30$) and a sample of females ($n_{\text{female}} = 30$). Choose an α level and test the null hypothesis that sex of subject is independent of the truthfulness of response. What conclusion can be drawn? Compute the χ^2 both with and without the correction for continuity, and note the effect of the correction.

	MALE	FEMALE
TRUE	20	28
FALSE	10	2

11. What is the rule for determination of df for a chi square test of goodness of fit? Explain the rationale for the rule.

12. What is the rule for determination of df for a chi square test of independence? Explain the rationale for the rule.

13. What is the rule for determination of the expected frequency for any given cell of a chi square test of independence? Explain the basis for the rule.

14. What are the special circumstances in which the correction for continuity is used?

15. Relate the independence of variables to the independence of outcomes comprising an intersection of outcomes.

Prologue to Chapter 12

There is a sense in which Chapter 12, Measures of Correlation and Tests of Significance, is an attempt to "tie up some loose ends" relating to the general topic of correlation. Considering the large number of pages in Chapter 5, you may wonder why it is necessary to provide still further information regarding the general topic of correlation. Part of the answer to this question relates to the great importance of correlation in psychological research. And part of the answer relates to the fact that Chapter 5 treated only one measure of correlation, the Pearson product-moment correlation coefficient, and did not consider the important matter of tests of significance. The latter was omitted, of course, because the necessary groundwork in terms of sampling distributions, hypothesis testing, and so forth, had not yet been established. Now that such matters are behind us, the remaining loose ends can be tied up with dispatch.

12

Measures of Correlation and Tests of Significance

In Chapter 5, we discussed the single most important measure of correlation—the Pearson product–moment correlation coefficient. One purpose of the present chapter is to present some additional and less widely used measures of correlation. These are Spearman's rank–order coefficient, the point biserial coefficient, and the correlation ratio.

A second purpose of the present chapter is to present tests of significance for all of the above measures of correlation, including the Pearson product–moment correlation coefficient. Just as there are tests of significance for means and variances, there are also tests of significance for measures of correlation. These tests of significance will be limited to the one-sample case. In the one-sample case a statistic computed from a single sample is compared with a hypothesized population characteristic. Thus for any one particular measure of correlation, a sample coefficient will be compared with a hypothesized value for the population. In the typical case this value is zero.

SIGNIFICANCE OF THE PEARSON PRODUCT–MOMENT CORRELATION COEFFICIENT, r

t Test

Table 5.8 contains the calculation of the zero–order correlation between liking and dating in the Stroebe *et al.* experiment. The r is $+.80$ for the 120 pairs of scores. The two-tailed alternative hypothesis is that the ρ (pronounced rho), the symbol for population correlation, is not equal to zero:

$$H_1: \rho \neq 0$$

and the null hypothesis is the logical contradictory:

$$H_0: \rho = 0$$

In order to test this null hypothesis, the sampling distribution of t may be used. With $df = n - 2$:

(margin note: t-test for correlation in single sample)

$$t = \frac{r}{\sqrt{1 - r^2}} \sqrt{n - 2} \qquad (12\text{-}1)$$

where n is the number of paired observations, or the total frequency of the bivariate frequency distribution. Specifying α at .01 and noting that:

$$df = n - 2$$
$$= 120 - 2$$
$$= 118$$

the critical two-tailed t is approximately ± 2.617.
Substituting into formula 12-1:

$$t = \frac{.80}{\sqrt{1 - (.30)^2}} \sqrt{120 - 2}$$
$$= 14.48$$

Since the obtained t of 14.48 is larger than $+2.617$, the null hypothesis may be rejected.

Suppose, on the other hand, a one-tailed test had been used—a one-tailed test in which the correlation was expected to be greater than zero (i.e., positive):

$$H_1: \rho > 0$$
$$H_0: \rho \leq 0$$

In this case the critical value for $\alpha = .01$ is approximately $+2.358$. With $df = 118$, an *(obtained t)* r greater than approximately $+2.358$ would enable rejection of H_0.

Suppose, finally, the investigator had expected the correlation to be less than zero (i.e., negative):

$$H_1: \rho < 0$$
$$H_0: \rho \geq 0$$

In this case the critical value is approximately -2.358. An *(obtained t)* r less than -2.358 would enable rejection of H_0. In all other respects the procedure for a one-tailed test is identical to that for a two-tailed test. The t is calculated in exactly the same way in all cases.

Table of Significant Values for r

Table V in Appendix C contains the minimum values of r necessary to reject a two-tailed null hypothesis at the .10, .05, .02, and .01 levels of significance, and of a one-tailed null hypothesis at the .05, .025, .01, and .005 levels of significance. These values, which have been calculated for various degrees of freedom, eliminate the necessity of using formula 12–1. In the case of a two-tailed test, the values in Table V should be read $\pm$ (plus or minus); and in the case of a one-tailed test, the values should be read $+$ (plus) if a positive correlation is expected and $-$ (minus) if a negative correlation is expected.

In the present example, with $df = 118$, an r of approximately $\pm .25$ is required to reject the two-tailed null hypothesis at the .01 level of significance.

SPEARMAN'S RANK–ORDER CORRELATION COEFFICIENT, r'

A second measure of correlation is Spearman's rank–order correlation coefficient, symbolized r'. The rank–order coefficient is designed specifically for the situation in which the X and Y scores are converted to ranks and the desired correlation is between the ranks. In Table 12.1 the history and English exam scores for a class of 20 students have been converted to ranks. Thus, the history ranks range from 1 to 20, and the English ranks range from 1 to 20.

Formula

The Spearman rank–order coefficient is defined as follows:

$$r' = 1 - \frac{6 \sum d^2}{n(n^2 - 1)} \qquad (12\text{–}2)$$

Where: d symbolizes the difference between paired ranks
n symbolizes the number of paired ranks.

> The Spearman rank–order coefficient is actually a special case of the Pearson product–moment coefficient. The rank-order coefficient is the special case that arises when the two variables are the first n consecutive untied integers. Stated somewhat differently, formula 12–2 for r' and formula 5–15 for r will yield the same result when applied to untied ranks. When some of the ranks are tied, as illustrated below, the two formulas will give somewhat different results. When the correlation between ranks is desired, formula 12–2 is preferred because of its greater simplicity.

TABLE 12.1. *Calculation of r' Between Score Ranks.*

STUDENT	HISTORY X	ENGLISH Y	RANK X	RANK Y	d	d^2
1	100	99	1	2.5	−1.5	2.25
2	99	100	2	1	1.0	1.00
3	98	99	3	2.5	0.5	0.25
4	95	90	4	6	−2.0	4.00
5	90	98	5	4	1.0	1.00
6	89	85	6	11	5.0	25.00
7	88	90	7	6	1.0	1.00
8	86	90	8	6	2.0	4.00
9	85	86	9	10	−1.0	1.00
10	83	87	10	9	1.0	1.00
11	82	81	11	14	−3.0	9.00
12	78	70	12	17	−5.0	25.00
13	75	66	13	19	−6.0	36.00
14	71	88	14	8	6.0	36.00
15	69	55	15	20	−5.0	25.00
16	65	74	16	15	1.0	1.00
17	59	84	17	12	5.0	25.00
18	53	72	18	16	2.0	4.00
19	35	83	19	13	6.0	36.00
20	31	68	20	18	2.0	4.00
						241.50

$$r' = 1 - \frac{6 \sum d^2}{n(n^2 - 1)}$$

$$= 1 - \frac{6(241.50)}{20(20^2 - 1)}$$

$$= 1 - \frac{1449}{7980}$$

$$= 1 - .18$$

$$= .82$$

Calculation

The calculation of r' is illustrated in Table 12.1. The initial step involves the ranking of the X scores and the ranking of the Y scores. The largest X, or history score, of 100 is assigned the rank of 1; the next largest score of 99 is assigned the rank of 2; and so on, to the lowest score, 31, which is assigned the rank of 20.

A similar procedure is used for the Y, or English, scores. In this instance, however, the situation is more complicated in view of the fact that some scores occur more than once, thus producing tied ranks. When there are tied ranks, the rule is to *assign each of the tied values the mean of the ranks which they jointly occupy*. The second and third scores, which are both 99, are assigned a rank of 2.5: (2 + 3) / 2 = 2.5. Similarly, the fifth, sixth, and seventh scores, which are all 90, are assigned a rank of 6: (5 + 6 + 7) / 3 = 6.

The fifth column, labeled d, is obtained by calculating *rank X − rank Y* for each set of paired ranks. In the sixth column, these values are squared and summed to give 241.50. When this sum, along with an n of 20, is inserted in formula 12–2, an r' of .82 is obtained.

Test of Significance

The statements of the alternative and null hypotheses parallel previous examples. Thus, for a two-tailed test:

$$H_1: \text{population value} \neq 0$$
$$H_0: \text{population value} = 0$$

For a one-tailed test in which a positive correlation is expected:

$$H_1: \text{population value} > 0$$
$$H_0: \text{population value} \leq 0$$

Table VI, Appendix C, gives the critical values for testing the null hypothesis that the population correlation equals zero. As before (see Chapter 9), such a null hypothesis is used for both two-tailed and one-tailed tests. In the case of a two-tailed test, the values in Table V should be read ± (plus or minus); and in the case of a one-tailed test, the values should be read + (plus) if a positive correlation is expected and − (minus) if a negative correlation is expected.

For the above example, an r' greater than +.450 or less than −.450 is required for rejection of the two-tailed H_0 with $\alpha = .05$ and $n = 20$. The obtained r' of +.82 is significant. The null hypothesis may be rejected.

POINT BISERIAL COEFFICIENT, r_{pb}

The point biserial coefficient, r_{pb}, is designed for the circumstance in which one variable consists of just two subsets, or measurement categories, and the other, a series of subsets. The variable with two subsets is customarily designated the X variable, and the variable with a series of subsets is the Y variable. The subsets for the X variable may be ordered quantitatively (as, for example, with the subsets pass and fail), or may simply be qualitatively different (as, for example, with the subsets male and female). The subsets for the Y variable, on the other hand, must be quantitatively different.

The point biserial coefficient is useful for the situation in which an item analysis is done on a test. An item analysis is done to increase a test's internal consistency. Such an analysis is done by examination of the correlation between responses to each single item and responses to the total test, or sum of all the items. Elimination of those items that do not correlate highly with the total test will increase the test's internal consistency. If the possible responses for each item fall into just two subsets, as, for example, correct–incorrect or agree–disagree, r_{pb} can be used to obtain the item–whole correlations. The two possible item responses are the X variable; and the possible sum of responses to all of the items is the Y variable.

Table 12.2 contains the data necessary to compute the item–whole correlation for the first item from a 12 item, multiple-choice test. Each of the items was scored 1 or 0 according to whether the item was answered correctly or incorrectly. Thus, the possible scores on the 12 item test ranged from 0 to 12. These test scores have been placed into subsets according to whether the

TABLE 12.2. Total Test Scores on a Twelve Item Test for Individuals Who Responded Correctly or Incorrectly to the First Item.

X_0		X_1	
INCORRECT		CORRECT	
Y_0	Y_0^2	Y_1	Y_1^2
10	100	3	9
5	25	3	9
7	49	4	16
4	16	10	100
8	64	8	64
3	9	6	36
5	25	5	25
2	4	5	25
		9	81
		10	100
		7	49
		5	25
Σ 44	292	75	539

individual who obtained the particular test score responded correctly, X_1, or incorrectly, X_0. Of the 20 people who took the test, 8 got the first item incorrect; 12 got the first item correct. The Y_0 column contains the total test scores for the individuals who got the first item incorrect, and the Y_1 column contains the total test scores for the individuals who got the first item correct.

Definitional Formula

The point biserial correlation coefficient is defined as:

$$r_{pb} = \sqrt{\frac{SS_{BG}}{SS_t}} \qquad (12\text{--}3)$$

Where: SS_{BG} is the between groups sum of squares for only two groups
SS_t is the total sum of squares

These sums of squares were introduced in Chapter 10 in connection with analysis of variance. The total sum of squares is an index of the total variability. In the present context total variability refers to variability among all the Y scores. The between groups sum of squares is an index of the variability between groups of subjects. In the present context the between groups variability refers to the variability between the Y scores in one subset of X, and the Y scores in the other subset of X.

Recall that r, the Pearson product–moment correlation coefficient, varies from -1 to $+1$. Formula 12–3 for r_{pb}, however, yields a value that varies from 0 to $+1$. Thus, formula 12–3 informs us as to the magnitude of a relation but not its direction or sign. To facilitate communication, it is sometimes convenient to assign a positive or negative sign to the value of r_{pb}. The direction of a relation can be ascertained by determining whether the mean of the Y scores in the X_0 subset is larger or smaller than the mean of the Y scores in the X_1 subset. If the mean of the X_0 subset is smaller, the direction is positive; if larger, the direction is negative.

> In Chapter 5, it was explained that r^2, the square of the Pearson product–moment coefficient, is the proportion of the total variability that is due to the relationship between X and Y. This is also true of r_{pb}^2. The higher r_{pb}^2 is, the greater the proportion of the total variability that is due to the difference between the Y scores in the X_0 subset and the Y scores in the X_1 subset. This parallel between r and r_{pb} makes it evident that r_{pb} is a special case of r, or more exactly, of $|r|$ (the absolute value of r). The point biserial coefficient is equal to $|r|$ with the restriction that the X variable has only two subsets.

Computational Formula

Formula 10–2 is the computational formula for SS_t:

$$SS_t = \sum X^2 - \frac{(\sum X)^2}{n}$$

and formula 10–7 is the computational formula for SS_{BG}:

$$SS_{BG} = \frac{(\sum X_1)^2}{n_1} + \frac{(\sum X_2)^2}{n_2} + \cdots + \frac{(\sum X_k)^2}{n_k} - \frac{(\sum X)^2}{n}$$

Converting X to Y, changing the subscripts to 0 and 1, and making formula 10–7 specific to two groups gives the computational formula for r_{pb}:

$$r_{pb} = \sqrt{\frac{\frac{(\sum Y_0)^2}{n_0} + \frac{(\sum Y_1)^2}{n_1} - \frac{(\sum Y)^2}{n}}{\sum Y^2 - \frac{(\sum Y)^2}{n}}} \tag{12-4}$$

Calculation

Substituting the values from Table 12.2 into formula 12–4:

$$r_{pb} = \sqrt{\frac{\frac{(44)^2}{8} + \frac{(75)^2}{12} - \frac{(44+75)^2}{20}}{(292+539) - \frac{(44+75)^2}{20}}}$$

$$= \sqrt{\frac{\frac{1936}{8} + \frac{5625}{12} - \frac{(119)^2}{20}}{831 - \frac{(119)^2}{20}}}$$

$$= \sqrt{\frac{242 + 468.75 - 708.05}{831 - 708.05}}$$

$$= \sqrt{\frac{2.70}{122.95}}$$

$$= \sqrt{.022} = .15$$

The obtained correlation of .15 is obviously very low; and it possibly indicates that the internal consistency of the test could be improved by elimination of this item. Ideally, the item–whole correlation for all of the items should be computed. If such a set of correlations ranges from high to low, elimination of the few items with low correlations will improve the internal consistency of the test.

Test of Significance

In order to test the null hypothesis that the population correlation is zero, we use formula 10–12:

$$F = \frac{MS_{BG}}{MS_{WG}}$$

Formula 10–12 has $k - 1$ and $n_1 + n_2 + \cdots + n_k - k$ degrees of freedom. Since, in the present context, k equals 2 subsets (labeled 0 and 1), there are 1 and $n_0 + n_1 - 2$ degrees of freedom.

Formula 10–12 is the ratio of the mean square between groups to the mean square within groups. As indicated in Chapter 10, a mean square is a sum of squares divided by its degrees of freedom. Thus, according to formula 10–8:

$$MS_{BG} = \frac{SS_{BG}}{k - 1}$$

where k is the number of groups or subsets—in the case of r_{pb}, 2. The formula for SS_{BG} was given above under the square root radical in the numerator of formula 12–4. What about MS_{WG}? According to formula 10–11:

$$MS_{WG} = \frac{SS_{WG}}{n_1 + n_2 + \cdots + n_k - k}$$

where k is again 2 and the subscripts are 0 and 1. Finally SS_{WG} may be obtained by modifying formula 10–3, $SS_t = SS_{WG} + SS_{BG}$, so that:

$$SS_{WG} = SS_t - SS_{BG} \tag{12-5}$$

Table 12.3 summarizes the calculations for the previously obtained r_{pb} of .15. In calculating this r_{pb}, an SS_t of 112.95 and an SS_{BG} of 2.70 were obtained.

It was pointed out in Chapter 10 that F as defined by formula 10–12 is insensitive to the direction of the difference between (or among) means. In a similar manner the F from formula 10–12 is also insensitive to whether the mean of the Y scores in the X_0 subset is smaller or larger than the mean of the Y scores

TABLE 12.3. *Summary Table for the r_{pb} Test of Significance.*

SOURCE OF VARIANCE	SUM OF SQUARES	df	MEAN SQUARES	F
Between Groups	2.70	1	2.70	.40
Within Groups	120.25	18	6.68	
Total	122.95			

in the X_1 subset (and, thus, r_{pb} given a positive or negative sign). In view of this fact, it is pointless to state a directional, alternative hypothesis. Therefore:

$$H_1: \text{population value} \neq 0$$
$$H_0: \text{population value} = 0$$

In spite of the fact that the alternative hypothesis is two-tailed, only the right, or positive tail, of the theoretical F distribution is being used. This is due to the fact that F as defined by formula 10–12 is always the ratio of MS_{BG} to MS_{WG}, and never, MS_{WG} to MS_{BG}. Thus, in order to reject H_0, an F larger than the positive critical value must be obtained.

With 1 and 18 df and $\alpha = .05$, the critical value is $+4.41$. The obtained F of $+.40$ is not larger than $+4.41$; and H_0 cannot be rejected.

CORRELATION RATIO, η

The Pearson product–moment correlation coefficient is a measure of *linear* relation. Not all variables, however, need be linearly related. Age, for example, is curvilinearly related to most types of aptitude. Consider reaction time. In the younger age ranges, increasing age is associated with faster and faster reaction times. The function, however, eventually levels off and then declines through the middle to older age ranges. Thus, the overall relation is curvilinear. In such a situation, r would be an inappropriate measure of correlation. Rather, the correlation ratio, symbolized η (Greek eta), should be used.

Formula

The formula for η is a simple extrapolation of formula 12–3 for r_{pb}. The point biserial coefficient, like the product–moment coefficient, is a measure of linear relation. Convention and parsimony dictate that two points are connected by a straight line. Thus, the two subsets of Y scores, X_0 and X_1, can be related only linearly. If the X variable, however, has three or more subsets, non-linear relationships are possible. Thus, we can write:

$$\eta = \sqrt{\frac{SS_{BG}}{SS_t}} \quad (12\text{–}6)$$

Like r_{pb}, obtained from formula 12–3, η varies between 0 and $+1$.

When the k subsets of X are 2, η will be the same as r_{pb}. The correlation ratio, however, is designed for the situation in which k is 3 or more.

The magnitude of η reflects the extent to which the total variability among

Y is due to the variability among the X subsets. Any number of relations between X and Y may produce a large η. The relationship may be linear, curvilinear, or some more complex form. With a linear relationship, η will approximately equal $|r|$. The magnitude of η does not describe the type of relation between X and Y; rather, it reflects the magnitude that happens to exist for that relation. Since the concept of negative or positive slope is less applicable with non-linear relations, the sign of η, unlike the sign of r_{pb}, is not altered.

Computational Formula

The computational formula for the correlation ratio is as follows:

$$\eta = \sqrt{\frac{\frac{(\sum Y_0)^2}{n_0} + \frac{(\sum Y_1)^2}{n_1} + \cdots + \frac{(\sum Y_k)^2}{n_k} - \frac{(\sum Y)^2}{n}}{\sum Y^2 - \frac{(\sum Y)^2}{n}}} \tag{12-7}$$

Formula 12-7 is obtained by substituting the computational formulas for SS_t (10-2) and SS_{BG} (10-7) into formula 12-6. As was the case with formula 12-4 for r_{pb}, Xs have been replaced with Ys, and the subscripts have been altered.

Comparison of formula 12-7 for η and formula 12-4 for r_{pb} indicates that the two formulas are identical when the X variable has only two subsets. Under this special circumstance, the point biserial correlation is the same as the correlation ratio.

Calculation

Table 12.4 gives the scores on a simple test of motor coordination for five different age groups (5, 10, 20, 40, and 80 yrs.).

TABLE 12.4. *Scores for Each of Five Age Groups on a Test of Motor Coordination.*

X_0 5 YRS.		X_1 10 YRS.		X_2 20 YRS.		X_3 40 YRS.		X_4 80 YRS.	
Y_0	Y_0^2	Y_1	Y_1^2	Y_2	Y_2^2	Y_3	Y_3^2	Y_4	Y_4^2
1	1	6	36	8	64	7	49	6	36
3	9	4	16	6	36	8	64	4	16
2	4	3	9	7	49	6	36	2	4
2	4	5	25	7	49	7	49	5	25
2	4	4	16	8	64	6	36	2	4
$\sum$ 10	22	22	102	36	262	34	234	19	85

Applying formula 12–7 to these data gives:

$$\eta = \sqrt{\frac{\frac{(10)^2}{5} + \frac{(22)^2}{5} + \frac{(36)^2}{5} + \frac{(34)^2}{5} + \frac{(19)^2}{5} + \frac{(10 + 22 + 36 + 34 + 19)^2}{5 + 5 + 5 + 5 + 5}}{22 + 102 + 262 + 234 + 85 - \frac{(10 + 22 + 36 + 34 + 19)^2}{5 + 5 + 5 + 5 + 5}}}$$

$$= \sqrt{\frac{\frac{100}{5} + \frac{484}{5} + \frac{1296}{5} + \frac{1156}{5} + \frac{361}{5} - \frac{(121)^2}{25}}{705 - \frac{(121)^2}{25}}}$$

$$= \sqrt{\frac{\frac{3397}{5} - \frac{14641}{25}}{705 - \frac{14641}{25}}}$$

$$= \sqrt{\frac{679.4 - 585.64}{705 - 585.64}}$$

$$= \sqrt{\frac{93.76}{119.36}} = \sqrt{.7855} = .89$$

Test of Significance

Except for the fact that there are more than two subsets of Y scores, the test of significance for η is the same as for r_{pb}. The within sum of squares is obtained by formula 12–5, $SS_{WG} = SS_t - SS_{BG}$. Since an SS_t of 119.36 and an SS_{BG} of 93.76 were obtained in calculating η for the data in Table 12.4, $SS_{WG} = 119.36 - 93.76 = 25.60$.

Table 12.5 summarizes the calculation of the relevant F. Using formulas 10–8 and 10–10 obtains the MS_{BG} and MS_{WG}. The ratio of these mean squares gives an F of 18.31. Such an F enables rejection of the null hypothesis that the population correlation ratio is zero at either of the conventional significance levels (.05 or .01). With 4 and 20 as the values for df, the critical value for $\alpha = .05$ is

TABLE 12.5. *Summary Table for the η Test of Significance.*

SOURCE OF VARIANCE	SUM OF SQUARES	df	MEAN SQUARES	F
Between Groups	93.76	4	23.44	18.31
Within Groups	25.60	20	1.28	
Total	119.36			

+2.87; and the critical value for $\alpha = .01$ is $+4.43$. The obtained F of 18.31 is greater than both of these values.

The F used here is the same F that is used in analysis of variance. The purpose of analysis of variance, however, is somewhat different from the correlation ratio. Analysis of variance tests for the existence of effects. The correlation ratio describes the magnitude of effects.

EXERCISES

1. Identify the special circumstance for which each of the following measures of correlation is designed:

(a) Spearman rank–order coefficient
(b) point biserial coefficient
(c) correlation ratio

2. Determine if the null hypothesis can be rejected in each of the following instances:

(a) $r = +.50, n = 12$, one-tailed $H_0, \alpha = .05$, positive correlation expected
(b) $r = +.50, n = 12$, two-tailed $H_0, \alpha = .05$
(c) $r = -.50, n = 12$, one-tailed $H_0, \alpha = .05$, negative correlation expected
(d) $r = -.50, n = 12$, two-tailed $H_0, \alpha = .05$
(e) $r = +.25, n = 90$, one-tailed $H_0, \alpha = .01$, positive correlation expected
(f) $r = +.25, n = 90$, two-tailed $H_0, \alpha = .01$
(g) $r = -.25, n = 90$, one-tailed $H_0, \alpha = .01$, negative correlation expected
(h) $r = -.25, n = 90$, two-tailed $H_0, \alpha = .01$.

3. Two judges ranked six brands of frozen orange juice in terms of preference.

(a) Determine the degree of agreement between the judges by computing r'.

JUDGE #1	JUDGE #2
6	5
5	6
4	4
3	2
2	3
1	1

(b) Can the two-tailed H_0 be rejected at the .01 level of significance?

4. Suppose, alternatively, that the judges rated the six brands of frozen orange juice on a seven-point evaluative scale.

(a) Compute r'.

JUDGE #1	JUDGE #2
7	6
7	6
6	7
3	2
2	2
1	2

(b) Can the two-tailed H_0 be rejected at the .05 level of significance?

5. Eight students took a five-item, true-false test and obtained the following results:

	ITEMS				
STUDENTS	1	2	3	4	5
1	correct	correct	incorrect	correct	incorrect
2	correct	correct	correct	incorrect	incorrect
3	correct	correct	incorrect	incorrect	incorrect
4	correct	correct	correct	correct	incorrect
5	incorrect	correct	correct	correct	correct
6	correct	correct	correct	correct	correct
7	correct	correct	correct	correct	incorrect
8	incorrect	incorrect	incorrect	incorrect	incorrect

(a) Compute r_{pb} for each of the five items.
(b) How could the internal consistency of the test be improved?

6. Compute the correlation ratio, η, for the deprivation and food intake experiment described in Chapter 10. The data are provided in Table 10.1.

7. Compute η for the physical attractiveness and dating preference data in Table 10.3.

8. Calculate the Spearman rank–order correlation coefficient for the following set of ranks.

X	Y
1	4
2	5
3	3
4	1
5	2

9. Calculate the Pearson product–moment correlation coefficient for the ranks in Exercise 8. Compare the obtained r with the value of r' obtained above.

10. Under what special circumstances does the correlation ratio equal the point biserial coefficient?

11. What is an item analysis? What is its purpose?

Prologue to Chapter 13

A minority of statisticians object to the classical test statistics, such as t and F. The basis for their objection are the beliefs: first, that one or more of the assumptions of these tests are often violated, and, second, that such violations lead to erroneous conclusions an unacceptably large number of times. As alternatives to the classical tests, these statisticians urge the use of so-called distribution-free statistical tests. These tests make less restrictive assumptions regarding the nature of the population distribution; hence, the name "distribution free." Chapter 13 describes a few of the many distribution-free tests that have been developed. The chapter also describes some of the relative advantages and disadvantages of distribution-free tests as compared to the classical tests. Although the majority of statisticians currently prefer classical tests in most situations, we psychologists should not make the mistake of assuming that the last word has been said on this matter.

13

Distribution-Free Statistical Tests

In this final chapter, we will briefly consider distribution-free statistical tests. Distribution-free statistical tests do not require as many assumptions—in particular, assumptions about the population—as do the more commonly used t and F tests.

In Chapter 9, it was stated that the use of the theoretical t distribution as a model to describe the sampling distribution of ts that result from actual samples of scores is based on a number of assumptions. For the one-sample case, the theoretical t distribution is an accurate model as long as three assumptions hold:

1. The sample set (from which $\bar{X}$ and est. $\sigma_{\bar{x}}$ are computed) is selected by independent random sampling.
2. A normal distribution accurately describes the probability distribution for the population from which the sample set is selected.
3. The population mean is known or can be assumed (e.g., the population mean specified in the null hypothesis).

For the two-sample case there are four assumptions:

1. The sample sets are selected by independent random sampling.
2. A normal distribution accurately describes the population distribution.
3. The difference between population means, μ_1 "minus" μ_2, is known or can be assumed (e.g., assumed as specified in the null hypothesis).
4. The population distributions have equal variances, σ_1^2 equals σ_2^2 (homogeneity of variance assumptions).

The first three assumptions for the one- and two-sample cases differ only by virtue of the fact that in the latter case there are two samples rather than one, and two populations rather than one. The fourth, or homogeneity of variance assumption, does not apply in the one-sample case.

In Chapter 10, four assumptions were stated for the use of the theoretical F distribution as a model to describe the sampling distribution of Fs that result from actual samples of scores:

1. The k sample sets are selected by independent random sampling.
2. A normal distribution accurately describes the populations.
3. The populations have equal means, $\mu_1 = \mu_2 = \cdots = \mu_k$ (i.e., H_0 is true).
4. The k sample variances estimate the same population variance (i.e., homogeneity of variance exists).

In the special case in which $k = 2$, the assumptions for F are the same as for the two-sample t.

The student should be able to appreciate the fact that it would be desirable to have statistical tests that do not assume normality or homogeneity of variance. The statistical tests that are of this type are referred to as distribution-free tests. *A distribution-free statistical test is a test that makes no assumptions about the precise form of the population distribution.* Special notice should be taken of the word "precise" in this definition. While it is true that distribution-free tests do not assume normality in the population, it is not true that no assumption is made regarding the population distribution. As we shall see, it is frequently assumed that the population distribution(s) is continuous.

DISTRIBUTION-FREE AND NONPARAMETRIC STATISTICAL TESTS

It was indicated in Chapter 8 that the term "parameter" is generally used to refer to any summary characteristic of a probability distribution (whether it is a population distribution or a sampling distribution); but, most technically, a parameter is an arbitrary constant specified in the equation for a probability distribution. Thus, μ, the mean of the population (probability) distribution, is a parameter. For classical tests, such as t and F, such parameters are, of course, used in the stating of null hypotheses.

A *nonparametric statistical test* can be defined as *a test that makes no hypothesis about the value of a parameter in a population distribution.* In popular usage, the terms "nonparametric" and "distribution-free" are used more or less synonymously. In the majority of instances, statistical tests that are nonparametric are also distribution-free. There are instances, however, in which this is not the case. In Chapter 7, the binomial test was used for the testing of hypotheses regarding proportions. This statistical test is distribution-free in that no assumptions were made regarding the normality of the population distribution. On the other hand, the test is not nonparametric. The use of the binomial test requires the making of an assumption regarding the true proportion in the population, for example, that the proportion of above average readers in the population is .5, or that $p = .5$.

We have chosen to use the term, "distribution-free," rather than the term, "nonparametric," to describe the statistical tests in this chapter. As Bradley (1968) points out, the term distribution-free comes closer to describing the quality that makes such tests desirable. The fact that the investigator may not have to assume that a population parameter has a certain value is interesting. However, the main advantage of the tests described in this chapter flows more directly from the less restrictive assumptions regarding the population or populations.

USE OF NOMINAL AND ORDINAL CHARACTERISTICS OF OBTAINED DATA

Suppose that a test of reading ability is administered to a sample of children. If the obtained measures of each child's reading ability have interval–scale characteristics, each child's score contains information regarding the magnitude of the intervals between his or her score and other possible scores. Thus, the difference between reading scores of 70 and 80 is equal to the difference between 80 and 90. One of the more salient characteristics of distribution-free statistical tests is that they do not make use of this information concerning the magnitude of intervals. Information regarding the magnitude of intervals is ignored by either categorizing the data or ranking it. Some distribution-free tests require that the data be categorized into nominal scale sets, for example, above and below average readers. Recall, from Chapters 6 and 7, that this is the case with the binomial. Other distribution-free statistical tests require that scores be ranked from highest to lowest. Such a procedure was used for the Spearman rank–order correlation, discussed in Chapter 12.

Of course, in many situations the original scores may already exist in nominal or ordinal scale form so that no transformation is necessary. Possible examples are the frequency of blue and brown eyes in a given population, and the rank–order of finish in an athletic contest.

Since distribution-free statistics make use of only nominal or ordinal characteristics of data, it is typically not necessary to make elaborate assumptions regarding the population distribution or distributions. However, just because a test does not make use of interval magnitude information does not mean that it is distribution free. The chi square test (Chapter 11) uses only the nominal characteristics of data but is not distribution-free. It is not distribution-free because the theoretical continuous chi square distribution, which is used as a model for the obtained χ^2 (as defined in formula 11–1), is based on the normal distribution.

ASSUMPTIONS

Two assumptions always are made by distribution-free statistical tests. These assumptions are:

1. The sample set is selected by independent, random sampling.
2. The data to which the statistic is applied contains no zero difference scores or tied ranks.

The first of these assumptions is, of course, also required for the classical statistical tests. The second assumption is unique to distribution-free tests.
Distribution-free tests that are based on the nominal characteristics of obtained data require that there be no zero difference scores; and distribution-free statistical tests that are based on the ordinal characteristics of obtained data require that there be no tied ranks. What is a zero difference score? Suppose that an investigator were interested in determining whether a given set of students contained more above average readers than below average readers. The intent would be to classify each student either as an above average reader or a below average reader. It could happen, however, that one of the students in the sample set had a reading score equal to the average value. Thus, the investigator would have a zero difference score. He would not know whether to classify the student as above or below average. Tied ranks occur when the same scores occur in the two conditions. When two or more scores have the same value, the investigator has no way of correctly knowing how to rank these scores.
In the literature on distribution-free statistical tests, it is frequently stated that such tests assume that the population distribution is continuous. The intent of this assumption is to state the sufficient condition for the *theoretical* non–occurrence of zero difference scores or tied ranks. Since continuous populations contain an infinite number of cases, the probability is zero that an investigator will obtain two scores of exactly the same magnitude. In actual fact, however, zero difference scores or tied ranks can occur when the investigator samples from a continuous population. The reason is that, as was pointed

out in Chapter 1, the measurement of a continuous variable is always discrete. Given the imprecision of measurement some probability is always present that zero difference scores or tied ranks will occur.

TREATMENT OF ZERO DIFFERENCE SCORES AND TIED RANKS

One of the major problems connected with distribution-free tests concerns the treatment of zero difference scores and tied ranks. If it is assumed that the population distribution is continuously distributed, the zero difference scores or tied ranks are due to the imprecision or discreteness of measurement. Thus, one of the possible resolutions of the ambiguous data represents the true state of affairs. However, given that certain non-differing scores have been obtained, it is impossible to know what the one true resolution of the ambiguous data is.

The method or procedure that will be followed here is simply to *drop the non-differing scores from the sample set*. Non differing scores are eliminated, thus reducing n. This method *assumes* that the *ratio of measurably positive to measurably negative difference scores is the same as that of truly positive to truly negative difference scores.*

Bradley (1968) discusses a number of methods for treating zero difference scores and tied ranks, some of which we will briefly consider. Discussion of these methods will assume continuity of the population distribution.

METHOD 1: *Obtain Upper and Lower Probability Bounds*

Method 1 requires that the test statistic be calculated twice. One time, values are assigned to the non-differing scores in such a way as to increase maximally the probability of rejecting the null hypothesis. A second time, values are assigned to the non-differing scores in such a way as to decrease maximally the probability of rejecting the null hypothesis. If the associated probabilities of both the calculated values for the test statistic are greater than the level of significance (.05, for example), the null hypothesis cannot be rejected. On the other hand, if the associated probabilities of both calculated values for the test statistic are less than the level of significance, the null hypothesis can be rejected. Thus, this method requires that the investigator obtain upper and lower probability bounds.

An ambiguous situation arises when the level of significance falls between the upper and lower probability bounds. In this situation perhaps the investigator is justified in rejecting the null hypothesis, but he has no way of knowing.

METHOD 2: *Drop the Non-differing Scores from the Sample Set*

Method 2 handles the problem of non-differing scores by simply eliminating them from the sample set, and thus, reducing n. As pointed out above, this method assumes that the ratio of measurably positive to measurably negative difference scores is the same as that of truly positive to truly negative difference scores. If this assumption is correct, method 2 will result in a valid test.

METHOD 3: *Randomly Assign Difference Scores or Ranks*

According to method 3 pluses and minuses are randomly assigned to zero difference scores, and ranks are randomly assigned to equal observations. This method assumes that each of the possible resolutions of the ambiguous data is equally probable if the null hypothesis is true.

METHOD 4: *Assign Half of the Zero Difference Scores a Plus, and Half, a Minus, and Use Midranks for Tied Ranks*

According to method 4, the zero difference scores can be numbered; the even numbered ones can be given a plus, and the odd numbered ones, a minus. Midranks are the average of the set of ranks for the set of consecutively tied scores. The procedure for calculating midranks was illustrated for the Spearman rank–order correlation in Chapter 12. Method 4, like method 3, assumes that each of the possible resolutions of the ambiguous data is equally probable if the null hypothesis is true.

All four of the methods discussed above result in a loss of power relative to the situation in which the ambiguous cases are correctly resolved. Method 1 loses power when the true probability level lies between the two calculated probability bounds. Method 2 loses power by reducing n. Methods 3 and 4 lose power by forcing the statistic into the distribution assumed under the null hypothesis.

In the situation in which continuity of the population actually exists, Bradley (1968) prefers the methods in the rank–order in which they have been presented (1 through 4). Method 1 yields a probability statement that is true of the entire sample. Method 2 yields a probability statement that is true of the unambiguous part of the sample. And, methods 3 and 4 yield probability statements that are only estimates of the true probability level for the entire sample.

As Bradley points out, however, in the situation in which the continuity assumption does not hold, the upper and lower probability bounds cannot be considered exact. Rather, they are *estimates* of an interval containing the true values of the test statistic. Thus, when the population is discrete, method 1 becomes more vague. This, however, is not necessarily true of the remaining methods. Thus, method 2 is most generally preferable.

SIGN TEST FOR ONE SAMPLE

The first two distribution-free tests that we will consider, the sign test for one sample and the sign test for two dependent samples, both are adaptations of the binomial test. (The binomial test was previously discussed in Chapters 6 and 7.) The sign test for one sample is, as the name suggests, designed for the one-sample case; i.e., the case in which the results from a single sample are compared with some hypothesized value.

Illustrative Data

Suppose that an investigator is interested in determining whether or not the reading ability of rural children from the midwest is above or below the national average on a particular test of reading ability. This national average could be a mean or a median; and, of course, when the distribution is symmetrical the mean and median will be the same. However, in the context of distribution-free tests, it is perhaps most appropriate to use the median as a measure of average value or central tendency.

Suppose that a sample of 20 rural children from the midwest is randomly selected and that a reading test is administered. Hypothetical reading scores for 20 students are presented in Table 13.1. If the national median is 50, it is possible to calculate the difference between each student's score and the national median, $X - $ Mdn. These calculations are illustrated in the third column of Table 13.1. The sign test, however, does not utilize the information regarding the magnitude of the $X - $ Mdn. difference, but, more simply, whether that

TABLE 13.1. *Calculation of Signs for a Sample of Reading Scores.*

READING SCORES X	NATIONAL MDN.	$X - $ Mdn.
40	50	-10
45	50	-5
52	50	$+2$
55	50	$+5$
61	50	$+11$
50	50	0
61	50	$+11$
43	50	-7
43	50	-7
38	50	-12
65	50	$+15$
73	50	$+23$
48	50	-2
52	50	$+2$
49	50	-1
49	50	-1
51	50	$+1$
51	50	$+1$
54	50	$+4$
52	50	$+2$

Number $+$ = 11
Number $-$ = 8

difference is positive or negative. The sign test, thus, focuses on the sign of the difference. For the data in Table 13.1, there are 11 positive differences and 8 negative differences. One zero score was obtained and dropped from the analysis, thus reducing n from 20 to 19.

Stating the Alternative and Null Hypotheses

One way of stating the alternative and null hypotheses is in terms of the probability of obtaining above or below average readers from the population of rural midwestern children. Thus:

$$H_1: p \text{ of above or below average readers} \neq .5$$

and,

$$H_0: p \text{ of above or below average readers} = .5$$

Another way of stating the alternative and null hypotheses is in terms of the mean of the binomial, or two-valued, population. If the above average readers are assigned a score of 1 and below average readers are assigned a score of 0, the mean will equal .5—if there are an equal number of 1's and 0's (which is the case if the probability of either occurrence is .5). Thus:

$$H_1: \text{mean of the population} \neq .5$$

and,

$$H_0: \text{mean of the population} = .5$$

Means of binomial populations are parameters. Thus, although the sign test is relatively distribution free, it is not nonparametric. A value for the population mean is assumed.

Completing the Test

Setting α at .05, can we reject the null hypothesis that the mean of the population equals .5? Table 13.1 indicates that of the 19 signs 8 are negative and 11 are positive. Table VII in Appendix C gives the one-tailed cumulative probabilities for sample sizes (ns) between 5 and 25 and rs between 0 and 15, where r is the frequency of the less frequent sign. For the data in Table 13.1, r is 8, the frequency of negative signs; and n is, of course, 19. According to Table VII, the one-tailed probability of obtaining 8 out of 19 is .324. Two-tailed probabilities are obtained by doubling the tabled entries. The two-tailed probability is .648. Clearly, the null hypothesis cannot be rejected at the .05 level of significance.

For a one-tailed test the probabilities can be read directly from Table 13.1, assuming the results are in the direction expected by the alternative hypothesis. If they are not, the null hypothesis cannot be rejected.

Relation to the Binomial

The probability entries in Table VII were obtained by application of formula 6–11, the binomial formula:

$$\text{Prob.}\left(\frac{r}{n}\right) = \frac{n!}{r!(n-r)!} p^r q^{n-r}$$

Where: n is the sample size
   r is the frequency of the less frequently occurring set
   p is the probability specified in the null hypothesis
   q is $1 - p$

As was indicated in Chapter 6, application of this formula gives the probability of obtaining one particular frequency for r. In the case of the data in Table 13.1, this would be the probability of obtaining exactly 8 out of 19 negative signs. Since, however, our interest is typically in the probability of obtaining any of the extreme frequencies, it is necessary to calculate probabilities for all frequencies between 0 and the obtained frequency for the less frequent set. In the above example it would be necessary to calculate the probability of 0 out of 19, 1 of 19, 2 of 19,..., 8 of 19; and then, sum these probabilities. The probabilities in Table VII were obtained by such a procedure. The obvious value of the Table is that it eliminates the necessity of such laborious calculation.

SIGN TEST FOR TWO DEPENDENT SAMPLES

Dependent sampling occurs when repeated measures are taken on the same subjects, or when the subjects in the different conditions are matched on one or more properties. An investigator, for example, may try two different educational training techniques on sets of subjects that were matched for initial aptitude. Or, an investigator may inject one set of rats with a drug, and their litter mates with a placebo. All of these procedures have the potential for producing dependency between the pairs of matched observations; i.e., a relationship such that the occurrence of a particular value for one member of a matched pair increases the probability for the occurrence of a particular value for the other matched pair.

In Chapter 9, it was shown how the one sample t could be applied to the case of two dependent samples by treating the differences between matched pairs as X scores. Application of the sign test to dependent samples follows an analogous procedure. It is assumed that the difference scores are obtained by random and independent sampling.

Illustrative Data

Suppose that a sample set of subjects takes a simulated driving test under two conditions, normal and intoxicated. Order of conditions is counterbalanced so that half take the test in the normal–intoxicated order, and half in the intoxicated–normal order. The two tests are taken 24 hours apart. Scores on the hypothetical test are given in Table 13.2. This table also illustrates the calculation of difference scores, and the summing of signs for the difference scores.

Stating the Alternative and Null Hypotheses

As was the case for the sign test for one sample, the null hypothesis can be stated either as a probability or as a mean. In this case, however, the two-valued

TABLE 13.2. *Calculation of Signs for Dependent Samples of Simulated Driving Scores.*

SUBJECT	NORMAL X	INTOXICATED Y	$X - Y$
1	80	50	+30
2	92	81	+11
3	90	84	+6
4	97	84	+13
5	73	73	0
6	78	79	−1
7	78	70	+8
8	81	50	+31
9	84	46	+38
10	83	45	+38
11	87	49	+38
12	78	39	+39
13	79	40	+39
14	80	82	−2
15	76	66	+10

Number + = 12
Number − = 2

population relates to $X - Y$ difference scores rather than to $X -$ Mdn. difference scores.

Since intoxication is expected to impair performance on the driving test, a one-tailed test is appropriate. Thus, in terms of probability:

$$H_1: p \text{ of intoxication impairing performance} > .5$$

and,

$$H_0: p \text{ of intoxication impairing performance} \leq .5$$

Assigning a 1 to improved performance (a plus $X - Y$ difference score) and a 0 to impaired performance (a minus $X - Y$ difference score) allows for a statement of the alternative and null hypotheses in terms of the mean of the binomial population; thus:

$$H_1: \text{mean of population} < .5$$

and,

$$H_0: \text{mean of population} \geq .5$$

Completing the Test

Setting α at .05, can we reject the null hypothesis that the mean of the population equals .5? Table 13.2 indicates that 12 out of 15 difference scores are positive, and 2 are negative. The one zero-difference score is dropped from the sample, giving an n of 14. In this case, r, the frequency of the less frequent sign, is 2. According to Appendix C, Table VII, the one-tailed probability of obtaining 2 out of 14 is .006. With an α level of .05, the null hypothesis that the mean of the population equals .5 can be rejected. Thus, by implication the one-tailed null hypothesis that the mean of the population is equal to or greater than .5 can also be rejected. If it is improbable that the mean of the population equals .5, it is even more improbable that the mean of the population is greater than .5.

WILCOXON RANK-SUM TEST

The Wilcoxon rank-sum test is a test for the identity of two populations based upon two independent samples. It is a distribution-free alternative to the t test for two independent samples.

If the student has any previous acquaintance with distribution-free tests, he or she may have heard of the Mann–Whitney U Test, one of the more popular distribution-free tests. The Mann–Whitney U Test is functionally equivalent to the Wilcoxon rank–sum test. The Wilcoxon test is presented here simply because it is somewhat simpler to compute.

Illustrative Data

Suppose that an investigator was interested in determining the relative efficacy of high-fear and low-fear arousing communications in producing a change in attitude toward cigarette smoking among teenagers. The low-fear communication presents various statistical and scientific data linking cigarette smoking to lung cancer, heart disease, and emphyzema. The high-fear communication presents the same factual information together with colored slides of various cancerous body parts, and offers pointed "this could happen to you" references. Attitude toward cigarette smoking was measured sometime before and immediately after exposure to one of the two persuasive communications. Before–after difference scores, or attitude change scores, for each subject are presented in Table 13.3. The higher the difference score is, the greater the amount of attitude change. Fourteen teenagers were randomly assigned to each of the two conditions.

TABLE 13.3. *Attitude Change Toward Cigarette Smoking for Subjects Exposed to High-Fear and Low-Fear Communications.*

HIGH FEAR		LOW FEAR	
X_1	RANK	X_2	RANK
−4.0	1	−3.0	2.5
−1.3	6	−3.0	2.5
−0.5	—	−2.3	4
1.0	11	−2.1	5
1.5	—	−0.5	—
2.4	12	0.0	7.5
2.6	13.5	0.0	7.5
2.6	13.5	0.5	9
3.5	16	0.9	10
3.7	17	1.5	—
4.1	19	1.5	—
6.0	22	3.0	15
6.0	22	4.0	18
6.0	22	4.2	20

$\Sigma = 101.0$

Conversion to Ranks

The Wilcoxon rank–sum test requires that the total sample ($n_1 + n_2$) be ranked. In order to accomplish this, it is helpful if the observations within each condition are arranged in increasing order of magnitude. This has already been done in Table 13.3. Thus, the X_1 scores begin with -4.0 and end with 6.0. Similarly, the X_2 scores begin with -3.0 and end with 4.2.

In converting the total sample to ranks, begin by finding the lowest score in either sample; assign that score a rank of 1. Proceed in the same manner until all of the scores in both samples have been ranked. Two difficulties may be encountered when this general procedure is followed. First, there may be *tied ranks within the same sample*; and second, there may be *tied ranks between samples*. Tied ranks within the same sample occur when the same score appears more than once within the same sample. Such tied ranks are treated by averaging the ranks. In Table 13.3, rank 1 was given to the lowest score, -4.0. The next lowest score, -3.0, however, occurs twice within the low-fear condition. These two scores are assigned the average of ranks 2 and 3, 2.5.

Tied ranks between samples occur when the same score appears in both samples. In Table 13.3, both samples contain X scores of -0.5 and of 1.5. Consistent with the procedure discussed earlier, such tied ranks are handled by dropping these scores from the samples. Thus, n_1, for the high-fear condition, is reduced from 14 to 12; and n_2, for the low-fear condition, is reduced from 14 to 11.

Stating the Alternative and Null Hypotheses

The Wilcoxon rank–sum test is a test of the equality of two populations. It is not specifically a test that the populations have different means or medians, or variances, and so forth. It is a general test regarding the two populations. For that reason the Wilcoxon rank–sum test is nonparametric. The test is not based on any assumed value or range of values for a population parameter. Nevertheless, it is still true that the rank–sum test is most sensitive to population differences in central tendency; e.g., mean or median. If this were not the case, the rank–sum test would not be a distribution-free alternative to the t test for differences in population means.

The two-tailed alternative hypothesis for the rank–sum test is that the populations are not identical; and the null hypothesis is that the populations are identical:

$$H_1: population_1 \neq population_2$$
$$H_0: population_1 = population_2$$

Suppose, however, that before collecting the data in Table 13.3 the investigator expected that the high-fear condition would produce more attitude change than would the low-fear condition. In this instance a one-tailed test would be appropriate. If *population*$_1$ is the high-fear population and *population*$_2$ is the low-fear population:

$$H_1: population_1 > population_2$$
$$H_0: population_1 \leq population_2$$

Completing the Test

After setting α at some level, typically .05, the next step is to calculate W_n. *The statistic W_n is calculated by summing the ranks in the smaller of the two samples.* When the two samples are of equal size, the sum of ranks in either of the samples may be calculated. For the data in Table 13.3, the low-fear sample is smaller. The sum of the ranks, W_n, for this sample is 101.0.

Table VIII in Appendix C contains the lower-tail critical values of W_n. Thus, Table VIII allows for the direct testing of one-tailed null hypotheses that the population sampled by the smaller or equal sized sample is equal to or less than the population sampled by the other sample. This was the case for the above one-tailed null hypothesis. Table VIII lists the smaller sample size, n, along the top, and the larger sample, m, down the first column. For the data in Table 13.3, these values are 11 and 12. For 11 and 12 observations, the critical value of W_n at the .05 level of significance is 104. Since the obtained W_n, 101, is *smaller*, the null hypothesis may be rejected, and the alternative hypothesis accepted. High-fear communications produce more attitude change than low-fear communications.

The student should be aware of the fact that for the Wilcoxon rank–sum test the calculated value must be *smaller* than the tabled entry. A calculated statistic that is smaller than the tabled entry means that the statistic is further into the tail of the sampling distribution than is the critical value. For t and F, the calculated statistics had to be *larger* than the absolute value of the tabled entry in order to be significant. For t and F, the larger the absolute value of the statistic is, the more extreme is that statistic. In all cases the underlying principle is to obtain a statistic that is more extreme, or further into the tail of the sampling distribution, than is the critical tabled entry.

Table VIII contains only the critical values for the lower tail of the sampling distribution of W_n. Lower tail values are appropriate when the sample that is expected to have the lower sum of ranks, and does, in fact, have the lower sum of ranks, also has the smaller n. This was the case for the above example. It is quite possible, however, for the sample that is expected to have the lower sum of ranks, and does, in fact, have the lower sum of ranks, to have the larger n.

This would have been the case if the low-fear sample had a larger n than the high-fear sample. The statistic W_n is still calculated for the smaller sample. In this situation, however, the larger the sum of ranks is, the greater the difference in the expected direction; i.e., the critical value lies in the upper tail of the sampling distribution of W_n. Since Table VIII, however, contains values for only the lower tail, a transformation is necessary. To do an upper-tail test, calculate:

$$2\bar{W} - W_n$$

and reject the null hypothesis if the obtained value is smaller than the appropriate tabled entry. The statistic, W_n, is still calculated on the smaller sample, and the values for $2\bar{W}$ are contained in the right-hand column of Table VIII. To do a two-tailed test, use either W_n or $2\bar{W} - W_n$, whichever is smaller; and double the α values in Table VIII. Thus, .025 is the two-tailed .05, and .005 is the two-tailed .01.

Sampling Distribution of W_n

As previously indicated, Table VIII contains the critical values for the lower tail of the sampling distribution of W_n. But, how is the sampling distribution of W_n constructed? The sampling distribution is obtained by calculating W_n for each of the possible pairs of samples obtainable from the total of $n_1 + n_2$, or $n + m$, observations. Different sampling distributions are, of course, calculated for different sample sizes.

Suppose, for example, that $n_1 = 2$ and $n_2 = 5$, giving a total n of 7 and a set of ranks from 1 to 7. The statistic W_n is calculated for each of the possible pairs of 2 ranks selected from the total of 7 ranks. When these values of W_n are placed in a distribution with relative frequency on the vertical axis and the values of W_n on the horizontal axis, the sampling distribution of W_n for 2 and 5 observations has been constructed. According to Table VIII, the one-tailed .05 critical value of W_n for such a sampling distribution is 3. Such a value can only occur when the two smallest ranks, 1 and 2, fall in the smaller sample.

WILCOXON SIGNED-RANK TEST

The Wilcoxon signed-rank test is a distribution-free test for two dependent samples. It is an alternative to the t test for dependent samples, as well as the sign test for dependent samples. A salient difference between the two Wilcoxon tests and the two sign tests is that the former make use of the ordinal characteristics of the measured outcomes, while the latter sign tests do not.

Illustrative Data

As illustrative data we can use the simulated driving scores previously discussed in connection with the sign test for two dependent samples (Table 13.2). These scores, which represent 15 subjects' performance under both normal and intoxicated conditions, have been repeated in Table 13.4. Scores in the normal and intoxicated conditions are symbolized X and Y rather than X_1 and X_2 because the two samples are dependent.

Table 13.4 illustrates the calculation of signed ranks for the $X - Y$ scores. The procedure involves ranking the absolute values of $X - Y$, $|X - Y|$. The smallest absolute difference receives the first rank. The one instance in which $X - Y$ equals 0 is dropped, reducing n from 15 to 14. Tied ranks are averaged. Finally, each rank is given the sign of the $X - Y$ difference score.

Stating the Alternative and Null Hypotheses

If the population of X treatment and the population of Y treatment are both symmetric, the null hypothesis is a hypothesis regarding the common axis of symmetry; i.e., regarding a difference in median or mean value. If the population of X treatment and the population of Y treatment are asymmetric, the null

TABLE 13.4. *Calculation of Signed Ranks for Dependent Samples of Simulated Driving Scores.*

SUBJECT	NORMAL X	INTOXICATED Y	$X - Y$	SIGN OF $X - Y$	RANK OF $\|X - Y\|$	SIGNED RANK
1	80	50	+30	+	8	+8
2	92	81	+11	+	5	+5
3	90	84	+6	+	7	+7
4	97	84	+13	+	6	+6
5	73	73	0			
6	78	79	−1	−	1	−1
7	78	70	+8	+	3	+3
8	81	50	+31	+	9	+9
9	84	46	+38	+	11	+11
10	83	45	+38	+	11	+11
11	87	49	+38	+	11	+11
12	78	39	+39	+	13.5	+13.5
13	79	40	+39	+	13.5	+13.5
14	80	82	−2	−	2	−2
15	76	66	+10	+	4	+4

$$W_+ = 102$$
$$|W_-| = 3$$

hypothesis is a general hypothesis regarding equality of the two populations. In this latter instance the null hypothesis relates to any population difference. This was the situation with regard to the Wilcoxon rank–sum test. If the populations are asymmetric, the null and alternative hypotheses can be stated as they were for that test. In stating the null and alternative hypotheses in this section, we will assume that populations are symmetric.

If $population_x$ is the normal population and $population_y$ is the intoxicated population, the one-tailed alternative hypothesis is:

$$H_1: mdn.\ of\ population_x > mdn.\ of\ population_y$$

and, the null hypothesis is:

$$H_0: mdn.\ of\ population_x \leq mdn.\ of\ population_y$$

Completing the Test

After setting α at some level, typically .05, the next step is to calculate the sum of positive ranks and the sum of negative ranks. As is indicated in Table 13.4, the sum of the positive ranks, W_+, is 102; and the sum of the negative ranks, W_-, is -3. In the case of negative ranks, however, what is needed is the absolute value of the sum $|W_-|$, which is 3.

Table IX in Appendix C gives the lower tail probabilities for W_+, the sum of positive-signed ranks. The W_+ values in Table IX are those that lie on either side of the stated α level. The discrete characteristic of the sampling distribution of W_+ prevents the tabling of exact cumulative probabilities for the various commonly used α levels (.05, .01, etc.).

Table IX gives the lower-tail probabilities for W_+. For a one-tailed test that rejects H_0 if W_+ is too small, refer the calculated value of W_+ directly to Table IX. For a one-tailed test that rejects H_0 if W_+ is too large, calculate the sum of the absolute values of the negative-signed ranks, $|W_-|$, and refer this value to Table IX. For a two-tailed test, double the α levels in Table IX, and use either W_+ or $|W_-|$, whichever is smaller. In all cases reject H_0 if the calculated value is *smaller* than the tabled entry.

> What is the reason for the use of $|W_-|$? The sampling distribution is symmetrical, and extends from 0 to $n(n+1)/2$. The latter expression is the highest possible sum of ranks for a given n. Table IX gives only the lower-tail probabilities of W_+. Due to the fact that the sampling distribution is symmetrical, however, the upper-tail probability of W_+ is the same as the lower-tail probability of $[n(n+1)/2] - W_+$. And, this expression equals the absolute value of the sum of negative ranks, $|W_-|$:
>
> $$[n(n+1)/2] - W_+ = |W_-|$$

For the data in Table 13.4, $W_+ = 102$; $|W_-| = 3$, and $n = 14$. Substituting these values in the above equality:

$$[14(14 + 1)/2] - 102 = 3$$
$$105 - 102 = 3$$
$$3 = 3$$

Since the above one-tailed H_0 is rejected if W_+ is too large, Table IX is entered with $|W_-|$. As Table 13.4 indicates, this value is 3. According to Table IX, with an n of 14 the critical value for the .05 level is between 25 and 26. Since $|W_-|$ is smaller than 25, the H_0 can be rejected, and the H_1 accepted. Intoxication lowered the driving scores.

ADVANTAGES AND DISADVANTAGES OF DISTRIBUTION-FREE TESTS

Now that a number of distribution-free statistical tests have been described, it is important to consider the advantages and disadvantages of such tests.

Ease and Speed of Calculation

The mathematical calculations required by distribution-free tests are frequently fairly simple. Thus, in the majority of instances distribution-free tests can be completed more easily and quickly than can classical tests. Before the advent of the high-speed computer, this advantage was, of course, more compelling than it is today.

Generality of Application

As indicated above, the main advantage of distribution-free tests is that they are relatively free of assumptions regarding the population distribution. How much of an advantage is it not to assume a normally distributed population or populations, and homogeneity of variance? The answer to this question depends in part on the frequency of non-normal population distributions, and in part on the effect of violating this assumption upon the validity of the classical test. No absolute answer can be given to a query concerning the frequency of non-normal distributions. However, it is known that many distributions are non-normal. Certainly, we should not regard non-normal distributions as atypical and "abnormal."

The answer to a query concerning the effect of violation of the normality assumption and the homogeneity of variance assumption upon the validity of distribution-fixed tests is, unfortunately, complicated. In some instances violation of these assumptions appears to make little difference in the probability of committing a Type I error; in some instances a moderate difference, and in some instances a sizeable difference. One important matter having a bearing on this problem is sample size (n). When $n \leq 10$, the violation of assumptions generally has the most devastating effect. However, in part due to the effect described by the central limit theorem, when $n > 30$, violation of assumptions *may* have little effect. Recall that the central limit theorem states that regardless of the form or shape of the population distribution, the sampling distribution of means approaches a normal distribution as sample size increases.

> How is it possible to know or find out about the effect of violating the normality assumption on some statistical test, the one-sample t for example? The most typical method is empirical. First, an actual non-normal population of scores is constructed. Second, a large number of samples of size n are randomly drawn (in these days by a computer) from the population. Third, after stating a null hypothesis assuming a certain value for μ, formula 9-1 is used to compute ts for each of the randomly drawn samples. Fourth and finally, the distribution of empirically obtained ts is compared with the theoretical t distribution. The greater the degree of overlap is, the less the effect of the violation of the normality assumption. Particular attention is, of course, paid to the degree of overlap in the tails. It may be, for example, that more than .05 of the empirically distributed ts fall beyond the critical t at the .05 interval of the theoretical t distribution. The term for describing the degree of overlap is *robustness*. In general, a *test statistic is said to be robust if its empirically obtained sampling distribution is not appreciably affected by violation of assumptions.*
>
> For the two-sample t, the procedure is analogous to that for the one-sample t. In the two-sample case, however, it is necessary to construct two populations, at least one of which is typically non-normal. Aside from the non-normality of the constructed population(s), an additional consideration relates to whether the populations have the same or "homogeneous" variances. In any event, after the populations have been constructed, samples are drawn from the two populations, and formula 9-12 is used to compute ts for each pair of samples. This formula, of course, requires that an assumption be made about the magnitude of the difference between μ_1 and μ_2. Usually, this difference is assumed to be zero. A further difference from the one-sample case relates to whether the sample ns are, or are not equal; and if unequal, what the exact magnitude of the inequality is.
>
> In many studies of t, χ^2, and F, it has been found that the amount of *non*-overlap between the empirically obtained distribution of the test statistic and the theoretical distribution of the test statistic varies all the way from negligible to unacceptably large. As indicated earlier, robustness frequently increases with an increase in sample size or degrees of freedom. Beyond this, however, there are additional circumstances affecting robustness, and such differences have lent themselves to various interpretations. We will content ourselves with the summary interpretation given by Scheffé (1959). First, in the case of inferences concerning means, non-normality has little effect. Second, heterogeneity, or inequality, of variances has little effect if the sample sets have equal ns, but sizeable effects if the sample sets have unequal ns. Third, inferences concerning means can be seriously affected if the means and variances are correlated across sample sets; i.e., the sample sets with large means tend to have large variances, and the sample sets with small means tend to have small variances.

Efficiency

In Chapter 8, the power of a statistical test was defined as the probability of rejecting a false null hypothesis. Since β is the probability of failing to reject a false null hypothesis; i.e., committing a Type II error, power is $1 - \beta$.

Since investigators wish to reject false null hypotheses, power is obviously desirable. As was indicated in Chapter 8, power is determined by a number of factors, such as sample size and the level of significance. In the context of the present chapter, however, it is more relevant to note that power is also determined by the type of statistical test that is used. In certain conditions a given statistical test (such as a t test) may be more powerful than some other test (such as the binomial test). The concept of *efficiency* refers to the relative power of two statistical tests when both tests have the same null and alternative hypotheses at the same level of significance. Given these conditions, the more powerful test is said to be more efficient.

One index of efficiency is based on the sample size required of the less powerful test to enable that test to equal the power of the more powerful test. Thus, the less powerful test may require 25 observations to attain the power of another test that has only 20 observations. The index is the ratio of these two sample sizes $(25/20 = .80)$. Under the particular conditions of the comparison, when the more powerful test is based on 20 observations, the less powerful test has an efficiency of .80.

Under conditions in which all of the assumptions of the classical or distribution-fixed tests (such as the t test) are met, distribution-free tests have somewhat lower efficiencies (efficiencies less than 1.00). With a small sample size, however, the efficiency of distribution-free tests may be only slightly less than 1.00. Generally, as sample size increases, the efficiency of distribution-free tests declines.

On the other hand, under conditions in which the assumptions of the distribution-fixed tests are not met and the less stringent assumptions of the distribution-free test are met, distribution-free tests often are more efficient than distribution-fixed tests. Thus, it is not correct to assume that distribution-fixed tests are always more efficient than distribution-free tests.

Higher-Order Interactions

In Chapter 10, two-factor analysis of variance was described. Two-factor analysis of variance allows for the testing of what is called a double interaction—an interaction between the two factors. It was also indicated in Chapter 10 that there are still more complex analyses of variance involving three, four, five, or any number of factors. As the number of factors increases, it becomes possible to test so-called higher-order interactions. Thus, with three factors it is possible to test a triple interaction; and with four factors, a quadruple interaction. In many instances such interactions may be of great importance to investigators.

In the present context, it is important to notice that no generally acceptable distribution-free tests are available for higher-order interactions. This does not mean, of course, that such tests will never be developed. At present, however, the nonexistence of acceptable distribution-free tests is one of the main advantages of the classical analysis of variance.

HISTORICAL PERSPECTIVE ON DISTRIBUTION-FREE TESTS

As indicated in Chapter 8, the normal curve was discovered in 1733 by De Moivre as the limiting form of the binomial distribution. The normal distribution was independently discovered about a half century later by Laplace. A third "discovery" was made still later by Gauss. Laplace and Gauss were both astronomer–mathematicians who derived the normal distribution as a mathematical description of distribution of errors of astronomical observations. It was believed that each observational error was contributed to by a large number of elementary errors—different physiological conditions of the observer, vibrations of the telescope, variations in the atmosphere, and so forth. Since it was assumed that each error was independent of all others, each observational error could be regarded as the algebraic sum of the contributing elementary errors. Thus, as the number of elementary errors approaches infinity, the distribution of errors of observations approaches a normal distribution.

Laplace and Gauss are two of the most illustrious figures in the history of mathematics. In view of their considerable prestige, it is understandable why their theoretical positions regarding the frequency of error should have been taken seriously. This was particularly true in view of the fact that empirical distributions of errors of astronomical observations were fit remarkably well by the theoretical normal distribution.

Quetelet, another mathematician–astronomer, discovered that the "Law of Frequency of Error" worked reasonably well for "anthropological" measures of soldiers in the Belgian army (weight, height, chest measures, etc.), as well as for other measures. As a result of the work of Laplace, Gauss, Quetelet, and others, it became fashionable to expect normal distributions. To quote Bradley, "The Error distribution began to fit, and to explain, almost everything; and soon it was regarded as a population archetype, that which was Gaussian was considered normal and that which was non-Gaussian was regarded as abnormal" (1968, p. 4).

The excitement and awe inspiring qualities of the Law of the Frequency of Error have been characterized as the "Normal Mystique." Unfortunately, all this enthusiasm for the normal distribution blinded many investigators to much of the non-normality that existed.

...the Gaussian error law came to act as a veritable Procrustean bed to which all possible measurement should be made to fit. The belief in authority so typical of modern German learning, which has also spread to America, was too great to question the supposed generality of the law discovered by the great Gauss. Statisticians could not conciliate themselves with the thought of the possible presence of "skew" frequency curves, although numerous data offered complete defiance to the Gaussian dogma and exhibited a markedly skew frequency distribution. Supposedly great authorities argued naïvely that the reason the data did not fit the curve of Gauss was that the observations were not numerous enough to eliminate the presence of skewness. In other words, skewness was regarded as a byproduct of sampling, and was believed could be made to disappear completely if we could take an infinite number of observations (Arne Fisher, 1923, p. 181).

It was in this general cultural context that Fisher, "Student," and others began to develop the classical or distribution-fixed tests in the early part of the twentieth century. As Bradley (1968) observes, it is not surprising that, even though they were aware of the non-normality of some empirical distributions, they developed tests which assumed normally distributed populations.

One possible beginning date for the history of distribution-free statistical tests is 1710. This is the date of a publication by John Arbuthnott in which an application of the Sign test was used as: "An argument for Divine Providence, taken from the constant Regularity observ'd in the Births of both Sexes." Savage (1953), however, puts the "true beginning" of distribution-free tests at 1936. It was at this time that distribution-free statistics took on the form of a semi-separate discipline, and that the developers of such tests became fully cognizant that their tests were relatively distribution free and did not require the sometimes unjustified assumption of normality.

Contemporary statisticians do not completely agree regarding the relative merit of distribution-free and distribution-fixed statistical tests. A small minority are somewhat wary of distribution-fixed tests and argue for the greater importance of distribution-free tests. A large majority of contemporary statisticians, however, do not agree with this point of view. While they admit that distribution-free tests may be of value in certain situations, they regard distribution-fixed tests as of much greater importance.

Future developments in statistics will undoubtedly contribute to a better understanding of the relative merits of distribution-free and distribution-fixed tests. Statistics obviously is not a static field. There is every reason to believe that statisticians will continue to provide psychologists with methodological tools of ever increasing sophistication.

EXERCISES

1. Define the following terms:

(a) distribution-free statistical test
(b) nonparametric statistical test
(c) nominal scale
(d) ordinal scale
(e) efficiency
(f) binomial population

2. Use the sign test to test the null hypothesis that the p of above or below average readers equals .5 for the data in Table 13.1, and the national medians listed below. Let $\alpha = .05$.

(a) 35 (b) 40
(c) 45 (d) 47
(e) 55 (f) 57
(g) 60 (h) 65
(i) 70

3. Using the sign test, and the same data and national medians given in Exercise 2, test the null hypothesis that p of an above average reader $\leq .5$. Let $\alpha = .05$.

4. Use the sign test to test the null hypothesis that p of intoxication impairing performance equals .5, after adding the following constants to the Y scores in Table 13.2. Let $\alpha = .05$.

(a) −10 (b) −5
(c) 5   (d) 10

5. Using the sign test, and the data and constants of Exercise 4, test the null hypothesis that the p of intoxication impairing performance $\leq .5$. Let $\alpha = .05$.

6. Suppose that the data in Table 13.3 were collected from two dependent, rather than two independent, samples. For example, suppose every subject in the high-fear condition was matched in terms of chronic anxiety and suggestibility with some other subject in the low-fear condition. Thus, the high-fear subject who had an attitude change score of −4.0 was matched with the low-fear subject who had an attitude change score of −3.0; the −1.3 subject with the −3.0 subject, and so forth. Using the sign test, test the null hypothesis that p of high fear producing more change than low fear equals .5. Let $\alpha = .05$.

7. Suppose that the data in Table 13.2 were obtained from two independent samples. Using the Wilcoxon rank–sum, and the .05 level of significance, test:

(a) the null hypothesis that the normal population equals the intoxicated population,
(b) the null hypothesis that the normal population is less than or equal to the intoxicated population.

8. Why is the Wilcoxon rank–sum test nonparametric?

9. What is the circumstance determining the use of the Wilcoxon rank–sum test as opposed to the Wilcoxon signed–rank test?

10. Use the data in Table 9.6 and the Wilcoxon signed–rank test to test:

(a) the null hypothesis that the median of the 2 hours deprivation population equals the median of the 0 hours deprivation population
(b) the median of the 2 hours deprivation population is less than or equal to the median of 0 hours deprivation population. Choose an appropriate α level.

11. State appropriately the two above null hypotheses for the situation in which the populations are asymmetric.

12. In the above situation why might an experimenter prefer the Wilcoxon signed–rank test as opposed to the t test?

13. What assumptions are made by distribution-free statistical tests?

14. Why is the sign test both distribution–free and parametric?

15. In what way is the calculation of the sign test for dependent samples analogous to the calculation of the t test for dependent samples? What is the rationale for such a calculational procedure?

16. What is one of the main advantages of the classical analysis of variance over distribution-free statistical tests?

Appendix A

Elementary Set Theory

Many concepts presented in this book are implicitly set concepts, or at least, understandable in terms of set theory. Set theory is used as a basis for discussion of such topics as probability theory, measurement, sampling, and statistical relations.

Elementary set notions have been used for two reasons. First, set concepts enable a clear, readily understandable definition and explanation of many statistical concepts. This is true, in part, because some statistical concepts are specific examples of the more general, higher order, set concepts. Second, each year more undergraduate students are likely to be acquainted with set theory. In a few years students will probably not be reading this appendix, but instead, will proceed through the text understanding the set terminology without explanation and definition.

CONCEPTS AND OPERATIONS OF SET THEORY

Set theory is a mathematical system, and, as such, is completely abstract, having no definition in terms of real life objects and events. We make use of set theory as a model of reality; we identify the basic concepts of set theory with real life objects and events, and then use these set concepts to talk about the nature of real life objects, events, and relationships.

Every mathematical system begins with a group of *operations*, *postulates*, and *basic concepts* that are said to be "given" as part of the system. That is, they form the basic definition of the system and, as such, are unquestioned and unproven. One can choose to accept or reject the system as useful; e.g., as a model of reality, but, once one decides to work within the system the basic concepts, the operations, and the postulates are accepted as axiomatic.

In this section the basic concepts and the operations of set theory will be defined. Since set theory will be used in only a limited way, the postulates will not be presented. The interested student is referred to Hamilton and Landin, *Set Theory: The Structure of Arithmetic* (1961).

Concepts of Set Theory: Sets and Elements

The concept of *set* is very simple: *A set is a well-defined collection of elements.* By well-defined is meant that a set is clearly specified so that one can readily decide whether or not a given element belongs to the set. For example, the following sets are well-defined: a collection of all fruits, a collection of all whole numbers between and inclusive of 1 and 10, and a collection of all students at a given university.

The use of the term *element* also needs clarification. *An element is a general term for anything that exists in a set.* The elements comprising a set may be completely abstract, that is, unidentified with real life objects, events, and properties. An example would be the set of whole numbers between and inclusive of 1 and 10. The numbers, per se, are completely abstract. They contain no reference to concrete objects. On the other hand, the elements comprising the set may be concrete objects, such as all students currently enrolled in statistics classes in a given university, or all families in a given community with a total income of less than $3,000. In still other instances, the elements may not be concrete objects per se, but rather, properties of concrete objects. In psychology the objects are typically human or nonhuman subjects, and the properties include such qualities as IQ, weight, introversion, and so forth. Psychologists often deal with sets of real numbers that represent various quantities of a measured property, rather than with the actual objects or subjects.

Finally, it should be mentioned that the elements may be only logical possibilities. Frequently, in an experiment the set of elements of concern is the set of all possible outcomes to an experiment. The intent is to examine an obtained outcome with reference to the complete set of possible outcomes. An outcome to an experiment does not occur until the experiment is run—it is not real until the experiment is run. However, the set of all outcomes (elements) that are logically possible can be specified, given the experimental procedures. And, these logically possible outcomes comprise a set.

Two methods are used to specify the elements of a given set: (1) *listing*; that is, complete enumeration of all elements and (2) *stating a rule* that defines a procedure for deciding which elements are included and which are excluded from the set. The set A of all whole numbers between 1 and 10 would be *listed* as $A = \{2, 3, 4, 5, 6, 7, 8, 9\}$ and specified by a *rule* such as $A = \{a$ such that a is a whole number and $1 < a < 10\}$. The latter rule provides a procedure for specifying each element of A. Generally, capital letters, such as A, will be used to symbolize a set; and the corresponding small letters will be used to symbolize the elements of the set.

Several special sets play a fundamental role in set theory and, therefore, need definition. These include *universal set*, *empty set*, and *subset*.

A universal set, symbolized W, is defined as the set of all elements under consideration. In psychological research this might be the total set of subjects one is studying, or the set of all possible outcomes to a given experiment. In other words, the universal set includes all of the elements one wants to talk about or refer to in a given problem or experiment. In psychological research a universal set is frequently called a *population* or *population set*.

The empty set, symbolized as E, is simply a set that has no elements.

A set whose entire membership belongs to a larger "parent" set is called a subset. For example, if B is a subset of A, then all elements in B are also members of A. If A is the set of all female students at a given university and B is a set of all female students with red hair at the same university, then B is a subset of A. In this text the terms *category*, *class*, *sample set*, *outcome*, and *event* are frequently used as substitutes for the term *subset*.

Operations of Set Theory

Operations that form given sets into new sets will now be considered. These operations will be illustrated through the use of drawings called Venn diagrams. The universal set will be represented as a rectangle and the various subsets as circles within the rectangle. Figure A–1 illustrates a universal set, W, and two subsets, A and B.

UNION. First consider the *union* of sets. *Given two sets, A and B, the union of A and B is a new set comprised of those elements belonging to A or to B, or to both A and B.* The union operator is symbolized $\cup$; therefore, the union of A and B is symbolized $A \cup B$. Using Venn diagrams, $A \cup B$ is described in Fig. A–2 by the shaded area. As an example, if A is the set of whole numbers $\{1, 4, 6, 8\}$ and B is the set of whole numbers $\{4, 8, 9, 12\}$, then $A \cup B$ is the set of whole numbers $\{1, 4, 6, 8, 9, 12\}$.

Suppose the sets do not overlap. In this case there are no elements in both A and B, and $A \cup B$ consists of just the elements in A or B. Thus, if A is the set of whole numbers $\{1, 6\}$ and B is the set of whole numbers $\{9, 12\}$, $A \cup B$ is the set of whole numbers $\{1, 6, 9, 12\}$.

INTERSECTION. A second operation is the *intersection* of sets. Using sets A and B for illustrative purposes, *the intersection of A and B is defined as a new set*

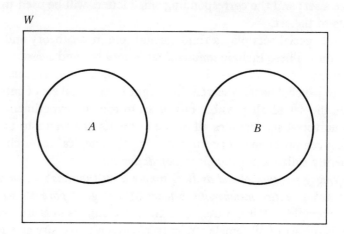

FIGURE A–1. *A universal set and two (mutually exclusive) subsets.*

comprised of only elements belonging to both A and B. The intersection operator is symbolized as ∩, and the intersection of A and B is symbolized as $A \cap B$. Using Venn diagrams, $A \cap B$ is described in Fig. A–3 by the shaded area. If A is a set of anxiety scores obtained from subjects under a low-stress condition, $A = \{4, 3, 4, 6, 8\}$, and B is a set of anxiety scores obtained from subjects under a high-stress condition, $B = \{6, 8, 10, 10, 12\}$; then, $A \cap B$ is the set of scores belonging to both A and B: $A \cap B = \{6, 8\}$.

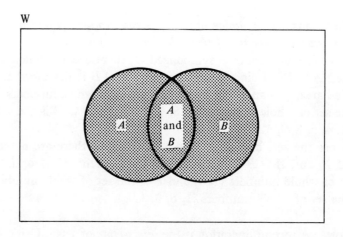

FIGURE A–2. *The union of two sets.*

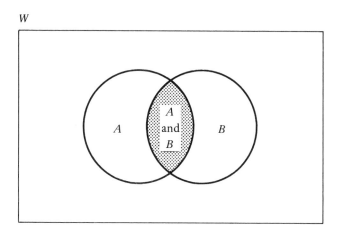

FIGURE A-3. *The intersection of two sets.*

When the intersection of two sets is an empty set, that is, when it has no elements, the two sets are said to be mutually exclusive. Again, using sets A and B, this situation can be symbolized as $A \cap B = E$. Figure A-1 presents $A \cap B = E$.

An example of two mutually exclusive sets would be the set of all women enrolled in statistics at a given university, and the set of all men enrolled in statistics at the same university. No persons belong to both sets. On the other hand, the set of all persons having blue eyes and the set of all women would not be mutually exclusive sets; for there are some elements, namely, blue-eyed women, which belong to both sets. The concept of *mutually exclusive* sets will be very useful in studying measurement, frequency distributions, and probability distributions.

COMPLEMENT. A third operation is the *complement* of a set. *Taking the complement of any set,* e.g., set A, *is accomplished by forming a new set comprised of all elements in W that do not belong to set A.* The *complement* of set A is symbolized $\bar{A}$. In Fig. A-4, the shaded area illustrates the *complement* of A.

From Fig. A-4, it can be seen that A and $\bar{A}$ are mutually exclusive sets; i.e., $A \cap \bar{A} = E$, and that $A \cup \bar{A}$ equals the universal set; i.e., $A \cup \bar{A} = W$. If the total population of concern, the universal set, includes all the laborers in a given cotton mill, and set A is comprised of all the laborers in that mill who are married; then, the complement of A, $\bar{A}$, is the set of all laborers in that mill who are not married.

DIFFERENCE. Finally, consider the *difference* between two sets, i.e., $A - B$.

Concepts and Operations of Set Theory

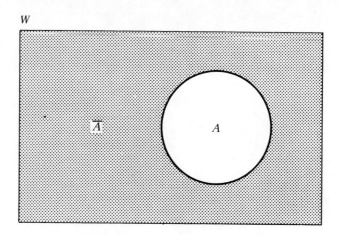

FIGURE A–4. *The complement of a set.*

This difference is illustrated by the shaded area in Fig. A–5. *When B is subtracted from A one is left with all of the elements in A that are not also elements in B.* As an example, suppose the universal set, or population set of concern in an experiment, is the undergraduate students currently enrolled in introductory psychology classes at a given university. A is the subset of these students who scored high on a test of conformity, and B is a subset of these students who

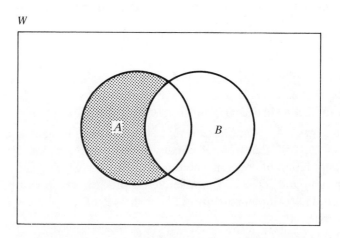

FIGURE A–5. *The difference between two sets.*

Elementary Set Theory

scored high on a test of verbal intelligence. The set $A - B$ includes all of the students who scored high on the conformity test and low on the intelligence test.

The difference operation is closely allied to the intersection operation, for it can be shown that $A - B = A \cap \bar{B}$. The shaded area in Fig. A–5 is both A "minus" B and the intersection of A with the complement of B. The complement of B, $\bar{B}$, consists of the elements in W that are not in B. And the intersection of A and $\bar{B}$ is the shaded area in Fig. A–5.

PRODUCT SET AND TWO SUBSETS

A special type of set, a *product set*, and two special types of subsets, *relations* and *functions*, will now be considered.

Product Set

Suppose each element in set A is paired with each element in set B so as to form a new set whose elements are all possible ordered pairs (a, b), where a is an element of A and b is an element of B. This new set, symbolized $A \times B$, is a product set. Notice that a product set is defined such that the pairs are ordered. Thus, the product set, $A \times B$, whose members are (a, b), is not the same as the product set, $B \times A$, whose members are (b, a).

Consider an example. One well-known set that can be viewed as a product set is a regular deck of 52 playing cards. Let A be a set comprised of the symbols club, diamond, heart, and spade; while set B is comprised of the symbols ace, 2, 3, 4, 5, 6, 7, 8, 9, 10, jack, queen, and king. Forming the product set $B \times A$ gives a set of all possible pairs (b, a); e.g., (king, club), (queen, diamond), (king, diamond), etc. In total there are 52 such pairs (13×4) with each pair being a pair of symbols found on one of the 52 cards in the regular playing deck.

One frequently used product set is the product of a set with itself, symbolized $R \times R$. For example, if $R = 1, 2, 3$, then $R \times R = \{(1, 1), (1, 2), (1, 3), (2, 1), (2, 2), (2, 3), (3, 1), (3, 2), (3, 3)\}$. This new set can be graphically represented using coordinate axes with the R values in the first position in the pairs being designated by the symbol X and placed along the horizontal axis, and the R values in the second position in the pairs being designated by the symbol Y and placed along the vertical axis. The $R \times R$ set of pairs listed above is graphically described in Fig. A–6. From this figure it can be seen that, because each position has a different meaning (the first position referring to the X axis and the second position to the Y axis), order in each pair has meaning; and thus, there are nine different pairs (3×3). Pairs such as $(2, 1)$ and $(1, 2)$, and $(3, 1)$ and $(1, 3)$ are distinctly different elements in this $R \times R$ set.

Relation and Function Subsets

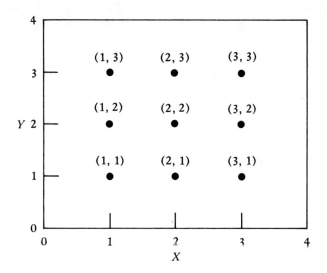

FIGURE A-6. *A product set.*

The concept of a product set provides the context for the introduction of two additional concepts, both of which are basic to any science—*relation* and *function*. *A relation is a subset of a product set.* It is a specified subset of one or more pairs, within a given product set. Using the example of a deck of 52 playing cards, any subset of 52 symbol pairs, such as the pairs on cards in a hand of five cards dealt in a poker game, (that is, two of diamonds, king of clubs, king of spades, queen of hearts, six of clubs), is a relation. Likewise, any subset of the $R \times R$ product referred to above; e.g., (1, 1), (1, 2), and (1, 3), is a relation. This particular subset consists of the three elements in the first "column" of Fig. A-6.

A *function* (sometimes called *functional relation*) is a specific kind of relation. The set of all elements that occupy the left position in the ordered pairs comprising a relation is the *domain* of the relation; and the set of all elements occupying the right position is the *range* of the relation. *A function is a relation comprised of a set of ordered pairs where each element in the domain is paired with one and only one element in the range.* Using the $R \times R$ product defined above, the following relations are functions:

1. [(1, 1), (2, 2), (3, 3)],   2. [(1, 1), (2, 1), (3, 1)],   3. [(2, 1), (3, 2), (1, 2)]

In these examples, each element in the domain is associated with one and only one element in the range. However, notice that an element in the range may be paired with several elements in the domain, and the relation is still a function.

Elements in the domain are not repeated, while elements in the range may be repeated. The following relations are not functions:

1. [(3, 1), (3, 2), (3, 3)],   2. [(2, 1), (2, 3), (1, 2)],   3. [(2, 1), (3, 2), (3, 3)]

In each case at least one element in the domain is associated with more than one element in the range.

When the elements comprising the domain and range of a relation are real numbers, a particular functional relation sometimes can be specified by a *function rule*, where the symbol X is used to represent the elements comprising the domain, and the symbol Y is used to represent the elements comprising the range. Students of algebra are familiar with equations such as $Y = \frac{1}{2}X$, $Y = 2X + 3$, $Y = X^2 + 2X + 4$. In the context of set theory, these equations are called function rules. They specify which (X, Y) pairs belong to the function. The function is the actual set of pairs. Once we have specified the real numbers comprising the domain and the range, we can specify the pairs belonging to a particular function by using its function rule. For example, if the domain consists of $\{1, 2, 3\}$, the range consists of $\{1, 2, 3, 4, 5, 6\}$ and the function rule is $Y = 2X$; the function specified by this rule is $\{(1, 2), (2, 4), (3, 6)\}$. According to the function rule, $Y = 2X$; if $X = 1$, then $Y = 2$; if $X = 2$, then $Y = 4$; and if $X = 3$, then $Y = 6$. This function is described using coordinate axes in Fig. A-7.

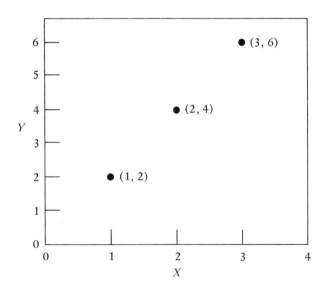

FIGURE A-7. *The function of* $Y = 2X$.

Product Set and Two Subsets

Relations and functions are sets; and as such, the elements (pairs) comprising them may be completely abstract or may represent real objects, e.g., humans or animals, or real properties of objects.

In psychological research the elements comprising the domain and range of a relation are often numbers representing amounts of quantified properties of humans or animals. As an example, let us identify the numbers comprising the domain and range of an $R \times R$ relation (where $R = \{(1, 2, 3)\}$ with measured properties of humans. The numbers along the horizontal axis represent quantities of anxiety induced by an experimental procedure, and the numbers along the vertical axis represent quantities of performance on a simple motor task. The subset of pairs that actually occurs as outcomes to our experiment (relating induced anxiety to performance) is a relation.

A major goal of psychological research is the discovery and verification of relations between properties of human and animal subjects. The relations sought are ones where values of one property are associated with different subsets of values of the other. Then, specifying a value of one property enables prediction of a limited set of paired values of the other. One gains some information about one property by knowing the value of another. The ideal state, of course, would be the discovery of a functional relation that would enable precise prediction of the value of one property, given the value of another.

Appendix B

An Introduction to Summation Algebra

Statistical procedures frequently require summing a set or sets of numbers representing measurements of a given property. Summation notation, e.g., $\sum X$, provides a shorthand expression for such summing operations.

Given a set of n numbers, symbolized as $X_1, X_2, X_3, \ldots, X_n$, their sum may be expressed as:

$$\sum_{i=1}^{n} X_i = X_1 + X_2 + \cdots + X_n$$

The notation $\sum$ symbolizes the summation operation, while $i=1$ and n, found below and above $\sum$, designate the range of quantities to be summed. They are called the summation limits. The subscript i can represent any one of the whole number subscripts (from 1 to n) that identify specific quantities in the set. Thus, $\sum_{i=1}^{n} X_i$ should be read, "Sum all X scores whose subscripts are between and inclusive of 1 and n."* *For a set of n scores, this means simply summing all of the scores in the set.* For example, for the set of six numbers, 5, 7, 8, 6, 2, 4:

$$\sum_{i=1}^{n} X_i = \sum_{i=1}^{6} X_i = 5 + 7 + 8 + 6 + 2 + 4 = 32$$

Since summation symbol $\sum$ is used in this text primarily to indicate the sum of an entire set of scores whose specification is clear from the context, summation limits and subscripts will be deleted. The reader should read an expression such as $\sum X$ as, "Sum the entire set of n scores where X represents any score in the set."

---

* Space limitations prevent writing $\sum_{i=1}^{n}$ as above, with n above and $i=1$ below $\sum$.

We also will have occasion to sum sets of squared quantities, products of quantities in two different sets, and sums and differences of quantities in two different sets. If each score in a set is squared and the squares then are summed, we symbolize this operation by $\sum X^2$:

$$\sum X^2 = X_1^2 + X_2^2 + \cdots + X_n^2$$

That is, $\sum X^2$ symbolizes the sum of a set of n squared quantities. The term $\sum X^2$ is to be distinguished from the term $(\sum X)^2$, which is the square of the sum of a set of n quantities:

$$(\sum X)^2 = (X_1 + X_2 + \cdots + X_n)^2$$

For the set of six quantities used above, 5, 7, 8, 6, 2, 4:

$$\sum X^2 = (5)^2 + (7)^2 + (8)^2 + (6)^2 + (2)^2 + (4)^2$$
$$= 25 + 49 + 64 + 36 + 4 + 16$$
$$= 194$$

and,

$$(\sum X)^2 = (5 + 7 + 8 + 6 + 2 + 4)^2$$
$$= (32)^2$$
$$= 1024$$

The psychologist frequently deals with two or more sets of scores where each set represents scores derived from measurement of the same subjects on a different property. Typically, different letters are used to represent the different sets of measurements. Scores derived from measurement of one property might be symbolized by X, scores for a second property by Y, scores for a third property Z, and so forth. Thus, for a set of n subjects, scores on the three properties would be X_1, Y_1, Z_1, for subject one; X_2, Y_2, Z_2, for subject two; $\ldots$; and X_n, Y_n, Z_n, for the nth subject.

The sum of products of scores representing amounts of two properties for a set of n subjects is given by:

$$\sum XY = X_1 Y_1 + X_2 Y_2 + \cdots + X_n Y_n$$

In Table B.1, the same subjects are represented by two sets of measures. The first column indicates subject number, the second each subject's score on the X variable, the third each subject's score on the Y variable, and the fourth the product, for each subject, of his score on variable X times his score on variable Y. The $\sum XY$ represented at the bottom of column four is the sum of products of scores for each subject.

TABLE B.1. *Summations Involving Measures of Two Properties.*

SUBJECT	X	Y	XY	X + Y	X − Y
1	5	6	30	11	−1
2	4	4	16	8	0
3	8	10	80	18	−2
4	3	2	6	5	1
5	9	8	72	17	1
	$\Sigma X = 29$	$\Sigma Y = 30$	$\Sigma XY = 204$	$\Sigma(X+Y) = 59$	$\Sigma(X-Y) = -1$

$$\text{Rule 2:} \quad \Sigma(X - Y) = \Sigma X - \Sigma Y$$
$$= 29 - 30$$
$$= -1$$

Likewise, the summation of the sums of, and differences between, scores on different properties are given by:

$$\Sigma(X + Y) = (X_1 + Y_1) + (X_2 + Y_2) + \cdots + (X_n + Y_n)$$

and,

$$\Sigma(X - Y) = (X_1 - Y_1) + (X_2 - Y_2) + \cdots + (X_n - Y_n)$$

For the quantities given in Table B.1, these sums are recorded in columns 5 and 6, respectively.

Notice the order of operations in Table B.1. In every case the summation follows a prior algebraic operation on the pairs of X and Y scores. Consider, further, the expression $\Sigma(X - Y)^2$:

$$\Sigma(X - Y)^2 = (X_1 - Y_1)^2 + (X_2 - Y_2)^2 + \cdots + (X_n - Y_n)^2$$

The expression $\Sigma(X - Y)^2$ is the summation of the squares of differences between scores. The sum of squared differences is calculated for the quantities in Table B.1 by subtracting, squaring, and then summing the $(X - Y)$ difference scores in column 6:

$$\Sigma(X - Y)^2 = (-1)^2 + (0)^2 + (-2)^2 + (1)^2 + (1)^2$$
$$= 7$$

Consider, on the other hand, the expression $[\Sigma(X - Y)]^2$. In this instance the summation precedes the squaring. For the numbers in Table B.1, column 6:

$$[\Sigma(X - Y)]^2 = (-1)^2 = +1$$

The order of operations is from the inside to outside—first subtract, then sum, and then square.

Certain rules exist regarding the algebraic manipulation of summations. Three of these rules are used in certain small print sections in this book concerned with algebraic proof of statistical concepts. The reader planning to study these proofs should understand the following summation rules.

RULE 1: *The summation of constants.* Notice that the sum of the numbers 3, 3, 3, 3, 3 can be obtained either by adding the five 3s, or by multiplying 5 times 3. This is true because 3 is a constant. The summation of any constant n times is equal to n times the constant. Symbolizing a constant as a, this rule can be written:

$$\sum a = na$$

This formula holds for any score that always remains the same or is a constant for a given distribution of scores. The arithmetic mean, for example, is a constant for a given set of scores. Since there is just one arithmetic mean for a set of scores:

$$\sum \bar{X} = n\bar{X}$$

where $\bar{X}$ symbolizes the arithmetic mean. In the above formula, $\sum \bar{X}$ signifies summing n $\bar{X}$s. Thus, if there are 5 scores with a mean of 3, five 3s are summed. Following this numerical example, $n\bar{X}$ signifies multiplying 5 times 3.

RULE 2: *The summation of sums or differences.* The summation of sums equals the sum of summations, and the summation of differences equals the difference of summations. Symbolically, this can be expressed as:

$$\sum (X + Y) = \sum X + \sum Y$$

and,

$$\sum (X - Y) = \sum X - \sum Y$$

where X and Y represent two different properties and $(X + Y)$ and $(X - Y)$ represent the sum and difference between a score on one property and a score on the other property for a given subject.

An illustration of the equivalence of the summation of differences and the difference of summations is given in Table B.1. Columns 2 and 3 give the scores on the two properties and the summations of each set of scores. The sixth column gives the differences between pairs of scores in the two sets and a summation of these differences. Notice that $\sum (X - Y) = -1$, and $\sum X - \sum Y = 29 - 30 = -1$. The student can work out a comparable example for the summation of sums.

RULE 3: *The summation of a constant times a variable.* Rule 3 is symbolically expressed as:

$$\sum aX = a \sum X$$

TABLE B.2. *Illustration of Summation Rule 3.*

X	a	aX
3	2	6
0	2	0
2	2	4
1	2	2
4	2	8
$\Sigma = 10$		$\Sigma = 20$

$$\Sigma(aX) = 20$$
$$a\Sigma X = (2)(10) = 20$$
$$\text{Therefore, } \Sigma(aX) = a\Sigma X = 20$$

Rule 3 indicates that a constant may be moved from one side of the summation sign to the other. Table B.2 illustrates the validity of this formula. Here it can be seen that ΣaX and $a \Sigma X$ both equal 20.

Summation rules 2 and 3 indicate that under certain, specified circumstances (addition and subtraction, and multiplication by a constant) the order of operations may be altered. Recall, however, that when multiplying by a variable, multiplying and then summing will not give the same result as summing and then multiplying $[\Sigma X^2 \neq (\Sigma X)^2]$.

EXERCISES

Given two sets of quantities:

X	Y
7	6
4	2
3	5
3	4
6	8

calculate the following summations:

1. ΣX
2. ΣX^2
3. $(\Sigma X)^2$
4. ΣY
5. ΣY^2
6. $(\Sigma Y)^2$
7. ΣXY
8. $\Sigma (X + Y)$
9. $\Sigma (X + Y)^2$
10. $[\Sigma (X + Y)]^2$
11. $\Sigma (X - Y)$
12. $\Sigma (X - Y)^2$
13. $[\Sigma (X - Y)]^2$

Appendix C

Tables

TABLE I. *Proportions of Area Under the Normal Curve*

TABLE II. *Critical Values of t*

TABLE III(a). *Critical Values of F for α Equal to .05 and .01*

TABLE III(b). *Critical Values of F for α Equal to .025 and .005*

TABLE IV. *Critical Values of Chi Square*

TABLE V. *Critical Values of the Correlation Coefficient*

TABLE VI. *Critical Values of r', the Spearman Rank–Order Coefficient*

TABLE VII. *Table of Probabilities Associated with Values as Small as Observed Values of r in the Sign Test*

TABLE VIII. *Critical Lower-Tail Values of W_n for Wilcoxon's Rank–Sum Test*

TABLE IX. *Critical and Quasi-Critical Lower-Tail Values of W_+ (and Their Probability Levels) for Wilcoxon's Signed–Rank Test*

TABLE I. Proportions of Area Under the Normal Curve.

B: P(actual value is btwn EV or 0 & Z)
C: P(actual value is ≥ Z or ≤ -Z)

(A) z	(B) area between mean and z	(C) area beyond z	(A) z	(B) area between mean and z	(C) area beyond z	(A) z	(B) area between mean and z	(C) area beyond z
0.00	.0000	.5000	0.55	.2088	.2912	1.10	.3643	.1357
0.01	.0040	.4960	0.56	.2123	.2877	1.11	.3665	.1335
0.02	.0080	.4920	0.57	.2157	.2843	1.12	.3686	.1314
0.03	.0120	.4880	0.58	.2190	.2810	1.13	.3708	.1292
0.04	.0160	.4840	0.59	.2224	.2776	1.14	.3729	.1271
0.05	.0199	.4801	0.60	.2257	.2743	1.15	.3749	.1251
0.06	.0239	.4761	0.61	.2291	.2709	1.16	.3770	.1230
0.07	.0279	.4721	0.62	.2324	.2676	1.17	.3790	.1210
0.08	.0319	.4681	0.63	.2357	.2643	1.18	.3810	.1190
0.09	.0359	.4641	0.64	.2389	.2611	1.19	.3830	.1170
0.10	.0398	.4602	0.65	.2422	.2578	1.20	.3849	.1151
0.11	.0438	.4562	0.66	.2454	.2546	1.21	.3869	.1131
0.12	.0478	.4522	0.67	.2486	.2514	1.22	.3888	.1112
0.13	.0517	.4483	0.68	.2517	.2483	1.23	.3907	.1093
0.14	.0557	.4443	0.69	.2549	.2451	1.24	.3925	.1075
0.15	.0596	.4404	0.70	.2580	.2420	1.25	.3944	.1056
0.16	.0636	.4364	0.71	.2611	.2389	1.26	.3962	.1038
0.17	.0675	.4325	0.72	.2642	.2358	1.27	.3980	.1020
0.18	.0714	.4286	0.73	.2673	.2327	1.28	.3997	.1003
0.19	.0753	.4247	0.74	.2704	.2296	1.29	.4015	.0985
0.20	.0793	.4207	0.75	.2734	.2266	1.30	.4032	.0968
0.21	.0832	.4168	0.76	.2764	.2236	1.31	.4049	.0951
0.22	.0871	.4129	0.77	.2794	.2206	1.32	.4066	.0934
0.23	.0910	.4090	0.78	.2823	.2177	1.33	.4082	.0918
0.24	.0948	.4052	0.79	.2852	.2148	1.34	.4099	.0901
0.25	.0987	.4013	0.80	.2881	.2119	1.35	.4115	.0885
0.26	.1026	.3974	0.81	.2910	.2090	1.36	.4131	.0869
0.27	.1064	.3936	0.82	.2939	.2061	1.37	.4147	.0853
0.28	.1103	.3897	0.83	.2967	.2033	1.38	.4162	.0838
0.29	.1141	.3859	0.84	.2995	.2005	1.39	.4177	.0823
0.30	.1179	.3821	0.85	.3023	.1977	1.40	.4192	.0808
0.31	.1217	.3783	0.86	.3051	.1949	1.41	.4207	.0793
0.32	.1255	.3745	0.87	.3078	.1922	1.42	.4222	.0778
0.33	.1293	.3707	0.88	.3106	.1894	1.43	.4236	.0764
0.34	.1331	.3669	0.89	.3133	.1867	1.44	.4251	.0749
0.35	.1368	.3632	0.90	.3159	.1841	1.45	.4265	.0735
0.36	.1406	.3594	0.91	.3186	.1814	1.46	.4279	.0721
0.37	.1443	.3557	0.92	.3212	.1788	1.47	.4292	.0708
0.38	.1480	.3520	0.93	.3238	.1762	1.48	.4306	.0694
0.39	.1517	.3483	0.94	.3264	.1736	1.49	.4319	.0681
0.40	.1554	.3446	0.95	.3289	.1711	1.50	.4332	.0668
0.41	.1591	.3409	0.96	.3315	.1685	1.51	.4345	.0655
0.42	.1628	.3372	0.97	.3340	.1660	1.52	.4357	.0643
0.43	.1664	.3336	0.98	.3365	.1635	1.53	.4370	.0630
0.44	.1700	.3300	0.99	.3389	.1611	1.54	.4382	.0618
0.45	.1736	.3264	1.00	.3413	.1587	1.55	.4394	.0606
0.46	.1772	.3228	1.01	.3438	.1562	1.56	.4406	.0594
0.47	.1808	.3192	1.02	.3461	.1539	1.57	.4418	.0582
0.48	.1844	.3156	1.03	.3485	.1515	1.58	.4429	.0571
0.49	.1879	.3121	1.04	.3508	.1492	1.59	.4441	.0559
0.50	.1915	.3085	1.05	.3531	.1469	1.60	.4452	.0548
0.51	.1950	.3050	1.06	.3554	.1446	1.61	.4463	.0537
0.52	.1985	.3015	1.07	.3577	.1423	1.62	.4474	.0526
0.53	.2019	.2981	1.08	.3599	.1401	1.63	.4484	.0516
0.54	.2054	.2946	1.09	.3621	.1379	1.64	.4495	.0505

SOURCE: Reprint of Table A from Runyon, R. P., and Haber, A. *Fundamentals of Behavioral Statistics.* Reading, Mass.: Addison-Wesley, 1971, pp. 290–291.

TABLE I. Proportions of Area Under the Normal Curve (Continued).

(A) z	(B) area between mean and z	(C) area beyond z	(A) z	(B) area between mean and z	(C) area beyond z	(A) z	(B) area between mean and z	(C) area beyond z
1.65	.4505	.0495	2.22	.4868	.0132	2.79	.4974	.0026
1.66	.4515	.0485	2.23	.4871	.0129	2.80	.4974	.0026
1.67	.4525	.0475	2.24	.4875	.0125	2.81	.4975	.0025
1.68	.4535	.0465	2.25	.4878	.0122	2.82	.4976	.0024
1.69	.4545	.0455	2.26	.4881	.0119	2.83	.4977	.0023
1.70	.4554	.0446	2.27	.4884	.0116	2.84	.4977	.0023
1.71	.4564	.0436	2.28	.4887	.0113	2.85	.4978	.0022
1.72	.4573	.0427	2.29	.4890	.0110	2.86	.4979	.0021
1.73	.4582	.0418	2.30	.4893	.0107	2.87	.4979	.0021
1.74	.4591	.0409	2.31	.4896	.0104	2.88	.4980	.0020
1.75	.4599	.0401	2.32	.4898	.0102	2.89	.4981	.0019
1.76	.4608	.0392	2.33	.4901	.0099	2.90	.4981	.0019
1.77	.4616	.0384	2.34	.4904	.0096	2.91	.4982	.0018
1.78	.4625	.0375	2.35	.4906	.0094	2.92	.4982	.0018
1.79	.4633	.0367	2.36	.4909	.0091	2.93	.4983	.0017
1.80	.4641	.0359	2.37	.4911	.0089	2.94	.4984	.0016
1.81	.4649	.0351	2.38	.4913	.0087	2.95	.4984	.0016
1.82	.4656	.0344	2.39	.4916	.0084	2.96	.4985	.0015
1.83	.4664	.0336	2.40	.4918	.0082	2.97	.4985	.0015
1.84	.4671	.0329	2.41	.4920	.0080	2.98	.4986	.0014
1.85	.4678	.0322	2.42	.4922	.0078	2.99	.4986	.0014
1.86	.4686	.0314	2.43	.4925	.0075	3.00	.4987	.0013
1.87	.4693	.0307	2.44	.4927	.0073	3.01	.4987	.0013
1.88	.4699	.0301	2.45	.4929	.0071	3.02	.4987	.0013
1.89	.4706	.0294	2.46	.4931	.0069	3.03	.4988	.0012
1.90	.4713	.0287	2.47	.4932	.0068	3.04	.4988	.0012
1.91	.4719	.0281	2.48	.4934	.0066	3.05	.4989	.0011
1.92	.4726	.0274	2.49	.4936	.0064	3.06	.4989	.0011
1.93	.4732	.0268	2.50	.4938	.0062	3.07	.4989	.0011
1.94	.4738	.0262	2.51	.4940	.0060	3.08	.4990	.0010
1.95	.4744	.0256	2.52	.4941	.0059	3.09	.4990	.0010
1.96	.4750	.0250	2.53	.4943	.0057	3.10	.4990	.0010
1.97	.4756	.0244	2.54	.4945	.0055	3.11	.4991	.0009
1.98	.4761	.0239	2.55	.4946	.0054	3.12	.4991	.0009
1.99	.4767	.0233	2.56	.4948	.0052	3.13	.4991	.0009
2.00	.4772	.0228	2.57	.4949	.0051	3.14	.4992	.0008
2.01	.4778	.0222	2.58	.4951	.0049	3.15	.4992	.0008
2.02	.4783	.0217	2.59	.4952	.0048	3.16	.4992	.0008
2.03	.4788	.0212	2.60	.4953	.0047	3.17	.4992	.0008
2.04	.4793	.0207	2.61	.4955	.0045	3.18	.4993	.0007
2.05	.4798	.0202	2.62	.4956	.0044	3.19	.4993	.0007
2.06	.4803	.0197	2.63	.4957	.0043	3.20	.4993	.0007
2.07	.4808	.0192	2.64	.4959	.0041	3.21	.4993	.0007
2.08	.4812	.0188	2.65	.4960	.0040	3.22	.4994	.0006
2.09	.4817	.0183	2.66	.4961	.0039	3.23	.4994	.0006
2.10	.4821	.0179	2.67	.4962	.0038	3.24	.4994	.0006
2.11	.4826	.0174	2.68	.4963	.0037	3.25	.4994	.0006
2.12	.4830	.0170	2.69	.4964	.0036	3.30	.4995	.0005
2.13	.4834	.0166	2.70	.4965	.0035	3.35	.4996	.0004
2.14	.4838	.0162	2.71	.4966	.0034	3.40	.4997	.0003
2.15	.4842	.0158	2.72	.4967	.0033	3.45	.4997	.0003
2.16	.4846	.0154	2.73	.4968	.0032	3.50	.4998	.0002
2.17	.4850	.0150	2.74	.4969	.0031	3.60	.4998	.0002
2.18	.4854	.0146	2.75	.4970	.0030	3.70	.4999	.0001
2.19	.4857	.0143	2.76	.4971	.0029	3.80	.4999	.0001
2.20	.4861	.0139	2.77	.4972	.0028	3.90	.49995	.00005
2.21	.4864	.0136	2.78	.4973	.0027	4.00	.49997	.00003

TABLE II. Critical Values of t.

For any given df, the table shows the values of t corresponding to various levels of probability.
Obtained t is significant at a given level if it is equal to or *greater than* the value shown in the table.

	Level of significance for one-tailed test					
	.10	.05	.025	.01	.005	.0005
	Level of significance for two-tailed test					
df	.20	.10	.05	.02	.01	.001
1	3.078	6.314	12.706	31.821	63.657	636.619
2	1.886	2.920	4.303	6.965	9.925	31.598
3	1.638	2.353	3.182	4.541	5.841	12.941
4	1.533	2.132	2.776	3.747	4.604	8.610
5	1.476	2.015	2.571	3.365	4.032	6.859
6	1.440	1.943	2.447	3.143	3.707	5.959
7	1.415	1.895	2.365	2.998	3.499	5.405
8	1.397	1.860	2.306	2.896	3.355	5.041
9	1.383	1.833	2.262	2.821	3.250	4.781
10	1.372	1.812	2.228	2.764	3.169	4.587
11	1.363	1.796	2.201	2.718	3.106	4.437
12	1.356	1.782	2.179	2.681	3.055	4.318
13	1.360	1.771	2.160	2.650	3.012	4.221
14	1.345	1.761	2.145	2.624	2.977	4.140
15	1.341	1.753	2.131	2.602	2.947	4.073
16	1.337	1.746	2.120	2.583	2.921	4.015
17	1.333	1.740	2.110	2.567	2.898	3.965
18	1.330	1.734	2.101	2.552	2.878	3.922
19	1.328	1.729	2.093	2.539	2.861	3.883
20	1.325	1.725	2.086	2.528	2.845	3.850
21	1.323	1.721	2.080	2.518	2.831	3.819
22	1.321	1.717	2.074	2.508	2.819	3.792
23	1.319	1.714	2.069	2.500	2.807	3.767
24	1.318	1.711	2.064	2.492	2.797	3.745
25	1.316	1.708	2.060	2.485	2.787	3.725
26	1.315	1.706	2.056	2.479	2.779	3.707
27	1.314	1.703	2.052	2.473	2.771	3.690
28	1.313	1.701	2.048	2.467	2.763	3.674
29	1.311	1.699	2.045	2.462	2.756	3.659
30	1.310	1.697	2.042	2.457	2.750	3.646
40	1.303	1.684	2.021	2.423	2.704	3.551
60	1.296	1.671	2.000	2.390	2.660	3.460
120	1.289	1.658	1.980	2.358	2.617	3.373
∞	1.282	1.645	1.960	2.326	2.576	3.291

SOURCE: Taken from Table III of Fisher, R. A., and Yates, F. *Statistical Tables for Biological, Agricultural and Medical Research.* 6th ed. Edinburgh: Oliver and Boyd, 1974. This abridgement is a reprint of Table C from Runyon, R. P., and Haber, A. *Fundamentals of Behavioral Statistics.* Reading, Mass.: Addison-Wesley, 1971, p. 293.

TABLE III(a). Critical Values of F for α Equal to .05 and .01.

The obtained F is significant at the .05 level (light row) and the .01 level (dark row) if it is equal to or greater than the value shown in the table for the distribution of F.

Degrees of Freedom in Numerator of F (df_1)

df_2	1	2	3	4	5	6	7	8	9	10	11	12	14	16	20	24	30	40	50	75	100	200	500	∞
1	161	200	216	225	230	234	237	239	241	242	243	244	245	246	248	249	250	251	252	253	253	254	254	254
	4052	4999	5403	5625	5764	5859	5928	5981	6022	6056	6082	6106	6142	6169	6208	6234	6258	6286	6302	6323	6334	6352	6361	6366
2	18.51	19.00	19.16	19.25	19.30	19.33	19.36	19.37	19.38	19.39	19.40	19.41	19.42	19.43	19.44	19.45	19.46	19.47	19.47	19.48	19.49	19.49	19.50	19.50
	98.49	99.01	99.17	99.25	99.30	99.33	99.34	99.36	99.38	99.40	99.41	99.42	99.43	99.44	99.45	99.46	99.47	99.48	99.48	99.49	99.49	99.49	99.50	99.50
3	10.13	9.55	9.28	9.12	9.01	8.94	8.88	8.84	8.81	8.78	8.76	8.74	8.71	8.69	8.66	8.64	8.62	8.60	8.58	8.57	8.56	8.54	8.54	8.53
	34.12	30.81	29.46	28.71	28.24	27.91	27.67	27.49	27.34	27.23	27.13	27.05	26.92	26.83	26.69	26.60	26.50	26.41	26.30	26.27	26.23	26.18	26.14	26.12
4	7.71	6.94	6.59	6.39	6.26	6.16	6.09	6.04	6.00	5.96	5.93	5.91	5.87	5.84	5.80	5.77	5.74	5.71	5.70	5.68	5.66	5.65	5.64	5.63
	21.20	18.00	16.69	15.98	15.52	15.21	14.98	14.80	14.66	14.54	14.45	14.37	14.24	14.15	14.02	13.93	13.83	13.74	13.69	13.61	13.57	13.52	13.48	13.46
5	6.61	5.79	5.41	5.19	5.05	4.95	4.88	4.82	4.78	4.74	4.70	4.68	4.64	4.60	4.56	4.53	4.50	4.46	4.44	4.42	4.40	4.38	4.37	4.36
	16.26	13.27	12.06	11.39	10.97	10.67	10.45	10.27	10.15	10.05	9.96	9.89	9.77	9.68	9.55	9.47	9.38	9.29	9.24	9.17	9.13	9.07	9.04	9.02
6	5.99	5.14	4.76	4.53	4.39	4.28	4.21	4.15	4.10	4.06	4.03	4.00	3.96	3.92	3.87	3.84	3.81	3.77	3.75	3.72	3.71	3.69	3.68	3.67
	13.74	10.92	9.78	9.15	8.75	8.47	8.26	8.10	7.98	7.87	7.79	7.72	7.60	7.52	7.39	7.31	7.23	7.14	7.09	7.02	6.99	6.94	6.90	6.88
7	5.59	4.74	4.35	4.12	3.97	3.87	3.79	3.73	3.68	3.63	3.60	3.57	3.52	3.49	3.44	3.41	3.38	3.34	3.32	3.29	3.28	3.25	3.24	3.23
	12.25	9.55	8.45	7.85	7.46	7.19	7.00	6.84	6.71	6.62	6.54	6.47	6.35	6.27	6.15	6.07	5.98	5.90	5.85	5.78	5.75	5.70	5.67	5.65
8	5.32	4.46	4.07	3.84	3.69	3.58	3.50	3.44	3.39	3.34	3.31	3.28	3.23	3.20	3.15	3.12	3.08	3.05	3.03	3.00	2.98	2.96	2.94	2.93
	11.26	8.65	7.59	7.01	6.63	6.37	6.19	6.03	5.91	5.82	5.74	5.67	5.56	5.48	5.36	5.28	5.20	5.11	5.06	5.00	4.96	4.91	4.88	4.86
9	5.12	4.26	3.86	3.63	3.48	3.37	3.29	3.23	3.18	3.13	3.10	3.07	3.02	2.98	2.93	2.90	2.86	2.82	2.80	2.77	2.76	2.73	2.72	2.71
	10.56	8.02	6.99	6.42	6.06	5.80	5.62	5.47	5.35	5.26	5.18	5.11	5.00	4.92	4.80	4.73	4.64	4.56	4.51	4.45	4.41	4.36	4.33	4.31
10	4.96	4.10	3.71	3.48	3.33	3.22	3.14	3.07	3.02	2.97	2.94	2.91	2.86	2.82	2.77	2.74	2.70	2.67	2.64	2.61	2.59	2.56	2.55	2.54
	10.04	7.56	6.55	5.99	5.64	5.39	5.21	5.06	4.95	4.85	4.78	4.71	4.60	4.52	4.41	4.33	4.25	4.17	4.12	4.05	4.01	3.96	3.93	3.91
11	4.84	3.98	3.59	3.36	3.20	3.09	3.01	2.95	2.90	2.86	2.82	2.79	2.74	2.70	2.65	2.61	2.57	2.53	2.50	2.47	2.45	2.42	2.41	2.40
	9.65	7.20	6.22	5.67	5.32	5.07	4.88	4.74	4.63	4.54	4.46	4.40	4.29	4.21	4.10	4.02	3.94	3.86	3.80	3.74	3.70	3.66	3.62	3.60
12	4.75	3.88	3.49	3.26	3.11	3.00	2.92	2.85	2.80	2.76	2.72	2.69	2.64	2.60	2.54	2.50	2.46	2.42	2.40	2.36	2.35	2.32	2.31	2.30
	9.33	6.93	5.95	5.41	5.06	4.82	4.65	4.50	4.39	4.30	4.22	4.16	4.05	3.98	3.86	3.78	3.70	3.61	3.56	3.49	3.46	3.41	3.38	3.36
13	4.67	3.80	3.41	3.18	3.02	2.92	2.84	2.77	2.72	2.67	2.63	2.60	2.55	2.51	2.46	2.42	2.38	2.34	2.32	2.28	2.26	2.24	2.22	2.21
	9.07	6.70	5.74	5.20	4.86	4.62	4.44	4.30	4.19	4.10	4.02	3.96	3.85	3.78	3.67	3.59	3.51	3.42	3.37	3.30	3.27	3.21	3.18	3.16
14	4.60	3.74	3.34	3.11	2.96	2.85	2.77	2.70	2.65	2.60	2.56	2.53	2.48	2.44	2.39	2.35	2.31	2.27	2.24	2.21	2.19	2.16	2.14	2.13
	8.86	6.51	5.56	5.03	4.69	4.46	4.28	4.14	4.03	3.94	3.86	3.80	3.70	3.62	3.51	3.43	3.34	3.26	3.21	3.14	3.11	3.06	3.02	3.00
15	4.54	3.68	3.29	3.06	2.90	2.79	2.70	2.64	2.59	2.55	2.51	2.48	2.43	2.39	2.33	2.29	2.25	2.21	2.18	2.15	2.12	2.10	2.08	2.07
	8.68	6.36	5.42	4.89	4.56	4.32	4.14	4.00	3.89	3.80	3.73	3.67	3.56	3.48	3.36	3.29	3.20	3.12	3.07	3.00	2.97	2.92	2.89	2.87

Degrees of Freedom in Denominator of F (df_2)

SOURCE: Taken from Snedecor, G. W., and Cochran, W. C. *Statistical Methods.* Ames, Iowa: Iowa State University Press, 1967. This table is a reprint of Table D of Runyon, R. P., and Haber, A. *Fundamentals of Behavioral Statistics.* Reading, Mass.: Addison-Wesley, 1971, pp. 294–297.

TABLE III(a). Critical Values of F for α Equal to .05 and .01 (Continued)

Degrees of Freedom in Numerator of F (df_1)

df_2	1	2	3	4	5	6	7	8	9	10	11	12	14	16	20	24	30	40	50	75	100	200	500	∞
16	4.49 8.53	3.63 6.23	3.24 5.29	3.01 4.77	2.85 4.44	2.74 4.20	2.66 4.03	2.59 3.89	2.54 3.78	2.49 3.69	2.45 3.61	2.42 3.55	2.37 3.45	2.33 3.37	2.28 3.25	2.24 3.18	2.20 3.10	2.16 3.01	2.13 2.96	2.09 2.89	2.07 2.86	2.04 2.80	2.02 2.77	2.01 2.75
17	4.45 8.40	3.59 6.11	3.20 5.18	2.96 4.67	2.81 4.34	2.70 4.10	2.62 3.93	2.55 3.79	2.50 3.68	2.45 3.59	2.41 3.52	2.38 3.45	2.33 3.35	2.29 3.27	2.23 3.16	2.19 3.08	2.15 3.00	2.11 2.92	2.08 2.86	2.04 2.79	2.02 2.76	1.99 2.70	1.97 2.67	1.96 2.65
18	4.41 8.28	3.55 6.01	3.16 5.09	2.93 4.58	2.77 4.25	2.66 4.01	2.58 3.85	2.51 3.71	2.46 3.60	2.41 3.51	2.37 3.44	2.34 3.37	2.29 3.27	2.25 3.19	2.19 3.07	2.15 3.00	2.11 2.91	2.07 2.83	2.04 2.78	2.00 2.71	1.98 2.68	1.95 2.62	1.93 2.59	1.92 2.57
19	4.38 8.18	3.52 5.93	3.13 5.01	2.90 4.50	2.74 4.17	2.63 3.94	2.55 3.77	2.48 3.63	2.43 3.52	2.38 3.43	2.34 3.36	2.31 3.30	2.26 3.19	2.21 3.12	2.15 3.00	2.11 2.92	2.07 2.84	2.02 2.76	2.00 2.70	1.96 2.63	1.94 2.60	1.91 2.54	1.90 2.51	1.88 2.49
20	4.35 8.10	3.49 5.85	3.10 4.94	2.87 4.43	2.71 4.10	2.60 3.87	2.52 3.71	2.45 3.56	2.40 3.45	2.35 3.37	2.31 3.30	2.28 3.23	2.23 3.13	2.18 3.05	2.12 2.94	2.08 2.86	2.04 2.77	1.99 2.69	1.96 2.63	1.92 2.56	1.90 2.53	1.87 2.47	1.85 2.44	1.84 2.42
21	4.32 8.02	3.47 5.78	3.07 4.87	2.84 4.37	2.68 4.04	2.57 3.81	2.49 3.65	2.42 3.51	2.37 3.40	2.32 3.31	2.28 3.24	2.25 3.17	2.20 3.07	2.15 2.99	2.09 2.88	2.05 2.80	2.00 2.72	1.96 2.63	1.93 2.58	1.89 2.51	1.87 2.47	1.84 2.42	1.82 2.38	1.81 2.36
22	4.30 7.94	3.44 5.72	3.05 4.82	2.82 4.31	2.66 3.99	2.55 3.76	2.47 3.59	2.40 3.45	2.35 3.35	2.30 3.26	2.26 3.18	2.23 3.12	2.18 3.02	2.13 2.94	2.07 2.83	2.03 2.75	1.98 2.67	1.93 2.58	1.91 2.53	1.87 2.46	1.84 2.42	1.81 2.37	1.80 2.33	1.78 2.31
23	4.28 7.88	3.42 5.66	3.03 4.76	2.80 4.26	2.64 3.94	2.53 3.71	2.45 3.54	2.38 3.41	2.32 3.30	2.28 3.21	2.24 3.14	2.20 3.07	2.14 2.97	2.10 2.89	2.04 2.78	2.00 2.70	1.96 2.62	1.91 2.53	1.88 2.48	1.84 2.41	1.82 2.37	1.79 2.32	1.77 2.28	1.76 2.26
24	4.26 7.82	3.40 5.61	3.01 4.72	2.78 4.22	2.62 3.90	2.51 3.67	2.43 3.50	2.36 3.36	2.30 3.25	2.26 3.17	2.22 3.09	2.18 3.03	2.13 2.93	2.09 2.85	2.02 2.74	1.98 2.66	1.94 2.58	1.89 2.49	1.86 2.44	1.82 2.36	1.80 2.33	1.76 2.27	1.74 2.23	1.73 2.21
25	4.24 7.77	3.38 5.57	2.99 4.68	2.76 4.18	2.60 3.86	2.49 3.63	2.41 3.46	2.34 3.32	2.28 3.21	2.24 3.13	2.20 3.05	2.16 2.99	2.11 2.89	2.06 2.81	2.00 2.70	1.96 2.62	1.92 2.54	1.87 2.45	1.84 2.40	1.80 2.32	1.77 2.29	1.74 2.23	1.72 2.19	1.71 2.17
26	4.22 7.72	3.37 5.53	2.89 4.64	2.74 4.14	2.59 3.82	2.47 3.59	2.39 3.42	2.32 3.29	2.27 3.17	2.22 3.09	2.18 3.02	2.15 2.96	2.10 2.86	2.05 2.77	1.99 2.66	1.95 2.58	1.90 2.50	1.85 2.41	1.82 2.36	1.78 2.28	1.76 2.25	1.72 2.19	1.70 2.15	1.69 2.13
27	4.21 7.68	3.35 5.49	2.96 4.60	2.73 4.11	2.57 3.79	2.46 3.56	2.37 3.39	2.30 3.26	2.25 3.14	2.20 3.06	2.16 2.98	2.13 2.93	2.08 2.83	2.03 2.74	1.97 2.63	1.93 2.55	1.88 2.47	1.84 2.38	1.80 2.33	1.76 2.25	1.74 2.21	1.71 2.16	1.68 2.12	1.67 2.10
28	4.20 7.64	3.34 5.45	2.95 4.57	2.71 4.07	2.56 3.76	2.44 3.53	2.36 3.36	2.29 3.23	2.24 3.11	2.19 3.03	2.15 2.95	2.12 2.90	2.06 2.80	2.02 2.71	1.96 2.60	1.91 2.52	1.87 2.44	1.81 2.35	1.78 2.30	1.75 2.22	1.72 2.18	1.69 2.13	1.67 2.09	1.65 2.06
29	4.18 7.60	3.33 5.52	2.93 4.54	2.70 4.04	2.54 3.73	2.43 3.50	2.35 3.32	2.28 3.20	2.22 3.08	2.18 3.00	2.14 2.92	2.10 2.87	2.05 2.77	2.00 2.68	1.94 2.57	1.90 2.49	1.85 2.41	1.80 2.32	1.77 2.27	1.73 2.19	1.71 2.15	1.68 2.10	1.65 2.06	1.64 2.03
30	4.17 7.56	3.32 5.39	2.92 4.51	2.69 4.02	2.53 3.70	2.42 3.47	2.34 3.30	2.27 3.17	2.21 3.06	2.16 2.98	2.12 2.90	2.09 2.84	2.04 2.74	1.99 2.66	1.93 2.55	1.89 2.47	1.84 2.38	1.79 2.29	1.76 2.24	1.72 2.16	1.69 2.13	1.66 2.07	1.64 2.03	1.62 2.01

Degrees of Freedom in Denominator of F (df_2)

TABLE III(a). *Critical Values of F for α Equal to .05 and .01 (Continued).*
0.05 (light row) and 0.01 (dark row) points for the distribution of F

Degrees of Freedom in Numerator of F (df_1)

df_2	1	2	3	4	5	6	7	8	9	10	11	12	14	16	20	24	30	40	50	75	100	200	500	∞
32	4.15 7.50	3.30 5.34	2.90 4.46	2.67 3.97	2.51 3.66	2.40 3.42	2.32 3.25	2.25 3.12	2.19 3.01	2.14 2.94	2.10 2.86	2.07 2.80	2.02 2.70	1.97 2.62	1.91 2.51	1.86 2.42	1.82 2.34	1.76 2.25	1.74 2.20	1.69 2.12	1.67 2.08	1.64 2.02	1.61 1.98	1.59 1.96
34	4.13 7.44	3.28 5.29	2.88 4.42	2.65 3.93	2.49 3.61	2.38 3.38	2.30 3.21	2.23 3.08	2.17 2.97	2.12 2.89	2.08 2.82	2.05 2.76	2.00 2.66	1.95 2.58	1.89 2.47	1.84 2.38	1.80 2.30	1.74 2.21	1.71 2.15	1.67 2.08	1.64 2.04	1.61 1.98	1.59 1.94	1.57 1.91
36	4.11 7.39	3.26 5.25	2.86 4.38	2.63 3.89	2.48 3.58	2.36 3.35	2.28 3.18	2.21 3.04	2.15 2.94	2.10 2.86	2.06 2.78	2.03 2.72	1.98 2.62	1.93 2.54	1.87 2.43	1.82 2.35	1.78 2.26	1.72 2.17	1.69 2.12	1.65 2.04	1.62 2.00	1.59 1.94	1.56 1.90	1.55 1.87
38	4.10 7.35	3.25 5.21	2.85 4.34	2.62 3.86	2.46 3.54	2.35 3.32	2.26 3.15	2.19 3.02	2.14 2.91	2.09 2.82	2.05 2.75	2.02 2.69	1.96 2.59	1.92 2.51	1.85 2.40	1.80 2.32	1.76 2.22	1.71 2.14	1.67 2.08	1.63 2.00	1.60 1.97	1.57 1.90	1.54 1.86	1.53 1.84
40	4.08 7.31	3.23 5.18	2.84 4.31	2.61 3.83	2.45 3.51	2.34 3.29	2.25 3.12	2.18 2.99	2.12 2.88	2.07 2.80	2.04 2.73	2.00 2.66	1.95 2.56	1.90 2.49	1.84 2.37	1.79 2.29	1.74 2.20	1.69 2.11	1.66 2.05	1.61 1.97	1.59 1.94	1.55 1.88	1.53 1.84	1.51 1.81
42	4.07 7.27	3.22 5.15	2.83 4.29	2.59 3.80	2.44 3.49	2.32 3.26	2.24 3.10	2.17 2.96	2.11 2.86	2.06 2.77	2.02 2.70	1.99 2.64	1.94 2.54	1.89 2.46	1.82 2.35	1.78 2.26	1.73 2.17	1.68 2.08	1.64 2.02	1.60 1.94	1.57 1.91	1.54 1.85	1.51 1.80	1.49 1.78
44	4.06 7.24	3.21 5.12	2.82 4.26	2.58 3.78	2.43 3.46	2.31 3.24	2.23 3.07	2.16 2.94	2.10 2.84	2.05 2.75	2.01 2.68	1.98 2.62	1.92 2.52	1.88 2.44	1.81 2.32	1.76 2.24	1.72 2.15	1.66 2.06	1.63 2.00	1.58 1.92	1.56 1.88	1.52 1.82	1.50 1.78	1.48 1.75
46	4.05 7.21	3.20 5.10	2.81 4.24	2.57 3.76	2.42 3.44	2.30 3.22	2.22 3.05	2.14 2.92	2.09 2.82	2.04 2.73	2.00 2.66	1.97 2.60	1.91 2.50	1.87 2.42	1.80 2.30	1.75 2.22	1.71 2.13	1.65 2.04	1.62 1.98	1.57 1.90	1.54 1.86	1.51 1.80	1.48 1.76	1.46 1.72
48	4.04 7.19	3.19 5.08	2.80 4.22	2.56 3.74	2.41 3.42	2.30 3.20	2.21 3.04	2.14 2.90	2.08 2.80	2.03 2.71	1.99 2.64	1.96 2.58	1.90 2.48	1.86 2.40	1.79 2.28	1.74 2.20	1.70 2.11	1.64 2.02	1.61 1.96	1.56 1.88	1.53 1.84	1.50 1.78	1.47 1.73	1.45 1.70
50	4.03 7.17	3.18 5.06	2.79 4.20	2.56 3.72	2.40 3.41	2.29 3.18	2.20 3.02	2.13 2.88	2.07 2.78	2.02 2.70	1.98 2.62	1.95 2.56	1.90 2.46	1.85 2.39	1.78 2.26	1.74 2.18	1.69 2.10	1.63 2.00	1.60 1.94	1.55 1.86	1.52 1.82	1.48 1.76	1.46 1.71	1.44 1.68
55	4.02 7.12	3.17 5.01	2.78 4.16	2.54 3.68	2.38 3.37	2.27 3.15	2.18 2.98	2.11 2.85	2.05 2.75	2.00 2.66	1.97 2.59	1.93 2.53	1.88 2.43	1.83 2.35	1.76 2.23	1.72 2.15	1.67 2.06	1.61 1.96	1.58 1.90	1.52 1.82	1.50 1.78	1.46 1.71	1.43 1.66	1.41 1.64
60	4.00 7.08	3.15 4.98	2.76 4.13	2.52 3.65	2.37 3.34	2.25 3.12	2.17 2.95	2.10 2.82	2.04 2.72	1.99 2.63	1.95 2.56	1.92 2.50	1.86 2.40	1.81 2.32	1.75 2.20	1.70 2.12	1.65 2.03	1.59 1.93	1.56 1.87	1.50 1.79	1.48 1.74	1.44 1.68	1.41 1.63	1.39 1.60
65	3.99 7.04	3.14 4.95	2.75 4.10	2.51 3.62	2.36 3.31	2.24 3.09	2.15 2.93	2.08 2.79	2.02 2.70	1.98 2.61	1.94 2.54	1.90 2.47	1.85 2.37	1.80 2.30	1.73 2.18	1.68 2.09	1.63 2.00	1.57 1.90	1.54 1.84	1.49 1.76	1.46 1.71	1.42 1.64	1.39 1.60	1.37 1.56
70	3.98 7.01	3.13 4.92	2.74 4.08	2.50 3.60	2.35 3.29	2.32 3.07	2.14 2.91	2.07 2.77	2.01 2.67	1.97 2.59	1.93 2.51	1.89 2.45	1.84 2.35	1.79 2.28	1.72 2.15	1.67 2.07	1.62 1.98	1.56 1.88	1.53 1.82	1.47 1.74	1.45 1.69	1.40 1.62	1.37 1.56	1.35 1.53
80	3.96 6.96	3.11 4.88	2.72 4.04	2.48 3.56	2.33 3.25	2.21 3.04	2.12 2.87	2.05 2.74	1.99 2.64	1.95 2.55	1.91 2.48	1.88 2.41	1.82 2.32	1.77 2.24	1.70 2.11	1.65 2.03	1.60 1.94	1.54 1.84	1.51 1.78	1.45 1.70	1.42 1.65	1.38 1.57	1.35 1.52	1.32 1.49

Degrees of Freedom in Denominator of F (df_2)

Appendix C

TABLE III(a). Critical Values of F for α Equal to .05 and .01 (Continued).

Degrees of Freedom in Numerator of F (df_1)

df_2																								
100	3.94 6.90	3.09 4.82	2.70 3.98	2.46 3.51	2.30 3.20	2.19 2.99	2.10 2.82	2.03 2.69	1.97 2.59	1.92 2.51	1.88 2.43	1.85 2.36	1.79 2.26	1.75 2.19	1.68 2.06	1.63 1.98	1.57 1.89	1.51 1.79	1.48 1.73	1.42 1.64	1.39 1.59	1.34 1.51	1.30 1.46	1.28 1.43
125	3.92 6.84	3.07 4.78	2.68 3.94	2.44 3.47	2.29 3.17	2.17 2.95	2.08 2.79	2.01 2.65	1.95 2.56	1.90 2.47	1.86 2.40	1.83 2.33	1.77 2.23	1.72 2.15	1.65 2.03	1.60 1.94	1.55 1.85	1.49 1.75	1.45 1.68	1.39 1.59	1.36 1.54	1.31 1.46	1.27 1.40	1.25 1.37
150	3.91 6.81	3.06 4.75	2.67 3.91	2.43 3.44	2.27 3.13	2.16 2.92	2.07 2.76	2.00 2.62	1.94 2.53	1.89 2.44	1.85 2.37	1.82 2.30	1.76 2.20	1.71 2.12	1.64 2.00	1.59 1.91	1.54 1.83	1.47 1.72	1.44 1.66	1.37 1.56	1.34 1.51	1.29 1.43	1.25 1.37	1.22 1.33
200	3.89 6.76	3.04 4.71	2.65 3.88	2.41 3.41	2.26 3.11	2.14 2.90	2.05 2.73	1.98 2.60	1.92 2.50	1.87 2.41	1.83 2.34	1.80 2.28	1.74 2.17	1.69 2.09	1.62 1.97	1.57 1.88	1.52 1.79	1.45 1.69	1.42 1.62	1.35 1.53	1.32 1.48	1.26 1.39	1.22 1.33	1.19 1.28
400	3.86 6.70	3.02 4.66	2.62 3.83	2.39 3.36	2.23 3.06	2.12 2.85	2.03 2.69	1.96 2.55	1.90 2.46	1.85 2.37	1.81 2.29	1.78 2.23	1.72 2.12	1.67 2.04	1.60 1.92	1.54 1.84	1.49 1.74	1.42 1.64	1.38 1.57	1.32 1.47	1.28 1.42	1.22 1.32	1.16 1.24	1.13 1.19
1000	3.85 6.66	3.00 4.62	2.61 3.80	2.38 3.34	2.22 3.04	2.10 2.82	2.02 2.66	1.95 2.53	1.89 2.43	1.84 2.34	1.80 2.26	1.76 2.20	1.70 2.09	1.65 2.01	1.58 1.89	1.53 1.81	1.47 1.71	1.41 1.61	1.36 1.54	1.30 1.44	1.26 1.38	1.19 1.28	1.13 1.19	1.08 1.11
∞	3.84 6.64	2.99 4.60	2.60 3.78	2.37 3.32	2.21 3.02	2.09 2.80	2.01 2.64	1.94 2.51	1.88 2.41	1.83 2.32	1.79 2.24	1.75 2.18	1.69 2.07	1.64 1.99	1.57 1.87	1.52 1.79	1.46 1.69	1.40 1.59	1.35 1.52	1.28 1.41	1.24 1.36	1.17 1.25	1.11 1.15	1.00 1.00

Degrees of Freedom in Denominator of F (df_2)

TABLE III(b). Critical Values of F for α Equal to .025 and .005.

The obtained F is significant at the .025 level (light row) and the .005 level (dark row) if it is equal to or greater than the value shown in the table for the distribution of F.

df_2 \ df_1	1	2	3	4	5	6	7	8	9	10	12	15	20	24	30	40	60	120	∞
1	648	800	864	900	922	937	948	957	963	969	977	985	993	997	1,001	1,006	1,010	1,014	1,018
	16,211	20,000	21,615	22,500	23,056	23,437	23,715	23,925	24,091	24,224	24,426	24,630	24,836	24,940	25,044	25,148	24,253	25,350	25,405
2	38.51	39.00	39.16	39.25	39.30	39.33	39.36	39.37	39.39	39.40	39.42	39.43	39.45	39.46	39.46	39.47	39.48	39.49	39.50
	198	199	199	199	199	199	199	199	199	199	199	199	199	199	199	199	199	199	200
3	17.44	16.04	15.44	15.10	14.88	14.74	14.62	14.54	14.47	14.42	14.34	14.25	14.17	14.12	14.08	14.04	13.99	13.95	13.90
	55.55	49.80	47.47	46.20	45.39	44.84	44.43	44.13	43.88	43.69	43.39	43.08	42.78	42.62	42.47	42.31	42.15	41.99	41.83
4	12.22	10.65	9.98	9.60	9.36	9.20	9.07	8.98	8.90	8.84	8.75	8.66	8.56	8.51	8.46	8.41	8.36	8.31	8.26
	31.33	26.28	24.26	23.16	22.46	21.98	21.62	21.35	21.14	20.97	20.70	20.44	20.17	20.03	19.89	19.75	19.61	19.47	19.32
5	10.01	8.43	7.76	7.39	7.15	6.98	6.85	6.76	6.68	6.62	6.52	6.43	6.33	6.28	6.23	6.18	6.12	6.07	6.02
	22.78	18.31	16.53	15.56	14.94	14.51	14.20	13.96	13.77	13.62	13.38	13.15	12.90	12.78	12.66	12.53	12.40	12.27	12.14
6	8.81	7.26	6.60	6.23	5.99	5.82	5.70	5.60	5.52	5.46	5.37	5.27	5.17	5.12	5.07	5.01	4.96	4.90	4.85
	18.64	14.54	12.92	12.03	11.46	11.07	10.79	10.57	10.39	10.25	10.03	9.81	9.59	9.47	9.36	9.24	9.12	9.00	8.88
7	8.07	6.54	5.89	5.52	5.29	5.12	4.99	4.90	4.82	4.76	4.67	4.57	4.47	4.42	4.36	4.31	4.25	4.20	4.14
	16.24	12.40	10.88	10.05	9.52	9.16	8.89	8.68	8.51	8.38	8.18	7.97	7.75	7.64	7.53	7.42	7.31	7.19	7.08
8	7.57	6.06	5.42	5.05	4.82	4.65	4.53	4.43	4.36	4.30	4.20	4.10	4.00	3.95	3.89	3.84	3.78	3.73	3.67
	14.69	11.04	9.60	8.81	8.30	7.95	7.69	7.50	7.34	7.21	7.01	6.81	6.61	6.50	6.40	6.29	6.18	6.06	5.95
9	7.21	5.71	5.08	4.72	4.48	4.32	4.20	4.10	4.03	3.96	3.87	3.77	3.67	3.61	3.56	3.51	3.45	3.39	3.33
	13.61	10.11	8.72	7.96	7.47	7.13	6.88	6.69	6.54	6.42	6.23	6.03	5.83	5.73	5.62	5.52	5.41	5.30	5.19
10	6.94	5.46	4.83	4.47	4.24	4.07	3.95	3.85	3.78	3.72	3.62	3.52	3.42	3.37	3.31	3.26	3.20	3.14	3.08
	12.83	9.43	8.08	7.34	6.87	6.54	6.30	6.12	5.97	5.85	5.66	5.47	5.27	5.17	5.07	4.97	4.86	4.75	4.64
11	6.72	5.26	4.63	4.28	4.04	3.88	3.76	3.66	3.59	3.53	3.43	3.33	3.23	3.17	3.12	3.06	3.00	2.94	2.88
	12.23	8.91	7.60	6.88	6.42	6.10	5.86	5.68	5.54	5.42	5.24	5.05	4.86	4.76	4.65	4.55	4.44	4.34	4.23
12	6.55	5.10	4.47	4.12	3.89	3.73	3.61	3.51	3.44	3.37	3.28	3.18	3.07	3.02	2.96	2.91	2.85	2.79	2.72
	11.75	8.51	7.23	6.52	6.07	5.76	5.52	5.35	5.20	5.09	4.91	4.72	4.53	4.43	4.33	4.23	4.12	4.01	3.90
13	6.41	4.97	4.35	4.00	3.77	3.60	3.48	3.39	3.31	3.25	3.15	3.05	2.95	2.89	2.84	2.78	2.72	2.66	2.60
	11.37	8.19	6.93	6.23	5.79	5.48	5.25	5.08	4.94	4.82	4.64	4.46	4.27	4.17	4.07	3.97	3.87	3.76	3.65
14	6.30	4.86	4.24	3.89	3.66	3.50	3.38	3.29	3.21	3.15	3.05	2.95	2.84	2.79	2.73	2.67	2.61	2.55	2.49
	11.06	7.92	6.68	6.00	5.56	5.26	5.03	4.86	4.72	4.60	4.43	4.25	4.06	3.96	3.86	3.76	3.66	3.55	3.44

Degrees of Freedom in Numerator of F (df_1)

Degrees of Freedom in Denominator of F (df_2)

Taken from Merrington, M., and Thompson, C. M. Tables of percentage points of the inverted beta (F) distribution, *Biometrika*, 1943, **33**, 73–78.

TABLE III(b). *Critical Values of F for α Equal to .025 and .005 (Continued).*

0.025 (light row) and 0.005 (dark row) points for the distribution of F

df_2 \ df_1	1	2	3	4	5	6	7	8	9	10	12	15	20	24	30	40	60	120	∞
15	6.20 10.80	4.76 7.70	4.15 6.48	3.80 5.80	3.58 5.37	3.41 5.07	3.29 4.85	3.20 4.67	3.12 4.54	3.06 4.42	2.96 4.25	2.86 4.07	2.76 3.88	2.70 3.79	2.64 3.69	2.58 3.58	2.52 3.48	2.46 3.37	2.40 3.26
16	6.12 10.58	4.69 7.51	4.08 6.30	3.73 5.64	3.50 5.21	3.34 4.91	3.22 4.69	3.12 4.52	3.05 4.38	2.99 4.27	2.89 4.10	2.79 3.92	2.68 3.73	2.63 3.64	2.57 3.54	2.51 3.44	2.45 3.33	2.38 3.22	2.32 3.11
17	6.04 10.38	4.62 7.35	4.01 6.16	3.66 5.50	3.44 5.07	3.28 4.78	3.16 4.56	3.06 4.39	2.98 4.25	2.92 4.14	2.82 3.97	2.72 3.79	2.62 3.61	2.56 3.51	2.50 3.41	2.44 3.31	2.38 3.21	2.32 3.10	2.25 2.98
18	5.98 10.22	4.56 7.21	3.95 6.03	3.61 5.37	3.38 4.96	3.22 4.66	3.10 4.44	3.01 4.28	2.93 4.14	2.87 4.03	2.77 3.86	2.67 3.68	2.56 3.50	2.50 3.40	2.44 3.30	2.38 3.20	2.32 3.10	2.26 2.99	2.19 2.87
19	5.92 10.07	4.51 7.09	3.90 5.92	3.56 5.27	3.33 4.85	3.17 4.56	3.05 4.34	2.96 4.18	2.88 4.04	2.82 3.93	2.72 3.76	2.62 3.59	2.51 3.40	2.45 3.31	2.39 3.21	2.33 3.11	2.27 3.00	2.20 2.89	2.13 2.78
20	5.87 9.94	4.46 6.99	3.86 5.82	3.51 5.17	3.29 4.76	3.13 4.47	3.01 4.26	2.91 4.09	2.84 3.96	2.77 3.85	2.68 3.68	2.57 3.50	2.46 3.32	2.41 3.22	2.35 3.12	2.29 3.02	2.22 2.92	2.16 2.81	2.09 2.69
21	5.83 9.83	4.42 6.89	3.82 5.73	3.48 5.09	3.25 4.68	3.09 4.39	2.97 4.18	2.87 4.01	2.80 3.88	2.73 3.77	2.64 3.60	2.53 3.43	2.42 3.24	2.37 3.15	2.31 3.05	2.25 2.95	2.18 2.84	2.11 2.73	2.04 2.61
22	5.79 9.73	4.38 6.81	3.78 5.65	3.44 5.02	3.22 4.61	3.05 4.32	2.93 4.11	2.84 3.94	2.76 3.81	2.70 3.70	2.60 3.54	2.50 3.36	2.39 3.18	2.33 3.08	2.27 2.98	2.21 2.88	2.14 2.77	2.08 2.66	2.00 2.55
23	5.75 9.63	4.35 6.73	3.75 5.58	3.41 4.95	3.18 4.54	3.02 4.26	2.90 4.05	2.81 3.88	2.73 3.75	2.67 3.64	2.57 3.47	2.47 3.30	2.36 3.12	2.30 3.02	2.24 2.92	2.18 2.82	2.11 2.71	2.04 2.60	1.97 2.48
24	5.72 9.55	4.32 6.66	3.72 5.52	3.38 4.89	3.15 4.49	2.99 4.20	2.87 3.99	2.78 3.83	2.70 3.69	2.64 3.59	2.54 3.42	2.44 3.25	2.33 3.06	2.27 2.97	2.21 2.87	2.15 2.77	2.08 2.66	2.01 2.55	1.94 2.43
25	5.69 9.48	4.29 6.60	3.69 5.46	3.35 4.84	3.13 4.43	2.97 4.15	2.85 3.94	2.75 3.78	2.68 3.64	2.61 3.54	2.51 3.37	2.41 3.20	2.30 3.01	2.24 2.92	2.18 2.82	2.12 2.72	2.05 2.61	1.98 2.50	1.91 2.38
26	5.66 9.41	4.27 6.54	3.67 5.41	3.33 4.79	3.10 4.38	2.94 4.10	2.82 3.89	2.73 3.73	2.65 3.60	2.59 3.49	2.49 3.33	2.39 3.15	2.28 2.97	2.22 2.87	2.16 2.77	2.09 2.67	2.03 2.56	1.95 2.45	1.88 2.33
27	5.63 9.34	4.24 6.49	3.65 5.36	3.31 4.74	3.08 4.34	2.92 4.06	2.80 3.85	2.71 3.69	2.63 3.56	2.57 3.45	2.47 3.28	2.36 3.11	2.25 2.93	2.19 2.83	2.13 2.73	2.07 2.63	2.00 2.52	1.93 2.41	1.85 2.29
28	5.61 9.28	4.22 6.44	3.63 5.32	3.29 4.70	3.06 4.30	2.90 4.02	2.78 3.81	2.69 3.65	2.61 3.52	2.55 3.41	2.45 3.25	2.34 3.07	2.23 2.89	2.17 2.79	2.11 2.69	2.05 2.59	1.98 2.48	1.91 2.37	1.83 2.25

Degrees of Freedom in Numerator of F (df_1)

Degrees of Freedom in Denominator of F (df_2)

TABLE III(b). Critical Values of F for α Equal to .025 and .005 (Continued).

0.025 (light row) and 0.005 (dark row) points for the distribution of F

df_2	\ df_1	1	2	3	4	5	6	7	8	9	10	12	15	20	24	30	40	60	120	∞
29		5.59	4.20	3.61	3.27	3.04	2.88	2.76	2.67	2.59	2.53	2.43	2.32	2.21	2.15	2.09	2.03	1.96	1.89	1.81
		9.23	6.40	5.28	4.66	4.26	3.98	3.77	3.61	3.48	3.38	3.21	3.04	2.86	2.76	2.66	2.56	2.45	2.33	2.21
30		5.57	4.18	3.59	3.25	3.03	2.87	2.75	2.65	2.57	2.51	2.41	2.31	2.20	2.14	2.07	2.01	1.94	1.87	1.79
		9.18	6.35	5.24	4.62	4.23	3.95	3.74	3.58	3.45	3.34	3.18	3.01	2.82	2.73	2.63	2.52	2.42	2.30	2.18
40		5.42	4.05	3.46	3.13	2.90	2.74	2.62	2.53	2.45	2.39	2.29	2.18	2.07	2.01	1.94	1.88	1.80	1.72	1.64
		8.83	6.07	4.98	4.37	3.99	3.71	3.51	3.35	3.22	3.12	2.95	2.78	2.60	2.50	2.40	2.30	2.18	2.06	1.93
60		5.29	3.93	3.34	3.01	2.79	2.63	2.51	2.41	2.33	2.27	2.17	2.06	1.94	1.88	1.82	1.74	1.67	1.58	1.48
		8.49	5.80	4.73	4.14	3.76	3.49	3.29	3.13	3.01	2.90	2.74	2.57	2.39	2.29	2.19	2.08	1.96	1.83	1.69
120		5.15	3.80	3.23	2.89	2.67	2.52	2.39	2.30	2.22	2.16	2.05	1.94	1.82	1.76	1.69	1.61	1.53	1.43	1.31
		8.18	5.54	4.50	3.92	3.55	3.28	3.09	2.93	2.81	2.71	2.54	2.37	2.19	2.09	1.98	1.87	1.75	1.61	1.43
∞		5.02	3.69	3.12	2.79	2.57	2.41	2.29	2.19	2.11	2.05	1.94	1.83	1.71	1.64	1.57	1.48	1.39	1.27	1.00
		7.88	5.30	4.28	3.72	3.35	3.09	2.90	2.74	2.62	2.52	2.36	2.19	2.00	1.90	1.79	1.67	1.53	1.36	1.00

Degrees of Freedom in Numerator of F (df_1)

Degrees of Freedom in Denominator of F (df_2)

TABLE IV. Critical Values of Chi Square.

Level of significance

df	.20	.10	.05	.02	.01	.001
1	1.64	2.71	3.84	5.41	6.64	10.83
2	3.22	4.60	5.99	7.82	9.21	13.82
3	4.64	6.25	7.82	9.84	11.34	16.27
4	5.99	7.78	9.49	11.67	13.28	18.46
5	7.29	9.24	11.07	13.39	15.09	20.52
6	8.56	10.64	12.59	15.03	16.81	22.46
7	9.80	12.02	14.07	16.62	18.48	24.32
8	11.03	13.36	15.51	18.17	20.09	26.12
9	12.24	14.68	16.92	19.68	21.67	27.88
10	13.44	15.99	18.31	21.16	23.21	29.59
11	14.63	17.28	19.68	22.62	24.72	31.26
12	15.81	18.55	21.03	24.05	26.22	32.91
13	16.98	19.81	22.36	25.47	27.69	34.53
14	18.15	21.06	23.68	26.87	29.14	36.12
15	19.31	22.31	25.00	28.26	30.58	37.70
16	20.46	23.54	26.30	29.63	32.00	39.29
17	21.62	24.77	27.59	31.00	33.41	40.75
18	22.76	25.99	28.87	32.35	34.80	42.31
19	23.90	27.20	30.14	33.69	36.19	43.82
20	25.04	28.41	31.41	35.02	37.57	45.32
21	26.17	29.62	32.67	36.34	38.93	46.80
22	27.30	30.81	33.92	37.66	40.29	48.27
23	28.43	32.01	35.17	38.97	41.64	49.73
24	29.55	33.20	36.42	40.27	42.98	51.18
25	30.68	34.38	37.65	41.57	44.31	52.62
26	31.80	35.56	38.88	42.86	45.64	54.05
27	32.91	36.74	40.11	44.14	46.96	55.48
28	34.03	37.92	41.34	45.42	48.28	56.89
29	35.14	39.09	42.69	46.69	49.59	58.30
30	36.25	40.26	43.77	47.96	50.89	59.70

For df greater than 30, the value obtained from the expression $\sqrt{2\chi^2} - \sqrt{2df - 1}$ may be used as a t-ratio.

Taken from Table IV of Fisher, R. A., and Yates, F. *Statistical Tables for Biological, Agricultural and Medical Research.* Edinburgh: Oliver and Boyd, 1974, 6th ed. This abridgement is a reprint of Table B of Runyon, R. P., and Haber, A. *Fundamentals of Behavioral Statistics.* Reading, Mass.: Addison-Wesley, 1967, p. 252.

TABLE V. Critical Values of the Correlation Coefficient.   *for Pearson r*

df	Level of significance for one-tailed test			
	.05	.025	.01	.005
	Level of significance for two-tailed test			
	.10	.05	.02	.01
1	.988	.997	.9995	.9999
2	.900	.950	.980	.990
3	.805	.878	.934	.959
4	.729	.811	.882	.917
5	.669	.754	.833	.874
6	.622	.707	.789	.834
7	.582	.666	.750	.798
8	.549	.632	.716	.765
9	.521	.602	.685	.735
10	.497	.576	.658	.708
11	.476	.553	.634	.684
12	.458	.532	.612	.661
13	.441	.514	.592	.641
14	.426	.497	.574	.623
15	.412	.482	.558	.606
16	.400	.468	.542	.590
17	.389	.456	.528	.575
18	.378	.444	.516	.561
19	.369	.433	.503	.549
20	.360	.423	.492	.537
21	.352	.413	.482	.526
22	.344	.404	.472	.515
23	.337	.396	.462	.505
24	.330	.388	.453	.496
25	.323	.381	.445	.487
26	.317	.374	.437	.479
27	.311	.367	.480	.471
28	.306	.361	.423	.468
29	.301	.355	.416	.456
30	.296	.349	.409	.449
35	.275	.325	.381	.418
40	.257	.304	.358	.393
45	.243	.288	.338	.372
50	.231	.273	.322	.354
60	.211	.250	.295	.325
70	.195	.232	.274	.303
80	.183	.217	.256	.283
90	.173	.205	.242	.267
100	.164	.195	.230	.254

Taken from Table VII of Fisher, R. A., and Yates, F. *Statistical Tables for Biological, Agricultural, and Medical Research.* Edinburgh: Oliver and Boyd, 1974, 6th ed.

TABLE VI. Critical Values of r′, the Spearman Rank–Order Coefficient.

	Level of significance for one-tailed test			
	.05	.025	.01	.005
	Level of significance for two-tailed test			
n*	.10	.05	.02	.01
5	.900	1.000	1.000	--
6	.829	.886	.943	1.000
7	.714	.786	.893	.929
8	.643	.738	.833	.881
9	.600	.683	.783	.833
10	.564	.648	.746	.794
12	.506	.591	.712	.777
14	.456	.544	.645	.715
16	.425	.506	.601	.665
18	.399	.475	.564	.625
20	.377	.450	.534	.591
22	.359	.428	.508	.562
24	.343	.409	.485	.537
26	.329	.392	.465	.515
28	.317	.377	.448	.496
30	.306	.364	.432	.478

*n = number of pairs

SOURCE: Taken from Table V of Olds, E. G. The 5 percent significance levels of sums of squares of rank differences and a correction. *Ann. Math. Statis.*, 1949, **20**: 117–118; and from Table V of Olds, E. G. Distribution of the sum of squares of rank differences for small numbers of individuals. *Ann. Math. Statis.*, 1938, **9**: 133–148. This abridgement is a reprint of Table G of Runyon, R. P., and Haber, A. *Fundamentals of Behavioral Statistics.* Reading, Mass.: Addison-Wesley, 1971, p. 301.

TABLE VII. *Table of Probabilities Associated With Values as Small as Observed Values of r in the Sign Test.*

Given in the body of this table are one-tailed probabilities under H_0 for the sign test when $p = q = \tfrac{1}{2}$. To save space, decimal points are omitted in the body of the table.

N \ m	0	1	2	3	4	5	6	7	8	9	10	11	12	13	14	15
5	031	188	500	812	969	*										
6	016	109	344	656	891	984	*									
7	008	062	227	500	773	938	992	*								
8	004	035	145	363	637	855	965	996	*							
9	002	020	090	254	500	746	910	980	998	*						
10	001	011	055	172	377	623	828	945	989	999	*					
11		006	033	113	274	500	726	887	967	994	*	*				
12		003	019	073	194	387	613	806	927	981	997	*	*			
13		002	011	046	133	291	500	709	867	954	989	998	*	*		
14		001	006	029	090	212	395	605	788	910	971	994	999	*	*	
15			004	018	059	151	304	500	696	849	941	982	996	*	*	*
16			002	011	038	105	227	402	598	773	895	962	989	998	*	*
17			001	006	025	072	166	315	500	685	834	928	975	994	999	*
18			001	004	015	048	119	240	407	593	760	881	952	985	996	999
19				002	010	032	084	180	324	500	676	820	916	968	990	998
20				001	006	021	058	132	252	412	588	748	868	942	979	994
21				001	004	013	039	095	192	332	500	668	808	905	961	987
22					002	008	026	067	143	262	416	584	738	857	933	974
23					001	005	017	047	105	202	339	500	661	798	895	953
24					001	003	011	032	076	154	271	419	581	729	846	924
25						002	007	022	054	115	212	345	500	655	788	885

† 1.0, or approximately 1.0.

Taken from Table IV, B, of Walker, H., and Lev, J. *Statistical Inference*. New York: Holt, 1953, p. 458.

TABLE VIII. Critical Lower-Tail Values of W_n for Wilcoxon's Rank-Sum Test.

n	\multicolumn{6}{c}{$m=1$}		\multicolumn{6}{c}{$m=2$}	n
	0·001	0·005	0·010	0·025	0·05	0·10	$2\overline{W}$	0·001	0·005	0·010	0·025	0·05	0·10	$2\overline{W}$	
2							4						—	10	2
3							5						3	12	3
4							6					—	3	14	4
5							7					3	4	16	5
6							8					3	4	18	6
7							9				—	3	4	20	7
8						—	10				3	4	5	22	8
9						1	11				3	4	5	24	9
10						1	12				3	4	6	26	10
11						1	13				3	4	6	28	11
12						1	14			—	4	5	7	30	12
13						1	15			3	4	5	7	32	13
14						1	16			3	4	6	8	34	14
15						1	17			3	4	6	8	36	15
16						1	18			3	4	6	8	38	16
17						1	19			3	5	6	9	40	17
18					—	1	20		—	3	5	7	9	42	18
19					1	2	21		3	4	5	7	10	44	19
20					1	2	22		3	4	5	7	10	46	20
21					1	2	23		3	4	6	8	11	48	21
22					1	2	24		3	4	6	8	11	50	22
23					1	2	25		3	4	6	8	12	52	23
24					1	2	26		3	4	6	9	12	54	24
25	—	—	—	—	1	2	27	—	3	4	6	9	12	56	25

Taken from Table I in Verdooren, L. R. Extended tables of critical values for Wilcoxon's test statistic. *Biometrika*, 1963, **50**, 177–186.

TABLE VIII. *Critical Lower-Tail Values of W_n for Wilcoxon's Rank-Sum Test (Continued).*

m	0.001	0.005	n = 3 0.010	0.025	0.05	0.10	$2\overline{W}$	0.001	0.005	n = 4 0.010	0.025	0.05	0.10	$2\overline{W}$	m
3					6	7	21								
4				—	6	7	24			—	10	11	13	36	4
5				6	7	8	27			10	11	12	14	40	5
6			—	7	8	9	30		10	11	12	13	15	44	6
7			6	7	8	10	33		10	11	13	14	16	48	7
8		—	6	8	9	11	36		11	12	14	15	17	52	8
9		6	7	8	10	11	39	—	11	13	14	16	19	56	9
10		6	7	9	10	12	42	10	12	13	15	17	20	60	10
11		6	7	9	11	13	45	10	12	14	16	18	21	64	11
12		7	8	10	11	14	48	10	13	15	17	19	22	68	12
13		7	8	10	12	15	51	11	13	15	18	20	23	72	13
14		7	8	11	13	16	54	11	14	16	19	21	25	76	14
15		8	9	11	13	16	57	11	15	17	20	22	26	80	15
16	—	8	9	12	14	17	60	12	15	17	21	24	27	84	16
17	6	8	10	12	15	18	63	12	16	18	21	25	28	88	17
18	6	8	10	13	15	19	66	13	16	19	22	26	30	92	18
19	6	9	10	13	16	20	69	13	17	19	23	27	31	96	19
20	6	9	11	14	17	21	72	13	18	20	24	28	32	100	20
21	7	9	11	14	17	21	75	14	18	21	25	29	33	104	21
22	7	10	12	15	18	22	78	14	19	21	26	30	35	108	22
23	7	10	12	15	19	23	81	14	19	22	27	31	36	112	23
24	7	10	12	16	19	24	84	15	20	23	27	32	38	116	24
25	7	11	13	16	20	25	87	15	20	23	28	33	38	120	25

m	0.001	0.005	n = 5 0.010	0.025	0.05	0.10	$2\overline{W}$	0.001	0.005	n = 6 0.010	0.025	0.05	0.10	$2\overline{W}$	m
5		15	16	17	19	20	55								
6		16	17	18	20	22	60		23	24	26	28	30	78	6
7	—	16	18	20	21	23	65	21	24	25	27	29	32	84	7
8	15	17	19	21	23	25	70	22	25	27	29	31	34	90	8
9	16	18	20	22	24	27	75	23	26	28	31	33	36	96	9
10	16	19	21	23	26	28	80	24	27	29	32	35	38	102	10
11	17	20	22	24	27	30	85	25	28	30	34	37	40	108	11
12	17	21	23	26	28	32	90	25	30	32	35	38	42	114	12
13	18	22	24	27	30	33	95	26	31	33	37	40	44	120	13
14	18	22	25	28	31	35	100	27	32	34	38	42	46	126	14
15	19	23	26	29	33	37	105	28	33	36	40	44	48	132	15
16	20	24	27	30	34	38	110	29	34	37	42	46	50	138	16
17	20	25	28	32	35	40	115	30	36	39	43	47	52	144	17
18	21	26	29	33	37	42	120	31	37	40	45	49	55	150	18
19	22	27	30	34	38	43	125	32	38	41	46	51	57	156	19
20	22	28	31	35	40	45	130	33	39	43	48	53	59	162	20
21	23	29	32	37	41	47	135	33	40	44	50	55	61	168	21
22	23	29	33	38	43	48	140	34	42	45	51	57	63	174	22
23	24	30	34	39	44	50	145	35	43	47	53	58	65	180	23
24	25	31	35	40	45	51	150	36	44	48	54	60	67	186	24
25	25	32	36	42	47	53	155	37	45	50	56	62	69	192	25

TABLE VIII. *Critical Lower-Tail Values of W_n for Wilcoxon's Rank-Sum Test (Continued).*

m	\multicolumn{7}{c}{n = 7}	\multicolumn{7}{c}{n = 8}	m
	0.001	0.005	0.010	0.025	0.05	0.10	$2\overline{W}$	0.001	0.005	0.010	0.025	0.05	0.10	$2\overline{W}$	
7	29	32	34	36	39	41	105								
8	30	34	35	38	41	44	112	40	43	45	49	51	55	136	8
9	31	35	37	40	43	46	119	41	45	47	51	54	58	144	9
10	33	37	39	42	45	49	126	42	47	49	53	56	60	152	10
11	34	38	40	44	47	51	133	44	49	51	55	59	63	160	11
12	35	40	42	46	49	54	140	45	51	53	58	62	66	168	12
13	36	41	44	48	52	56	147	47	53	56	60	64	69	176	13
14	37	43	45	50	54	59	154	48	54	58	62	67	72	184	14
15	38	44	47	52	56	61	161	50	56	60	65	69	75	192	15
16	39	46	49	54	58	64	168	51	58	62	67	72	78	200	16
17	41	47	51	56	61	66	175	53	60	64	70	75	81	208	17
18	42	49	52	58	63	69	182	54	62	66	72	77	84	216	18
19	43	50	54	60	65	71	189	56	64	68	74	80	37	224	19
20	44	52	56	62	67	74	196	57	66	70	77	83	90	232	20
21	46	53	58	64	69	76	203	59	68	72	79	85	92	240	21
22	47	55	59	66	72	79	210	60	70	74	81	88	95	248	22
23	48	57	61	68	74	81	217	62	71	76	84	90	98	256	23
24	49	58	63	70	76	84	224	64	73	78	86	93	101	264	24
25	50	60	64	72	78	86	231	65	75	81	89	96	104	272	25

m	\multicolumn{7}{c}{n = 9}	\multicolumn{7}{c}{n = 10}	m
	0.001	0.005	0.010	0.025	0.05	0.10	$2\overline{W}$	0.001	0.005	0.010	0.025	0.05	0.10	$2\overline{W}$	
9	52	56	59	62	66	70	171								
10	53	58	61	65	69	73	180	65	71	74	78	82	87	210	10
11	55	61	63	68	72	76	189	67	73	77	81	86	91	220	11
12	57	63	66	71	75	80	198	69	76	79	84	89	94	230	12
13	59	65	68	73	78	83	207	72	79	82	88	92	98	240	13
14	60	67	71	76	81	86	216	74	81	85	91	96	102	250	14
15	62	69	73	79	84	90	225	76	84	88	94	99	106	260	15
16	64	72	76	82	87	93	234	78	86	91	97	103	109	270	16
17	66	74	78	84	90	97	243	80	89	93	100	106	113	280	17
18	68	76	81	87	93	100	252	82	92	96	103	110	117	290	18
19	70	78	83	90	96	103	261	84	94	99	107	113	121	300	19
20	71	81	85	93	99	107	270	87	97	102	110	117	125	310	20
21	73	83	88	95	102	110	279	89	99	105	113	120	128	320	21
22	75	85	90	98	105	113	288	91	102	108	116	123	132	330	22
23	77	88	93	101	108	117	297	93	105	110	119	127	136	340	23
24	79	90	95	104	111	120	306	95	107	113	122	130	140	350	24
25	81	92	98	107	114	123	315	98	110	116	126	134	144	360	25

TABLE VIII. Critical Lower-Tail Values of W_n for Wilcoxon's Rank-Sum Test (Continued).

			$n = 11$							$n = 12$					
m	0.001	0.005	0.010	0.025	0.05	0.10	$2\overline{W}$	0.001	0.005	0.010	0.025	0.05	0.10	$2\overline{W}$	m
11	81	87	91	96	100	106	253								
12	83	90	94	99	104	110	264	98	105	109	115	120	127	300	12
13	86	93	97	103	108	114	275	101	109	113	119	125	131	312	13
14	88	96	100	106	112	118	286	103	112	116	123	129	136	324	14
15	90	99	103	110	116	123	297	106	115	120	127	133	141	336	15
16	93	102	107	113	120	127	308	109	119	124	131	138	145	348	16
17	95	105	110	117	123	131	319	112	122	127	135	142	150	360	17
18	98	108	113	121	127	135	330	115	125	131	139	146	155	372	18
19	100	111	116	124	131	139	341	118	129	134	143	150	159	384	19
20	103	114	119	128	135	144	352	120	132	138	147	155	164	396	20
21	106	117	123	131	139	148	363	123	136	142	151	159	169	408	21
22	108	120	126	135	143	152	374	126	139	145	155	163	173	420	22
23	111	123	129	139	147	156	385	129	142	149	159	168	178	432	23
24	113	126	132	142	151	161	396	132	146	153	163	172	183	444	24
25	116	129	136	146	155	165	407	135	149	156	167	176	187	456	25

			$n = 13$							$n = 14$					
m	0.001	0.005	0.010	0.025	0.05	0.10	$2\overline{W}$	0.001	0.005	0.010	0.025	0.05	0.10	$2\overline{W}$	m
13	117	125	130	136	142	149	351								
14	120	129	134	141	147	154	364	137	147	152	160	166	174	406	14
15	123	133	138	145	152	159	377	141	151	156	164	171	179	420	15
16	126	136	142	150	156	165	390	144	155	161	169	176	185	434	16
17	129	140	146	154	161	170	403	148	159	165	174	182	190	448	17
18	133	144	150	158	166	175	416	151	163	170	179	187	196	462	18
19	136	148	154	163	171	180	429	155	168	174	183	192	202	476	19
20	139	151	158	167	175	185	442	159	172	178	188	197	207	490	20
21	142	155	162	171	180	190	455	162	176	183	193	202	213	504	21
22	145	159	166	176	185	195	468	166	180	187	198	207	218	518	22
23	149	163	170	180	189	200	481	169	184	192	203	212	224	532	23
24	152	166	174	185	194	205	494	173	188	196	207	218	229	546	24
25	155	170	178	189	199	211	507	177	192	200	212	223	235	560	25

			$n = 15$							$n = 16$					
m	0.001	0.005	0.010	0.025	0.05	0.10	$2\overline{W}$	0.001	0.005	0.010	0.025	0.05	0.10	$2\overline{W}$	m
15	160	171	176	184	192	200	465								
16	163	175	181	190	197	206	480	184	196	202	211	219	229	528	16
17	167	180	186	195	203	212	495	188	201	207	217	225	235	544	17
18	171	184	190	200	208	218	510	192	206	212	222	231	242	560	18
19	175	189	195	205	214	224	525	196	210	218	228	237	248	576	19
20	179	193	200	210	220	230	540	201	215	223	234	243	255	592	20
21	183	198	205	216	225	236	555	205	220	228	239	249	261	608	21
22	187	202	210	221	231	242	570	209	225	233	245	255	267	624	22
23	191	207	214	226	236	248	585	214	230	238	251	261	274	640	23
24	195	211	219	231	242	254	600	218	235	244	256	267	280	656	24
25	199	216	224	237	248	260	615	222	240	249	262	273	287	672	25

Appendix C   419

TABLE VIII. Critical Lower-Tail Values of W_n for Wilcoxon's Rank-Sum Test (Continued).

m	|	$n = 17$							|	$n = 18$						m
	0.001	0.005	0.010	0.025	0.05	0.10	$2\overline{W}$		0.001	0.005	0.010	0.025	0.05	0.10	$2\overline{W}$	
17	210	223	230	240	249	259	595									
18	214	228	235	246	255	266	612		237	252	259	270	280	291	666	18
19	219	234	241	252	262	273	629		242	258	265	277	287	299	684	19
20	223	239	246	258	268	280	646		247	263	271	283	294	306	702	20
21	228	244	252	264	274	287	663		252	269	277	290	301	313	720	21
22	233	249	258	270	281	294	680		257	275	283	296	307	321	738	22
23	238	255	263	276	287	300	697		262	280	289	303	314	328	756	23
24	242	260	269	282	294	307	714		267	286	295	309	321	335	774	24
25	247	265	275	288	300	314	731		273	292	301	316	328	343	792	25

m	|	$n = 19$							|	$n = 20$						m
	0.001	0.005	0.010	0.025	0.05	0.10	$2\overline{W}$		0.001	0.005	0.010	0.025	0.05	0.10	$2\overline{W}$	
19	267	283	291	303	313	325	741									
20	272	289	297	309	320	333	760		298	315	324	337	348	361	820	20
21	277	295	303	316	328	341	779		304	322	331	344	356	370	840	21
22	283	301	310	323	335	349	798		309	328	337	351	364	378	860	22
23	288	307	316	330	342	357	817		315	335	344	359	371	386	880	23
24	294	313	323	337	350	364	836		321	341	351	366	379	394	900	24
25	299	319	329	344	357	372	855		327	348	358	373	387	403	920	25

m	|	$n = 21$							|	$n = 22$						m
	0.001	0.005	0.010	0.025	0.05	0.10	$2\overline{W}$		0.001	0.005	0.010	0.025	0.05	0.10	$2\overline{W}$	
21	331	349	359	373	385	399	903									
22	337	356	366	381	393	408	924		365	386	396	411	424	439	990	22
23	343	363	373	388	401	417	945		372	393	403	419	432	448	1012	23
24	349	370	381	396	410	425	966		379	400	411	427	441	457	1034	24
25	356	377	388	404	418	434	987		385	408	419	435	450	467	1056	25

m	|	$n = 23$							|	$n = 24$						m
	0.001	0.005	0.010	0.025	0.05	0.10	$2\overline{W}$		0.001	0.005	0.010	0.025	0.05	0.10	$2\overline{W}$	
23	402	424	434	451	465	481	1081									
24	409	431	443	459	474	491	1104		440	464	475	492	507	525	1176	24
25	416	439	451	468	483	500	1127		448	472	484	501	517	535	1200	25

m	|	$n = 25$						
	0.001	0.005	0.010	0.025	0.05	0.10	$2\overline{W}$	
25	480	505	517	536	552	570	1275	

TABLE IX. *Critical and Quasi-Critical Lower-Tail Values of W_+ (and Their Probability Levels) for Wilcoxon's Signed-Rank Test.*

n	$\alpha = .05$		$\alpha = .025$		$\alpha = .01$		$\alpha = .005$	
5	0	.0313						
	1	.0625						
6	2	.0469	0	.0156				
	3	.0781	1	.0313				
7	3	.0391	2	.0234	0	.0078		
	4	.0547	3	.0391	1	.0156		
8	5	.0391	3	.0195	1	.0078	0	.0039
	6	.0547	4	.0273	2	.0117	1	.0078
9	8	.0488	5	.0195	3	.0098	1	.0039
	9	.0645	6	.0273	4	.0137	2	.0059
10	10	.0420	8	.0244	5	.0098	3	.0049
	11	.0527	9	.0322	6	.0137	4	.0068
11	13	.0415	10	.0210	7	.0093	5	.0049
	14	.0508	11	.0269	8	.0122	6	.0068
12	17	.0461	13	.0212	9	.0081	7	.0046
	18	.0549	14	.0261	10	.0105	8	.0061
13	21	.0471	17	.0239	12	.0085	9	.0040
	22	.0549	18	.0287	13	.0107	10	.0052
14	25	.0453	21	.0247	15	.0083	12	.0043
	26	.0520	22	.0290	16	.0101	13	.0054
15	30	.0473	25	.0240	19	.0090	15	.0042
	31	.0535	26	.0277	20	.0108	16	.0051
16	35	.0467	29	.0222	23	.0091	19	.0046
	36	.0523	30	.0253	24	.0107	20	.0055
17	41	.0492	34	.0224	27	.0087	23	.0047
	42	.0544	35	.0253	28	.0101	24	.0055
18	47	.0494	40	.0241	32	.0091	27	.0045
	48	.0542	41	.0269	33	.0104	28	.0052
19	53	.0478	46	.0247	37	.0090	32	.0047
	54	.0521	47	.0273	38	.0102	33	.0054
20	60	.0487	52	.0242	43	.0096	37	.0047
	61	.0527	53	.0266	44	.0107	38	.0053

Taken from Table II in Wilcoxon, F., Katti, S. K., and Wilcox, R. A. *Critical Values and Probability Levels for the Wilcoxon Rank-Sum Test and the Wilcoxon Signed-Rank Test*, American Cyanamid Company (Lederle Laboratories Division, Pearl River, N.Y.) and Florida State University (Department of Statistics, Tallahasse, Fla.), August 1963.

TABLE IX. Critical and Quasi-Critical Lower-Tail Values of W_+ (and Their Probability Levels) for Wilcoxon's Signed-Rank Test (Continued).

n	$\alpha = .05$		$\alpha = .025$		$\alpha = .01$		$\alpha = .005$	
21	67	.0479	58	.0230	49	.0097	42	.0045
	68	.0516	59	.0251	50	.0108	43	.0051
22	75	.0492	65	.0231	55	.0095	48	.0046
	76	.0527	66	.0250	56	.0104	49	.0052
23	83	.0490	73	.0242	62	.0098	54	.0046
	84	.0523	74	.0261	63	.0107	55	.0051
24	91	.0475	81	.0245	69	.0097	61	.0048
	92	.0505	82	.0263	70	.0106	62	.0053
25	100	.0479	89	.0241	76	.0094	68	.0048
	101	.0507	90	.0258	77	.0101	69	.0053
26	110	.0497	98	.0247	84	.0095	75	.0047
	111	.0524	99	.0263	85	.0102	76	.0051
27	119	.0477	107	.0246	92	.0093	83	.0048
	120	.0502	108	.0260	93	.0100	84	.0052
28	130	.0496	116	.0239	101	.0096	91	.0048
	131	.0521	117	.0252	102	.0102	92	.0051
29	140	.0482	126	.0240	110	.0095	100	.0049
	141	.0504	127	.0253	111	.0101	101	.0053
30	151	.0481	137	.0249	120	.0098	109	.0050
	152	.0502	138	.0261	121	.0104	110	.0053
31	163	.0491	147	.0239	130	.0099	118	.0049
	164	.0512	148	.0251	131	.0105	119	.0052
32	175	.0492	159	.0249	140	.0097	128	.0050
	176	.0512	160	.0260	141	.0103	129	.0053
33	187	.0485	170	.0242	151	.0099	138	.0049
	188	.0503	171	.0253	152	.0104	139	.0052
34	200	.0488	182	.0242	162	.0098	148	.0048
	201	.0506	183	.0252	163	.0103	149	.0051
35	213	.0484	195	.0247	173	.0096	159	.0048
	214	.0501	196	.0257	174	.0100	160	.0051
36	227	.0489	208	.0248	185	.0096	171	.0050
	228	.0505	209	.0258	186	.0100	172	.0052
37	241	.0487	221	.0245	198	.0099	182	.0048
	242	.0503	222	.0254	199	.0103	183	.0050
38	256	.0493	235	.0247	211	.0099	194	.0048
	257	.0509	236	.0256	212	.0104	195	.0050
39	271	.0493	249	.0246	224	.0099	207	.0049
	272	.0507	250	.0254	225	.0103	208	.0051
40	286	.0486	264	.0249	238	.0100	220	.0049
	287	.0500	265	.0257	239	.0104	221	.0051

TABLE IX. Critical and Quasi-Critical Lower-Tail Values of W_+ (and Their Probability Levels) for Wilcoxon's Signed-Rank Test (Continued).

n	$\alpha = .05$		$\alpha = .025$		$\alpha = .01$		$\alpha = .005$	
41	302	.0488	279	.0248	252	.0100	233	.0048
	303	.0501	280	.0256	253	.0103	234	.0050
42	319	.0496	294	.0245	266	.0098	247	.0049
	320	.0509	295	.0252	267	.0102	248	.0051
43	336	.0498	310	.0245	281	.0098	261	.0048
	337	.0511	311	.0252	282	.0102	262	.0050
44	353	.0495	327	.0250	296	.0097	276	.0049
	354	.0507	328	.0257	297	.0101	277	.0051
45	371	.0498	343	.0244	312	.0098	291	.0049
	372	.0510	344	.0251	313	.0101	292	.0051
46	389	.0497	361	.0249	328	.0098	307	.0050
	390	.0508	362	.0256	329	.0101	308	.0052
47	407	.0490	378	.0245	345	.0099	322	.0048
	408	.0501	379	.0251	346	.0102	323	.0050
48	426	.0490	396	.0244	362	.0099	339	.0050
	427	.0500	397	.0251	363	.0102	340	.0051
49	446	.0495	415	.0247	379	.0098	355	.0049
	447	.0505	416	.0253	380	.0100	356	.0050
50	466	.0495	434	.0247	397	.0098	373	.0050
	467	.0506	435	.0253	398	.0101	374	.0051

References

BAKAN, D. The test of significance in psychological research. *Psychological Bulletin* 1966, **66**:423–437.

BARBER, T. X., and CALVERLY, D. S. An experimental study of "hypnotic" (auditory and visual) hallucinations. *Journal of Abnormal and Social Psychology*, 1964, **63**:13–20.

BLALOCK, H. M., JR. *Causal Inference in Non-experimental Research.* Chapel Hill: University of North Carolina Press, 1961.

BOISMONT, A. G. *On Hallucinations: A History and Explanation of Apparitions, Visions, Dreams, Ecstacy, Magnetism, and Somnambulism.* London: Henry Renshaw, 1859.

BOWERS, K. S. The effect of demands for honesty on reports of visual and auditory hallucinations. *International Journal of Clinical and Experimental Hypnosis*, 1967, **15**:31–36.

BRADLEY, J. V. *Distribution-Free Statistical Tests.* New Jersey: Prentice Hall, 1968.

DAVID, F. N. *Games, Gods, and Gambling.* New York: Hafner, 1962.

FISHER, A. *The Mathematical Theory of Probabilities.* Translated by C. Dickson and W. Bonynge. Vol. I. 2nd ed. New York: Macmillan, 1923.

FISHER, R. A. Application of "Student's" distribution. *Metron*, 1925, **5**:90–104.

GALTON, F. *Memories of My Life.* London: Methuen, 1908.

——— *Natural Inheritance.* London: Macmillan, 1889.

GREENWALD, A. G. Significance, non-significance, and interpretation of an ESP experiment. *Journal of Experimental Social Psychology*, 1975, **11**:180–191.

HAMILTON, N. T., and LANDIN, J. *Set Theory: The Structure of Arithmetic.* Boston: Allyn and Bacon, 1961.

HEATH, T. L. *A Manual of Greek Mathematics.* New York: Dover, 1930.

HUFF, D. *How to Lie with Statistics.* New York: W. W. Norton, 1954.

MEITZEN, A. *History, Theory and Technique of Statistics.* Philadelphia: American Academy of Political and Social Science, 1886.

MINTZ, S., and ALPERT, M. Imagery, vividness, reality testing, and schizophrenic hallucinations. *Journal of Abnormal Psychology*, 1972, **79**:310–316.

NEWMAN, J. R. A. FISHER, (1890–1962): An appreciation. *Science*, 1967, **156**: 1456–1462.

PEARSON, E. S. *Karl Pearson.* Cambridge: Cambridge University Press, 1938.

——— Student as statistician. *Biometrika*, 1938, **30**:210–250.

PEARSON, K. *The Grammar of Science.* 2nd ed. London: Adam and Charles Black, 1900.

——— Historical note on the normal curve of errors. *Biometrika*, 1924, **16**:402–404.

——— *Life, Letters, and Labours of Francis Galton.* Vol. IIIA. London: Cambridge University Press, 1930.

SAVAGE, I. R. Bibliography of nonparametric statistics and related topics. *Journal of the American Statistical Association*, 1953, **48**:844–906.

SCHEFFÉ, S. *The Analysis of Variance.* New York: Wiley, 1959.

SPANOS, N. P., and BARBER, T. X. "Hypnotic" experiences as inferred from subjective reports: auditory and visual hallucinations. *Journal of Experimental Research in Personality*, 1968, **3**:136–150.

STROEBE, W., INSKO, C. A., THOMPSON, V. D., and LAYTON, B. D. Effects of physical attractiveness, attitude similarity, and sex on interpersonal attraction. *Journal of Personality and Social Psychology*, 1971, **18**:79–91.

STUDENT. The probable error of a mean. *Biometrika*, 1908, **6**:1–25.

WALKER, H. M. *Studies in the History of Statistical Method.* Baltimore: Williams and Wilkins, 1929.

Exercise Answers

Chapter 1.

1. (a) p. 13 (b) p. 9 (c) p. 9 (d) p. 10 (e) pp. 6–7 (f) p. 11 (g) p. 11 (h) p. 11
   (i) p. 12 (j) p. 12 (k) p. 13 (l)
2. p. 7
3. p. 7
4. pp. 4–7
5. p. 11
6. p. 11
7. (a) continuous (b) discrete (c) continuous (d) discrete (e) discrete (f) discrete
   (g) discrete (h) discrete (i) discrete (j) discrete
8. pp. 11–12
9. p. 11
10. p. 3
13. p. 9
14. pp. 10–11
15. No
16. Yes
17. No
18. (a) No (b) No (c) No (d) No (e) Yes

Chapter 2.

1. (a) p. 18 (b) pp. 17–19 (c) p. 19 (d) p. 31 (e) p. 19 (f) pp. 19–20 (g) p. 21
   (h) p. 21 (i) p. 21 (j) p. 21 (k) p. 22 (l) p. 24 (m) p. 27
2. pp. 29–31
4. p. 19
5. pp. 31–33
6. p. 20

7. (a) 1.0 (b) 1.0 (c) 2.0 (d) 2.0 (e) 2.0 (f) 1.0 (g) 2.0 (h) 2.0 (i) 1.0 (j) 2.0
  (k) 2.0

8. (a)

Class Intervals	f(X)	Midpoints
95–104	2	99.5
85– 94	2	89.5
75– 84	4	79.5
65– 74	3	69.5
55– 64	1	59.5
45– 54	4	49.5
35– 44	1	39.5
25– 34	2	29.5
15– 24	0	19.5
5– 14	1	9.5
	20	

(c)

Real Class Limits	f(X)	f(X)/n	cp
94.5–104.5	2	.100	1.000
84.5– 94.5	2	.100	.900
74.5– 84.5	4	.200	.800
64.5– 74.5	3	.150	.600
54.5– 64.5	1	.050	.450
44.5– 54.5	4	.200	.400
34.5– 44.5	1	.050	.200
24.5– 34.5	2	.100	.150
14.5– 24.5	0	.000	.050
4.5– 14.5	1	.050	.050
	20		

(b) 87

14. No

Chapter 3.

1. (a) p. 39 (b) p. 48 (c) p. 48 (d) p. 56 (e) p. 52
2. pp. 40–45
3. p. 55
4. (a) p. 39 (b) p. 39 (c) p. 41 (d) p. 39 (e) p. 43 (f) p. 47 (g) p. 39 (h) p. 46 (i) p. 48 (j) p. 56
5. 61.55

6. (Cumulative Frequency used in Exercise 9)

X	f(X)	Xf(X)	Cumulative Frequency
95	2	190	20
88	1	88	18
87	1	87	17
83	1	83	16
80	1	80	15
76	2	152	14
69	1	69	12
67	1	67	11
65	1	65	10
60	1	60	9
48	3	144	8
45	1	45	5
35	1	35	4
29	2	58	3
8	1	8	1
	20	1231	

$$\bar{X} = \frac{1231}{20} = 61.55$$

7. $\bar{X} - 10 = \sum(X - 10)/20 = 1031/20 = 51.55 = 61.55 - 10$
8. Mdn. $= (67 + 65)/2 = 66$
   Mode $= 48$
9. $20(1/4) = 5, 5 - 4 = 1, 1/1$
   $Q_1 = 44.5 + 1/1 = 45.5$
   $20(3/4) = 15, 15 - 14 = 1, 1/1$
   $Q_3 = 79.5 + 1/1 = 80.5$
10. 55.83
11. (Cumulative Frequency, Q_3, Mdn, and Q_1 used in Exercise 13)

	X	f(X)	Xf(X)	Cumulative Frequency
	100	1	100	30
	99	1	99	29
	98	1	98	28
	80	1	80	27
	79	1	79	26
	75	1	75	25
	73	1	73	24
Q_3	70	1	70	23
	65	1	65	22
	60	1	60	21
	55	1	55	20
	52	1	52	19
	51	1	51	18
	50	1	50	17
	49	1	49	16
Mdn.	48	1	48	15
	47	3	141	14
	46	3	138	11
Q_1	45	2	90	8
	40	2	80	6
	39	1	39	4
	38	1	38	3
	25	1	25	2
	20	1	20	1
		30	1675	

$$\bar{X} = \frac{1675}{30} = 55.83$$

12. $\bar{X} = \dfrac{\sum Xf(X)/10}{n} = \dfrac{167.5}{30} = 5.583$

13. (a) $(30)(1/2) = 15, 15 - 14 = 1, 1/1$
    Mdn $= 47.5 + 1/1 = 48.5$
    (b) $(30)(1/4) = 7.5, 7.5 - 6 = 1.5, 1.5/2$
    $Q_1 = 44.5 + 1.5/2 = 45.25$
    (c) $(30)(3/4) = 22.5, 22.5 - 22 = .5, .5/1$
    $Q_3 = 69.5 + .5/1 = 70.0$
    (d) 46, 47

14. $\bar{X} = \dfrac{(10)(15) + (20)(25)}{10 + 20} = \dfrac{650}{30} = 21.67$

15. (a) 7.5 (b) 7 (c) 7.75 (d) 7.5 (e) 7.5 (f) 8 (g) 8 (h) 7 (i) 8

16. $\sqrt{(8)(2)} = 4$

17. $60(1/4) = 15, 15 - 6 = 9, 9/10$
    $Q_1 = 1.5 + 9/10 = 2.40$
    $60(3/4) = 45, 45 - 37 = 8, 8/13$
    $Q_3 = 3.5 + 8/13 = 4.12$

18. Case 2, Mdn $= 5.5$; Case 1, Mdn $= 6.0$

Chapter 4.

1. (a) p. 67 (b) p. 69 (c) p. 69 (d) p. 69 (e) p. 71 (f) p. 83
2. pp. 78–81
3. $s^2 = 77.72/7 = 11.10$
   $s = 3.33$
   $\hat{s}^2 = 77.72/6 = 12.95$
   $\hat{s} = 3.60$
4. $s^2 = 295.0/7 - 31.04 = 11.10$
   $s = 3.60$
   $\hat{s}^2 = (7)(295) - 1521/42 = 12.95$
   $\hat{s} = 3.33$
5. 9
6. $s^2 = 4/8 = .50$
   $s = .71$
   $\hat{s}^2 = 4/7 = .57$
   $\hat{s} = .75$
7. $s^2 = 36/8 - 4 = .5$
   $s = .71$
   $\hat{s}^2 = (8)(36) - 256/56 = .57$
   $\hat{s} = .75$
8. 2
9. $2.5 - 1.5/2 = .5$
10. $s^2 = 750/60 - 10.69 = 1.81$
    $s = 1.35$
    $\hat{s}^2 = (60)(750) - (196^2)/(60)(59) = 1.86$
    $s = 1.36$
11. $s^2 = 1.81$
    $s = 1.35$
12. pp. 68–69
13. p. 81
14. pp. 81–82
15. pp. 71–73
16. pp. 73–74
18. p. 79
19. pp. 74–75
20. p. 86

Chapter 5.

1. (a) pp. 89–90 (b) p. 93 (c) p. 94 (d) p. 97 (e) pp. 98–99 (f) p. 100 (g) p. 100
   (h) p. 104 (i) p. 118 (j) p. 119 (k) p. 126 (l) pp. 126–127
2. pp. 90–92
3. pp. 107–109
4. p. 110
5. .81, .64, .49, .36, .25, .16
6. .50

7. (a) $r = \dfrac{(10)(251) - (50)(50)}{\sqrt{[(10)(310) - (50)^2][(10)(310) - (50)^2]}} = .017$ (b) 5   (c) 5

   (d) $s_x = \sqrt{\dfrac{310}{10} - 25} = 2.45$ (e) $s_y = \sqrt{\dfrac{310}{10} - 25} = 2.45$

   (f) $\tilde{Y} = 5 + (.017)\left(\dfrac{2.45}{2.45}\right)(5 - 5) - 5$

8. (a) within rounding error of 0   (b) within rounding error of 0   (c) within rounding error of 1
   (d) within rounding error of 1

9. (a) $r = \dfrac{(5)(77) - (15)(31)}{\sqrt{[(5)(71) - (15)^2][(5)(219) - (31)^2]}} = -.606$ (b) 3   (c) 6.20

   (d) $s_x = \sqrt{\dfrac{71}{5} - 9} = 2.28$ (e) $s_y = \sqrt{\dfrac{219}{5} - (6.20)^2} = 2.32$

   (f) $\tilde{Y} = 6.20 + (-.606)\left(\dfrac{2.32}{2.28}\right)(4 - 3) = 5.58$ (g) $\tilde{Y} = 6.20 + (-.606)\left(\dfrac{2.32}{2.28}\right)(3 - 3) = 6.20$

10. (a) within rounding error of 0   (b) within rounding error of 0   (c) within rounding error of 1
    (d) within rounding error of 1

11. (a) $r_{xy \cdot z} = \dfrac{.8 - (.5)(.3)}{\sqrt{(1 - .25)(1 - .09)}} = .79$ (b) $r_{xy \cdot z} = \dfrac{.8 - (.5)(.2)}{\sqrt{(1 - .25)(1 - .04)}} = .83$

    (c) $r_{xy \cdot z} = \dfrac{.8 - (.5)(0)}{\sqrt{(1 - .25)(1 - 0)}} = .92$ (d) $r_{xy \cdot z} = \dfrac{.8 - (.3)(.5)}{\sqrt{(1 - .09)(1 - .25)}} = .79$

    (e) $r_{xy \cdot z} = \dfrac{.8 - (0)(.5)}{\sqrt{(1 - 0)(1 - .25)}} = .92$

12. $r = \dfrac{(10)(51) - (10)(50)}{\sqrt{[(10)(70) - (10)^2][(10)(310) - (50)^2]}} = .017$

13. pp. 107–109
14. (a)   (b)

15. Subtract 100 from each X score.
Subtract 1000 from each Y score.

$$r = \frac{(10)(510) - (76)(60)}{\sqrt{[(10)(672) - (76)^2][(10)(412) - (60)^2]}} = .77$$

16. pp. 125–129
17. p. 114
18. p. 117
19. p. 128
20. p. 127
21. pp. 129–130

Chapter 6.

1. (a) p. 137 (b) p. 139 (c) p. 139 (d) p. 141 (e) p. 159 (f) pp. 167–168 (g) pp. 143–144
(h) p. 144 (i) p. 144 (j) p. 145 (k) p. 145 (l) pp. 145–146 (m) p. 146 (n) p. 153
(o) p. 153 (p) p. 149 (q) p. 152 (r) p. 147

2. .9, .8, .7

3. .24, .20, .54

4. (a) $\left(\frac{1}{4}\right)^5 = .00098$ (b) $p\left(\frac{4}{5}\right) = \frac{5!}{4!(5-4)!}\left(\frac{1}{4}\right)^4\left(\frac{3}{4}\right)^1 = .01465$ (c) $\left(\frac{1}{4}\right)^4\left(\frac{3}{4}\right)^4 = .07910$

5. $\mu = 2(.2) + 3(.5) + 5(.3) = 3.4$
$\sigma^2 = (2 - 3.4)^2(.2) + (3 - 3.4)^2(.5) + (5 - 3.4)^2(.3) = 1.24$

6. $p\left(\frac{1}{1}\right) = \frac{550}{1400} = .393$

$p\left(\frac{3}{5}\right) = \frac{5!}{3!(5-3)!}(.393)^3(.607)^2 = .221$

7. $p\left(\frac{1}{6}\right) = \frac{6!}{1!(6-1)!}\left(\frac{1}{3}\right)^1\left(\frac{2}{3}\right)^5 = .263$

$p\left(\frac{5}{6}\right) = \frac{6!}{5!(6-5)!}\left(\frac{1}{3}\right)^5\left(\frac{2}{3}\right)^1 = .016$

$p\left(\frac{4}{6}\right) = \frac{6!}{4!(6-4)!}\left(\frac{1}{3}\right)^4\left(\frac{2}{3}\right)^2 = .082$

$p\left(\frac{5}{6}\right) = \phantom{\frac{6!}{5!(6-5)!}\left(\frac{1}{3}\right)^5\left(\frac{2}{3}\right)^1 =} .016$

$p\left(\frac{6}{6}\right) = \frac{6!}{6!(6-6)!}\left(\frac{1}{3}\right)^6\left(\frac{2}{3}\right)^0 = .001$

$\Sigma = .099$

8. (a) $p\left(\dfrac{2}{5}\right) = \dfrac{5!}{2!\,(3)!}\left(\dfrac{3}{5}\right)^2\left(\dfrac{2}{5}\right)^3 = .230$

(b) $p\left(\dfrac{3}{3}\right) = \dfrac{3!}{3!\,(1)!}\left(\dfrac{3}{10}\right)^3\left(\dfrac{7}{10}\right)^0 = .027$

(c) $p\left(\dfrac{1}{4}\right) = \dfrac{4!}{1!\,(3)!} = \left(\dfrac{4}{5}\right)^1\left(\dfrac{1}{5}\right)^3 = .0256$

9. (a) .50 (b) $\dfrac{5}{9}(.5) + \dfrac{5}{9}(.5) = .56$

10. $p\left(\dfrac{9}{10}\right) = \dfrac{10!}{9!\,(1)!}\left(\dfrac{3}{10}\right)^9\left(\dfrac{7}{10}\right)^1 = .00013778$

11. $p\left(\dfrac{10}{10}\right) = \dfrac{10!}{10!\,(1)!}\left(\dfrac{3}{10}\right)^{10}\left(\dfrac{7}{10}\right)^0 = .00000590$

$+\; p\left(\dfrac{9}{10}\right) \qquad\qquad\qquad\qquad = .00013778$

$\qquad\qquad\qquad\qquad\qquad\qquad\quad .00014368$

12. pp. 157–158

13. $p\left(\dfrac{5}{5}\right) = \dfrac{5!}{5!\,(0)!}\left(\dfrac{1}{2}\right)^5\left(\dfrac{1}{2}\right)^0 = .031$

14. (a) $p\left(\dfrac{0}{4}\right) = \dfrac{4!}{1!\,(4-0)!}\left(\dfrac{1}{4}\right)^0\left(\dfrac{3}{4}\right)^4 = .316$

(b) $p\left(\dfrac{1}{4}\right) = \dfrac{4!}{1!\,(4-1)!}\left(\dfrac{1}{4}\right)^1\left(\dfrac{3}{4}\right)^3 = .422$

(c) $p\left(\dfrac{2}{4}\right) = \dfrac{4!}{2!\,(4-2)!}\left(\dfrac{1}{4}\right)^2\left(\dfrac{3}{4}\right)^2 = .211$

(d) $p\left(\dfrac{3}{4}\right) = \dfrac{4!}{3!\,(4-3)!}\left(\dfrac{1}{4}\right)^3\left(\dfrac{3}{4}\right)^1 = .047$

(e) $p\left(\dfrac{4}{4}\right) = \dfrac{4!}{4!\,(0)!}\left(\dfrac{1}{4}\right)^4\left(\dfrac{3}{4}\right)^0 = .004$

sum (a) through (e) $= 1.000$

15. pp. 167–168

16. pp. 163–164

17. p. 142

18. pp. 140–141

19. As a definition, it is circular.   20. p. 138

Chapter 7.

1. (a) p. 175 (b) p. 175 (c) p. 175 (d) p. 176 (e) p. 184 (f) p. 187 (g) p. 194
   (h) pp. 196–197 (i) p. 175

2. pp. 187–191

3. Prob $(r/n) = \dfrac{n!}{r!\,(n-r)!}\,p^n q^{n-r}$

$p\left(\dfrac{0}{3}\right) = \dfrac{3!}{0!\,(3)!}\left(\dfrac{8}{10}\right)^0\left(\dfrac{2}{10}\right)^3 = (1)(1)(.008) = .008$

$p\left(\dfrac{1}{3}\right) = \dfrac{3!}{1!\,(2)!}\left(\dfrac{8}{10}\right)^1\left(\dfrac{2}{10}\right)^2 = (3)(.8)(.04) = .096$

$p\left(\dfrac{2}{3}\right) = \dfrac{3!}{2!\,(1)!}\left(\dfrac{8}{10}\right)^2\left(\dfrac{2}{10}\right)^1 = (3)(.64)(.2) = .384$

$p\left(\dfrac{3}{3}\right) = \dfrac{3!}{3!\,(0)!}\left(\dfrac{8}{10}\right)^3\left(\dfrac{2}{10}\right)^0 = (1)(.512)(1) = .512$

$\Sigma = 1.000$

Exercise Answers

4. (a) $r/n = 0/3 = .008$
$r/n = 1/3 = .096$
$\overline{.104}$

(b) $r/n = 0/3 = .008$
$r/n = 1/3 = .096$
$r/n = 2/3 = .384$
$\overline{.488}$

5. (a) $p\left(\dfrac{0}{4}\right) = \dfrac{4!}{0!\,(4)!}\left(\dfrac{6}{10}\right)^0\left(\dfrac{4}{10}\right)^4 = .026$

$p\left(\dfrac{1}{4}\right) = \dfrac{4!}{1!\,(3)!}\left(\dfrac{6}{10}\right)^1\left(\dfrac{4}{10}\right)^3 = .154$

$p\left(\dfrac{2}{4}\right) = \dfrac{4!}{2!\,(2)!}\left(\dfrac{6}{10}\right)^2\left(\dfrac{4}{10}\right)^2 = .346$

$p\left(\dfrac{3}{4}\right) = \dfrac{4!}{3!\,(1)!}\left(\dfrac{6}{10}\right)^3\left(\dfrac{4}{10}\right)^1 = .346$

$p\left(\dfrac{4}{4}\right) = \dfrac{4!}{4!\,(0)!}\left(\dfrac{6}{10}\right)^4\left(\dfrac{4}{10}\right)^0 = .130$

$\Sigma = 1.002$

(b) $p\left(\dfrac{0}{6}\right) = \dfrac{6!}{0!\,(6)!}\left(\dfrac{6}{10}\right)^0\left(\dfrac{4}{10}\right)^6 = .004$

$p\left(\dfrac{1}{6}\right) = \dfrac{6!}{1!\,(5)!}\left(\dfrac{6}{10}\right)^1\left(\dfrac{4}{10}\right)^5 = .037$

$p\left(\dfrac{2}{6}\right) = \dfrac{6!}{2!\,(4)!}\left(\dfrac{6}{10}\right)^2\left(\dfrac{4}{10}\right)^4 = .138$

$p\left(\dfrac{3}{6}\right) = \dfrac{6!}{3!\,(3)!}\left(\dfrac{6}{10}\right)^3\left(\dfrac{4}{10}\right)^3 = .276$

$p\left(\dfrac{4}{6}\right) = \dfrac{6!}{4!\,(2)!}\left(\dfrac{6}{10}\right)^4\left(\dfrac{4}{10}\right)^2 = .311$

$p\left(\dfrac{5}{6}\right) = \dfrac{6!}{5!\,(1)!}\left(\dfrac{6}{10}\right)^5\left(\dfrac{4}{10}\right)^1 = .187$

$p\left(\dfrac{6}{6}\right) = \dfrac{6!}{6!\,(0)!}\left(\dfrac{6}{10}\right)^6\left(\dfrac{4}{10}\right)^0 = .047$

$\Sigma = 1.000$

6. (d) for $n = 4$, $\sqrt{\dfrac{(.6)(.4)}{4}} = .245$

for $n = 6$, $\sqrt{\dfrac{(.6)(.4)}{6}} = .200$

7. (a) $\sigma_{\bar{x}} = \dfrac{4}{\sqrt{5}} = 1.789$ (b) $\sigma_{\bar{x}} = \dfrac{3.16}{\sqrt{2}} = 2.235$ (c) $\sigma_{\bar{x}}^2 = \dfrac{6}{7} = .857$ (d) $\sigma_{\bar{x}}^2 = \left(\dfrac{4}{\sqrt{15}}\right)^2 = 1.067$

(e) $\sigma = \sqrt{4} = 2$ (f) $\sigma_{\bar{x}} = \dfrac{6}{\sqrt{5}} = 2.68$ (g) $\sigma = (\sqrt{7})(\sqrt{3}) = 4.58$ (h) $\sigma_{\bar{x}}^2 = \dfrac{12}{15} = .8$

(i) $n = \left(\dfrac{4}{2}\right)^2 = 4$

8. (a) $p = .031$ (b) $p = .969$ **9.** p. 194   **13.** p. 187   **14.** pp. 194–196   **15.** p. 197

Chapter 8.

1. (a) p. 203 (b) p. 204 (c) pp. 204–205 (d) pp. 204–206 (e) p. 210 (f) p. 210 (g) p. 212 (h) p. 212 (i) p. 214 (j) p. 214 (k) p. 218 (l) p. 220 (m) p. 221

2. Using Appendix C, Table I:
   (a) .2119 (b) .0359 (c) $.1554 + .5000$ (0 to $-\infty$) $+ .2743 = .9297$
   (d) $.0793 + .5000$ (0 to $-\infty$) $= .5793$ (e) $.3849 - .2881 = .0968$
   (f) $.4918 - .0793 = .4125$ (g) $.1554 + .5000 + .0139 = .6693$

3. (a) $z = \dfrac{130 - 90}{30} = 1.33$ (b) $z_{\bar{x}} = \dfrac{20 - 45}{40/\sqrt{50}} = -4.42$ (c) $z_{\bar{x}} = \dfrac{3 - 2.22}{1.27} = .614$

4. (a) $z = \dfrac{9 - 8}{4} = .25,\ p = .4013$ (b) $z = \dfrac{2 - 4}{3} = -.67,\ p = .7486$

   (c) $z = \dfrac{8 - 5}{4} = .75,\ p = .2266$

5. (a) (one-tailed) $z = \dfrac{6 - 4}{4/\sqrt{9}} = 1.50$, do not reject H_0

   (b) (two-tailed) $z = \dfrac{12 - 8}{5/\sqrt{10}} = 2.53$, reject H_0

   (c) (one-tailed) $z = \dfrac{21 - 20}{13/\sqrt{25}} = .385$, do not reject H_0

   (d) (one-tailed) $z = \dfrac{30 - 45}{20/\sqrt{35}} = -4.437$, do not reject H_0

   (e) (two-tailed) $z = \dfrac{7.75 - 6}{10/\sqrt{100}} = 1.75$, do not reject H_0

6. $H_0: \mu = 12$ against $H_1: \mu \neq 12$

   $z = \dfrac{12.96 - 12}{4/\sqrt{100}} = 2.4$, reject H_0

7. pp. 201–202
8. p. 204
9. pp. 205–206
10. p. 211
11. p. 216
12. pp. 221–227
13. p. 229
14. pp. 217–218

Chapter 9.

1. (a) p. 240 (b) p. 241 (c) p. 247 (d) pp. 247–248 (e) p. 251 (f) p. 251 (g) p. 260
2. NOTE: Some probabilities are approximations from nearest t value in Table II.
   (a) $p = .025$ (b) $p = .05$ (c) $p = .10$ (d) $p \simeq .05$ (e) $1.00 - .01 = p \simeq .99$ (f) $p = .10$

3. (a) ±2.145 (b) +2.423 (c) between +4.032 and +6.859 (d) ±3.355 (e) −1.771
(f) ±2.423 (g) −4.541 (h) ±3.850

4. (a) $t = \dfrac{8 - 10}{2/\sqrt{20}} = -4.472$, two-tailed, $df = 19$, reject H_0

(b) $t = \dfrac{44 - 40}{3/\sqrt{15}} = 5.164$, one-tailed, $df = 14$, reject H_0

(c) $t = \dfrac{4 - 5}{2/\sqrt{25}} = 2.50$, one-tailed, $df = 24$, reject H_0

(d) $t = \dfrac{12 - 15}{\sqrt{15\left(\frac{1}{4} + \frac{1}{6}\right)}} = -1.199$, two-tailed, $df = 8$, do not reject H_0

(e) $t = \dfrac{12 - 48}{\sqrt{20\left(\frac{1}{10} + \frac{1}{10}\right)}} = -18.00$, one-tailed, $df = 18$, reject H_0

(f) $t = \dfrac{3 - 2}{\sqrt{2.81\left(\frac{1}{25} + \frac{1}{36}\right)}} = 2.291$, two-tailed, $df = 59$, do not reject H_0

(g) $t = \dfrac{-10 - 0}{5/\sqrt{6}} = -4.899$, one-tailed, $df = 5$, reject H_0

(h) $t = \dfrac{5.2 - 6}{4/\sqrt{16}} = -.800$, one-tailed, $df = 15$, do not reject H_0

(i) $t = \dfrac{125 - 100}{50/\sqrt{10}} = 1.581$, two-tailed, $df = 9$, do not reject H_0

(j) $t = \dfrac{20 - 17}{\sqrt{26\left(\frac{1}{10} + \frac{1}{10}\right)}} = 1.316$, two-tailed, $df = 18$, do not reject H_0

(k) $t = \dfrac{4 - 2}{\sqrt{5\left(\frac{1}{6} + \frac{1}{2}\right)}} = 1.095$, one-tailed, $df = 6$, do not reject H_0

(l) $t = \dfrac{4 - 0}{2/\sqrt{16}} = 8.000$, two-tailed, $df = 15$, reject H_0

(m) $t = \dfrac{55 - 45}{5/\sqrt{8}} = 5.657$, two-tailed, $df = 7$, reject H_0

5. (a) $F_{(29,19)} \simeq 2.45$, $F_{(19,29)} \simeq \dfrac{1}{2.32} = .43$, $F = \dfrac{10}{5} = 2$ or $\dfrac{5}{10} = .5$, do not reject H_0

(b) $F_{(4,4)} = 15.98$, $F = \frac{4}{2} = 2$, do not reject H_0

(c) $F_{(19,4)} \simeq 20.44$, $F_{(4,19)} = \frac{1}{5.27} = .19$, $F = \frac{6}{4} = 1.5$ or $\frac{4}{6} = .67$, do not reject H_0

(d) $F_{(19,4)} \simeq 14.15$, $F = 1.5$, do not reject H_0 (e) $F_{(30,24)} \simeq 1.94$, $F = 2$, reject H_0

6. $t = \dfrac{7 - 2.75}{\sqrt{3.794\left(\frac{1}{4} + \frac{1}{4}\right)}} = 3.08$, reject H_0, $F_{(3,3)} = 15.44$, $F_{(3,3)} = \dfrac{1}{15.44} = .0648$,

$F = \dfrac{6.67}{.917} = 7.27$ or $\dfrac{.917}{6.67} = .13748$, do not reject H_0

7. pp. 240–242

8. pp. 243, 253–254

9. pp. 259–260

10. p. 267

11. pp. 247–248

12. p. 247

13. p. 247

14. pp. 248–249

15. pp. 252–253, 261

Chapter 10.

1. (a) p. 278 (b) p. 282 (c) p. 282 (d) p. 283 (e) p. 285 (f) p. 285 (g) p. 287 (h) p. 287
   (i) p. 288 (j) pp. 296–297 (k) pp. 298–299

2.
Source	SS	df	MS	F
Between	75.45	2	37.72	7.91
Within	71.49	15	4.77	

Significant at $\alpha = .05$ or $.01$.

4.
Source	SS	df	MS	F
Between	4.10	3	1.37	0.85
Within	57.80	36	1.61	

Not significant at $\alpha = .05$ or $.01$.

3.
Source	SS	df	MS	F
Between	129.69	3	43.23	5.47
Within	94.75	12	7.90	

Significant at $\alpha = .05$, not significant at $\alpha = .01$.

5.
Source	SS	df	MS	F
Between	128.78	2	64.39	5.76
Within	167.66	15	11.18	

Significant at $\alpha = .05$, not significant at $\alpha = .01$.

6.
Source	SS	df	MS	F
A	16.00	1	16.00	2.82
B	0	1	0	0
A × B	64.00	1	64.00	11.29
Within	68.00	12	5.67	

Factors A and B not significant at $\alpha = .05$ or $.01$, interaction significant at $\alpha = .05$ or $.01$.

7.

Source	SS	df	MS	F
A	288.80	1	288.80	8.58
B	145.80	1	145.80	4.33
A × B	3.20	1	3.20	.10
Within	538.40	16	33.65	

Factor A significant at $\alpha = .05$ or $.01$; Factor B and interaction not significant at $\alpha = .05$ or $.01$.

8.

Source	SS	df	MS	F
A	.20	2	.10	.03
B	30.00	1	30.00	8.88
A × B	81.80	2	40.90	12.10
Within	81.20	24	3.38	

Factor A not significant at $\alpha = .05$ or $.01$; Factor B and interaction significant at $\alpha = .05$ or $.01$.

9.

Source	SS	df	MS	F
A	6.54	2	3.27	1.03
B	6.73	3	2.24	.71
A × B	181.46	6	30.24	9.54
Within	114.25	36	3.17	

Factors A and B not significant at $\alpha = .05$ or $.01$. Interaction significant at $\alpha = .05$ or $.01$.

11. pp. 284–285

12. pp. 289–290

13. p. 291

14. (a) main effect for B (b) main effect for A (c) A × B interaction (d) main effect for B (e) A × B interaction (f) main effect for A, main effect for B, A × B interaction (g) main effect for A, main effect for B, A × B interaction (h) main effect for A, main effect for B (i) main effect for A, main effect for B

16. pp. 316–317

Chapter 11.

1. (a) p. 323 (b) p. 325 (c) p. 326 (d) p. 325 (e) p. 329 (f) p. 335 (g) p. 336

2. $\chi^2 = \dfrac{(10.0 - 38.0)^2}{38.0} + \dfrac{(20.0 - 28.50)^2}{28.50} + \dfrac{(30.0 - 19.0)^2}{19.0} + \dfrac{(35.0 - 9.50)^2}{9.50}$

$= 97.98$, $df = 3$, significant at $\alpha = .05$ or $.01$, reject H_0

3. $\chi^2 = \dfrac{(40 - 60)^2}{60} + \dfrac{(40 - 30)^2}{30} + \dfrac{(20 - 10)^2}{10}$

$= 20.00$, $df = 2$, significant at $\alpha = .05$ or $.01$, reject H_0

4. $\chi^2 = \dfrac{(100.0 - 122.28)^2}{122.28} + \dfrac{(125.0 - 122.28)^2}{122.28} + \dfrac{(150.0 - 130.43)^2}{130.43} + \dfrac{(125.0 - 102.72)^2}{102.72}$

$+ \dfrac{(100.0 - 102.72)^2}{102.72} + \dfrac{(90.0 - 109.57)^2}{109.57}$

$= 15.46$, $df = 2$, significant at $\alpha = .05$ or $.01$, reject H_0

5. $\chi^2 = \dfrac{(20 - 23)^2}{23} + \dfrac{(30 - 27)^2}{27} + \dfrac{(26 - 23)^2}{23} + \dfrac{(24 - 27)^2}{27}$

$= 1.45$, $df = 1$, corrected for continuity $\chi^2 = 1.0062$, not significant at $\alpha = .05$ or $.01$, do not reject H_0

6. $\chi^2 = \dfrac{(20.0 - 15.0)^2}{15.0} + \dfrac{(20.0 - 17.50)^2}{17.50} + \dfrac{(15.0 - 22.50)^2}{22.50} + \dfrac{(10.0 - 15.0)^2}{15.0}$

$+ \dfrac{(15.0 - 17.50)^2}{17.50} + \dfrac{(30.0 - 22.50)^2}{22.50}$

$= 9.05$, $df = 2$, significant at $\alpha = .05$ or $.01$, reject H_0

7. $\chi^2 = \dfrac{(25 - 35)^2}{35} + \dfrac{(15 - 10)^2}{10} + \dfrac{(10 - 5)^2}{5}$

$= 10.36$, $df = 2$, significant at $\alpha = .01$, reject H_0

8. $\chi^2 = \dfrac{(50 - 40)^2}{40} + \dfrac{(30 - 40)^2}{40} + \dfrac{(30 - 40)^2}{40} + \dfrac{(35 - 40)^2}{40} + \dfrac{(55 - 40)^2}{40}$

$= 13.75$, $df = 4$, significant at $\alpha = .05$ or $.01$

9. $\chi^2 = \dfrac{(4.0 - 7.8)^2}{7.8} + \dfrac{(7.0 - 7.8)^2}{7.8} + \dfrac{(10.0 - 11.14)^2}{11.14} + \dfrac{(18.0 - 12.26)^2}{12.26}$

$+ \dfrac{(10.0 - 6.2)^2}{6.2} + \dfrac{(7.0 - 6.2)^2}{6.2} + \dfrac{(10.0 - 8.86)^2}{8.86} + \dfrac{(4.0 - 9.74)^2}{9.74}$

$= 10.70$, $df = 3$, significant at $\alpha = .05$ and not significant at $\alpha = .01$, reject H_0

10. $\chi^2 = \dfrac{(20 - 24)^2}{24} + \dfrac{(10 - 6)^2}{6} + \dfrac{(28 - 24)^2}{24} + \dfrac{(2 - 6)^2}{6}$

$= 6.67$, $df = 1$, corrected for continuity $\chi^2 = 5.10$, not significant at $\alpha = .05$ or $.01$, do not reject H_0

11. pp. 327–328

12. p. 335

13. p. 340

14. p. 346

15. p. 329

Chapter 12.

1. (a) p. 347 (b) pp. 349–350 (c) p. 354

2. (a) reject H_0 (b) do not reject H_0 (c) reject H_0 (d) do not reject H_0 (e) $t = 2.42$, reject H_0 (f) do not reject H_0 (g) $t = -2.42$, reject H_0 (h) do not reject H_0

3. (a) $r' = 1 - \dfrac{(6)(4)}{(6)(35)} = .886$ (b) do not reject H_0

4. (a) $r' = 1 - \dfrac{(6)(8)}{(6)(35)} = .771$ (b) do not reject H_0

5. (Item 1) $r_{pb} = \sqrt{\dfrac{\dfrac{(4)^2}{2} + \dfrac{(21)^2}{6} - \dfrac{(4+21)^2}{8}}{(16+79) - \dfrac{(4+21)^2}{8}}} = .447$

(Item 2) $r_{pb} = \sqrt{\dfrac{\dfrac{(0)^2}{1} + \dfrac{(25)^2}{7} - \dfrac{(0+25)^2}{8}}{(0+95) - \dfrac{(0+25)^2}{8}}} = .813$

(Items 3 and 4) $r_{pb} = \sqrt{\dfrac{\dfrac{(5)^2}{3} + \dfrac{(20)^2}{5} - \dfrac{(5+20)^2}{8}}{(13+82) - \dfrac{(5+20)^2}{8}}} = .778$

(Item 5) $r_{pb} = \sqrt{\dfrac{\dfrac{(16)^2}{6} + \dfrac{(9)^2}{2} - \dfrac{(16+9)^2}{8}}{(54+41) - \dfrac{(16+9)^2}{8}}} = .546$

6. $\eta = \sqrt{\dfrac{\dfrac{(16)^2}{10} + \dfrac{(28.7)^2}{10} + \dfrac{(52.8)^2}{10} + \dfrac{(66)^2}{10} - \dfrac{(16+28.7+52.8+66)^2}{40}}{835.83 - \dfrac{(16+28.7+52.8+66)^2}{40}}} = .959$

7. $\eta = \sqrt{\dfrac{\dfrac{(199)^2}{40} + \dfrac{(159)^2}{40} + \dfrac{(97)^2}{40} - \dfrac{(199+159+97)^2}{120}}{1131 + 745 + 311 - \dfrac{(199+159+97)^2}{120}}} = .535$

8. $r' = 1 - \dfrac{6(36)}{5(24)}$
    $= -.80$

9. $r = -.80$

10. p. 354

11. p. 350

Chapter 13.

1. (a) p. 362   (b) p. 362   (c) p. 363   (d) p. 363   (e) p. 380   (f) p. 368
2. (a) p = approx. .00, reject H_0 (b) p = approx. .00, reject H_0 (c) p = .020, reject H_0
    (d) p = .042, reject H_0 (e) p = .020, reject H_0 (f) p = .012, reject H_0 (g) p = .012, reject H_0
    (h) p = approx. .00, reject H_0 (i) p = approx. .00, reject H_0

3. (a) $p =$ approx. .00, reject H_0 (b) $p =$ approx. .00, reject H_0 (c) $p = .010$, reject H_0
   (d) $p = .021$, reject H_0 (e) results in wrong direction, do not reject H_0
   (f) results in wrong direction, do not reject H_0 (g) results in wrong direction, do not reject H_0
   (h) results in wrong direction, do not reject H_0 (i) results in wrong direction, do not reject H_0
4. (a) $p =$ approx. .00, reject H_0 (b) $p =$ approx. .00, reject H_0 (c) $p = .036$, reject H_0
   (d) $p = .424$, do not reject H_0
5. (a) $p =$ approx. .00, reject H_0 (b) $p =$ approx. .00, reject H_0 (c) $p = .018$, reject H_0
   (d) $p = .212$, do not reject H_0
6. $p = .002$, reject H_0
7. (a) $W_n(10) = 61$, $m = 11$
       Critical value for two-tailed $W_n = 81$, reject H_0
   (b) $W_n(10) = 61$, $m = 11$
       Critical value for one-tailed $W_n = 86$, reject H_0
8. p. 373
9. pp. 371, 375
10. (a) two-tailed, reject H_0 at $\alpha = .05$ (b) one-tailed, reject H_0 at $\alpha = .05$
11. pp. 376–377, 373–374
12. pp. 378–379
13. p. 364
14. p. 368
15. p. 370
16. pp. 380–381

Exercise Answers

Exercise Answers

APPENDIX B

1. 23
2. 119
3. 529
4. 25
5. 145
6. 625
7. 125
8. 48
9. 514
10. 2304
11. −2
12. 14
13. 4

Index

A

Absolute zero, 10
Addition rule, 149–152, 158
Alternative hypothesis, 210
Analysis of variance:
  between groups sum of squares, 283, 285–287, 293–295, 303–304, 312–313
  between means sum of squares, 285
  error variance, 279, 284–285, 289–290, 308
  F distribution, 290–291
  F statistic, 280, 290–291, 295–296, 307, 309, 314
  F test, 280, 291–292, 295–296, 307–309, 314
  Fisher, R. A., 318–319
  history, 318–319
  interaction, 296–300, 316–317, 380–381
  mean squares, 280, 287–289, 295, 306–307, 313–314
    degrees of freedom, 287–288, 295, 306–307, 313–314
  one-factor, 280–296
  overview, 278–280
  three-factor, 315–318
  total sum of squares, 282–283, 293, 301–302, 304, 310
  two-factor, 296–315
    main effects and interaction effects, 304–307, 313–314
  within groups sum of squares, 283–285, 293–295, 303–304, 310, 311
Arabic numerals, 13–14
Arithmetic mean, 23
  characteristics, 40–45
  coding, 44–45
  comparison with median and mode, 53–56
  of probability distribution, 163–164
  of sampling distribution of proportions, 181, 184, 187
  of the means of subsets, 45–46
  of ungrouped frequency distribution, 47–48
Axioms of probability, 138

B

Binomial formula, 157–159, 180–181, 183, 185–186, 369
Bivariate relation (*see* Frequency distribution)

C

Central Limit Theorem, 196–197
Central tendency, 29
  arithmetic mean (*see* Arithmetic mean)
  geometric mean (*see* Geometric mean)
  harmonic mean (*see* Harmonic mean)
  median (*see* Median)
  mode (*see* Mode)
Chi square distribution:
  assumptions, 324–325, 326–327, 379
  degrees of freedom, 324
Chi square statistic, 323
  computation, 325–326, 328, 335, 336–340

Chi square statistic (*Cont.*)
   correction for continuity, 336
   obtained and expected frequencies, 325–326, 328, 330–334, 337–340
Chi square test:
   goodness of fit, 325–328
   independence, 329–340
Class interval, 19–21
Class limits, 21–22
Classical model of probability, 139–140
Coding of arithmetic mean (*see* Arithmetic mean)
Coding of variance and standard deviation (*see* Variance and standard deviation)
Coefficient of determination, 117–120
Combinations, 149, 169
Complement, 389
Conditional probability, 154–155
Continuous variable (*see* Variables)
Correlation (*see also* Correlation ratio, Pearson product moment correlation coefficient, Point biserial coefficient, Prediction, Spearman's rank-order coefficient, and Slope):
   bivariate frequency distribution, 32–33
   causation, 127–130
   within cell, 114, 117, 127
   history, 130–134
   multiple, 126, 133
   partial, 126–127
   range, 125–126
   zero order, 114
Correlation ratio, 354–357
Critical interval(s), 212–215

D

Degrees of freedom, 73–74
   F distribution (*see* F distribution)
   t distribution (*see* t distribution)
Descriptive statistics, 12
   bivariate frequency distributions, 31–33
   characteristics of univariate frequency distribution, 29–31
   history, 33–36
Deviation score, 41
Difference (of two sets), 389–390
Discrete variable (*see* Variables)
Distribution-free statistical tests, 361–362
   advantages and disadvantages, 378–381
   assumptions, 364–365

   history, 381–382
   measurement, 363–364
   methods of treating zero difference scores and tied ranks, 365–366
   relation to nonparametric tests, 362–363
   sign test:
      one sample, 366–369
      two dependent samples, 369–371
   Wilcoxon rank-sum test, 371–375
   Wilcoxon signed-rank test, 375–378

E

Efficiency, 380
Equation for a straight line (*see* Function rule for a straight line)
Error:
   type I, 218–221
   type II, 218–221
      determinants, 221–228
Error variance:
   variability, 81, 86
Estimate of standard deviation (*see* Variance and standard deviation)
Estimate of variance (*see* Variance and standard deviation)
Experiment, 128–130, 139

F

F distribution, 266–268 (*see also* Analysis of variance):
   assumptions, 267, 379
   degrees of freedom, 267
   table, 267
F statistic:
   computation, 269–270
   formula, 269
F test:
   analysis of variance (*see* Analysis of variance)
   correlation ratio, 356–357
   equality of population variances, 265–270
   point biserial correlation, 353–354
Factorials, 147
Fisher, R. A., 273, 318–319, 381
Frequency distributions, 17–19
   bivariate, 31–33
      correlation (*see* Correlation)
      relation, 32–33
      scatterplot, 97–99

Frequency distributions (*Cont.*)
  relation to probability distribution, 163–167
  univariate, 19–31
    characteristics, 29–31
    graphed (*see* Graphs)
    grouped (*see* Grouped frequency distribution)
    marginal distributions, 32
    ungrouped (*see* Ungrouped frequency distribution)
Frequency polygon (*see* Graphs)
Function, 392–393
Function rule for a straight line, 93–96

G

Galton, F., 61–63, 86, 130–134, 233–234, 318–319
Geometric mean:
  correlation, 107–108
  formula, 56
  origin, 61
Graphs:
  conventions, 28
  cumulative proportion, 27–28
  frequency polygon, 23–26
  histogram, 22–23
  of central tendency, 56–58
  of nominal scale variables, 24–25
  that create false impressions, 58–60
Grouped frequency distribution, 19–21
  graphed (*see* Graphs)
  procedure for constructing, 20–21
  terminology, 21

H

Histogram (*see* Graphs)
History of statistics:
  analysis of variance, 318–319
  arabic numerals, 13–14
  central tendency, 60–63
  correlation, 130–134
  descriptive statistics, 33–36
  distribution-free tests, 381–382
  Fisher, R. A., 273, 318–319, 381
  Galton, F., 61–63, 86, 130–134, 233–234, 318–319
  normal distribution, 232–235, 381
  Pearson, K., 86, 131–134, 232, 272–273, 318–319
  probability, 168–172
  "student," 272–273, 381
  t test, 272–273
  variability, 86–87
Homogeneity of variance assumption, 251–252, 254, 259, 285, 291, 361–362, 379

I

Independence of samples, 248, 260, 329, 371, 375
Independence of variables, 329–330
Independent outcomes, 153–155, 329
Inferential statistics, 12, 137
Interaction (*see* Analysis of variance)
Intercept constant, 93–94
Intersection, 152–155, 387–389

K

Kurtosis, 31

L

Long-run relative frequency model of probability, 141–143

M

Magnitude of effect, 270–272
Mean (*see* Arithmetic mean, Geometric mean)
Mean squares (*see* Analysis of variance)
Measurement, 6–11
  illustrated, 7–9
  properties of organisms, 6–7, 11
  scales, 9–10
    interval, 10
    nominal, 9, 24–25
    ordinal, 9–10
    ratio, 10–11
Median, 48
  comparison with arithmetic mean and mode, 53–56
  computation, 49–52
  relation to semi-interquartile range, 83–85
Midpoints, 21–22
Mode, 33
  comparison with mean and median, 53–56

Multiple correlation, 126, 133
Multiplication rule, 152–155
Mutually exclusive outcomes, 139–140, 142, 149–150, 152, 389

N

Nonparametric tests, 362–363
Normal distributions:
    ascertaining probabilities, 206–209
    family, 204
    formula, 202–203
    history, 232–235
    parameters, 203–204
    relation to sampling distribution of proportions, 201–202
    standard normal distribution, 204–206
    standard normal transformation (*see* z scores)
Null hypothesis, 210, 217–218

O

One-tailed test, 214–215, 225, 227, 230–232

P

Partial correlation, 126–127
Pearson, K., 86, 131–134, 232, 272–273, 318–319
Pearson product–moment correlation coefficient:
    accountable variance and coefficient of determination, 117–120
    calculation and coding, 112–113
    formula and derivation, 110–112
    geometric mean, 108
    measure of linear correlation, 96
    measure of slope after z transformation, 108–110
    significance test, 345–347
Percentiles, 48
    median, 48–52
    quartiles, 48
        computed, 49–52
Permutations, 147–149, 155–159, 169
Point biserial coefficient, 349–354
Population:
    defined, 143–144
    discussed, 12
    distribution, 167–168

Power:
    defined, 221
    determinants, 221–228
Prediction:
    example problems, 123–125
    illustrative problem, 120–121
    multiple correlation, 126
    regression equations, 121–122
Probability:
    axioms, 138
    classical model, 139–140
    computation, 147–159
    conditional, 154–155
    definition, 138
    distributions, 159–168
    history, 168–172
    long-run, relative frequency model, 141–143
    numbers, 137
    procedures, 138–139
Probability distribution, 159
    continuous, 160–162
    discrete, 160
    mean, 163–164
    population, 167–168
    variance and standard deviation, 164–167
Procedures, 138–139
Product set, 32, 391–392
Properties of organisms, 6–7, 11

Q

Quartiles (*see* Percentiles)

R

Range, 67–68
    advantages and disadvantages, 68
    affect on correlation, 125–126
    illustrated, 18
    origin, 86
Regression of X on Y:
    formula, 105
    horizontal deviation, 104, 105
    rotation, 107
    slope expressed in deviation form, 105
Regression of Y on X:
    formula, 101
    rotation, 104
    slope expressed in deviation form, 102
    vertical deviation, 100, 105

Relation, 32–33, 392
Robustness, 379
Rounding numbers, 20

S

Sampling:
  defined, 143–144
  simple random, 144–145
  with and without replacement, 145–147
Sampling distributions, 175–176
  of chi square (*see* Chi square distribution)
  of differences between independent sample means, 250–251
  of means, 187–197
    computation of example, 189–191
    mean, 191–193
    shape, 196–197
    standard error, 193–196
  of proportions, 176–187
    computation, 177–187
    mean, 181, 184, 187
    p and q defined, 176
    sample proportion defined, 176–177
    standard error, 184–185
  of t (see t distribution)
Scales of measurement, 9–11 (*see* Measurement)
Semi-interquartile range, 83–85
  advantages and disadvantages, 86
  derived, 85
  origin, 86
  relation to median, 85
Set theory, 385–393
  concepts, 386–387
  operations, 387–391
  product set, 391–393
Sign test (*see* Distribution-free statistical test)
Significance level, 212–215, 223, 228–230
Skewness, 31
Slope:
  correlation index, 99–100
  difficulties as correlation index, 107–110
  straight line constant, 94–96
Slope constant, 93–96
Spearman rank-order correlation coefficient, 347–349
Standard deviation (*see* Variance and standard deviation)

Standard error of the mean, 193–196, 211, 239–240
Standard scores (*see* z scores)
Statistic, 13
  summary measures, 13
Statistics:
  descriptive, 12
  field, 1
  inferential, 12, 137
Straight-line function rule (*see* Function rule for a straight line)
"Student," 272–273, 381
Sum of squares, 69 (*see* Analysis of variance)
Summation algebra:
  notation, 395–397
  rules, 398–399

T

t distribution, 240–243
  assumptions, 243–244, 253–254, 259–260, 379
  comparison with standard normal distribution, 241–242
  degrees of freedom, 243, 252, 261
  sampling distribution of t, 241
  Student's t distribution, 241
  table, 241
t statistic, 240
  one-sample computation, 244–247
  t for differences between dependent sample means, 260–261
  t for differences between independent sample means, 252–253
t test:
  history, 272–273
  one-sample, 239–247
  Pearson product–moment correlation coefficient, 345–347
  two dependent samples, 260–264
  two independent samples, 248–259
Two-tailed test, 214–215, 225, 227, 230–232
Type I error (*see* Error)
Type II error (*see* Error)

U

Ungrouped frequency distribution:
  arithmetic mean of, 47–48, 164, 166
  illustrated, 19

Ungrouped frequency distribution: (Cont.)
   median of, 49–52
   terminology, 21
   variance and standard deviation, 81–83, 165–166
Union, 149–152, 387

V

Variability, 30, 67
   range, 67–68
   semi-interquartile range, 83–86
   variance and standard deviation, 68–83
Variables, 11–12
   continuous, 11–12
   discrete, 11–12
Variance and standard deviation, 68–71
   advantages and disadvantages, 74–75
   characteristics, 78–81
   coding, 79
   computational formulas, 75–78
   estimate, 71–74, 76, 78
   of probability distribution, 164–167
   ungrouped frequency distribution, 81–82

W

Wilcoxon rank-sum test, 371–375
Wilcoxon signed-rank test, 375–378

Z

z scores, 89–90
   characteristics, 90–92
   identity of regression line slopes, 110–112
Pearson product–moment correlation coefficient, 110–112
   standard normal transformation, 204–206, 215–217, 218